高等学校专业英语教材

机械工程专业英语教程
（第6版）

施 平 主编

电子工业出版社
Publishing House of Electronics Industry
北京·BEIJING

内 容 简 介

本书编者在前一版的基础上，汲取了多所大学在使用本书过程中提出的许多宝贵意见，对全书进行了修订和补充。本书的主要目的是使读者掌握机械工程专业英语术语及用法，培养和提高读者阅读和翻译专业英语文献资料的能力。本书的主要内容包括力学、机械零件与机构、机械工程材料、润滑与摩擦、机械制图、公差与配合、机械设计、机械制造、管理、现代制造技术、科技写作。全书由77篇课文组成，其中34篇课文有参考译文，9篇课文为阅读材料。本书提供电子课件等教学资源，读者可从华信教育资源网（www.hxedu.com.cn）免费下载。

本书既可以作为机械设计制造及自动化、机械工程及自动化、机电工程等专业的专业英语教材，也可以供从事机械工程工作的工程技术人员参考使用。

未经许可，不得以任何方式复制或抄袭本书之部分或全部内容。
版权所有，侵权必究。

图书在版编目（CIP）数据

机械工程专业英语教程 / 施平主编. -- 6 版.
北京：电子工业出版社，2024.7. -- ISBN 978-7-121-48363-9
Ⅰ.TH
中国国家版本馆 CIP 数据核字第 2024RX5970 号

责任编辑：秦淑灵　　文字编辑：徐　萍
印　　刷：三河市良远印务有限公司
装　　订：三河市良远印务有限公司
出版发行：电子工业出版社
　　　　　北京市海淀区万寿路 173 信箱　邮编：100036
开　　本：787×1092　1/16　印张：20　字数：666 千字
版　　次：2003 年 7 月第 1 版
　　　　　2024 年 7 月第 6 版
印　　次：2024 年 7 月第 1 次印刷
定　　价：55.00 元

凡所购买电子工业出版社图书有缺损问题，请向购买书店调换。若书店售缺，请与本社发行部联系，联系及邮购电话：(010)88254888，88258888。
质量投诉请发邮件至 zlts@phei.com.cn，盗版侵权举报请发邮件至 dbqq@phei.com.cn。
本书咨询联系方式：qinshl@phei.com.cn。

前　言

专业英语是大学英语教学的一个重要组成部分，是促进学生从英语学习过渡到实际应用的有效途径。教育部颁布的《大学英语教学大纲》明确规定专业英语为必修课程，要求通过四年不间断的大学英语学习，培养学生以英语为工具交流信息的能力。编者根据此精神编写了本书，以满足高等院校机械工程各专业学生的专业英语学习需求。

在此次再版前，编者汲取了多所大学在使用本书过程中提出的许多宝贵意见，对全书进行了修订和补充。本书所涉及的内容包括力学、机械零件与机构、机械工程材料、润滑与摩擦、机械制图、公差与配合、机械设计、机械制造、管理、现代制造技术、科技写作等方面。通过学习本书，学生不仅可以熟悉和掌握本专业常用的及与本专业有关的单词、词组及其用法，而且可以深化本专业的知识，从而为今后的学习和工作打下良好的基础。

全书由 77 篇课文组成，其中 34 篇课文有参考译文，9 篇课文为阅读材料。本书选材广泛，内容丰富，语言规范，难度适中，便于自学。为了方便教学，本书另配有电子课件，向采纳本书作为教材的教师免费提供（获取方式：登录电子工业出版社华信教育资源网 www.hxedu.com.cn 或电话联系 010-88254531 或邮件联系 qinshl@phei.com.cn 获得）。

本书由施平主编，参加编写和电子课件制作工作的有李越、胡明、乔世坤、田锐、施晓东，由贾艳敏担任主审。

由于编者水平有限，书中难免有不足和欠妥之处，恳请广大读者批评指正。

编　者

Contents

Lesson 1　Basic Concepts in Mechanics ……………………………………………………… (1)
Lesson 2　Forces and Their Effects ……………………………………………………………… (4)
Lesson 3　Overview of Engineering Mechanics ………………………………………………… (9)
Lesson 4　Shafts, Couplings, and Splines ……………………………………………………… (12)
Lesson 5　Shafts and Associated Parts ………………………………………………………… (16)
Lesson 6　Belts, Clutches, Brakes, and Chains ………………………………………………… (20)
Lesson 7　Fasteners and Springs ………………………………………………………………… (24)
Lesson 8　Rolling Bearings ……………………………………………………………………… (27)
Lesson 9　Turbine Engine Bearings for Ultra-High Temperatures …………………………… (30)
Lesson 10　Spindle Bearings (*Reading Material*) ……………………………………………… (34)
Lesson 11　Machine Tool Frames ………………………………………………………………… (37)
Lesson 12　Spur Gears …………………………………………………………………………… (41)
Lesson 13　Strength of Mechanical Elements …………………………………………………… (45)
Lesson 14　Physical Properties of Materials …………………………………………………… (48)
Lesson 15　Kinematics and Dynamics …………………………………………………………… (51)
Lesson 16　Basic Concepts of Mechanisms ……………………………………………………… (55)
Lesson 17　Material Selection …………………………………………………………………… (58)
Lesson 18　Selection of Materials ………………………………………………………………… (61)
Lesson 19　Gear Materials (*Reading Material*) ………………………………………………… (65)
Lesson 20　Friction, Wear, and Lubrication …………………………………………………… (69)
Lesson 21　Lubrication …………………………………………………………………………… (73)
Lesson 22　Introduction to Tribology …………………………………………………………… (77)
Lesson 23　Product Drawings …………………………………………………………………… (80)
Lesson 24　Sectional Views ……………………………………………………………………… (84)
Lesson 25　Computer Graphics (*Reading Material*) …………………………………………… (89)
Lesson 26　Dimensional Tolerance ……………………………………………………………… (92)
Lesson 27　Fundamentals of Manufacturing Accuracy ………………………………………… (95)
Lesson 28　Tolerances and Fits …………………………………………………………………… (99)
Lesson 29　Introduction to Mechanical Design ………………………………………………… (102)
Lesson 30　Engineering Design …………………………………………………………………… (105)
Lesson 31　Some Rules for Mechanical Design ………………………………………………… (109)
Lesson 32　Computer Applications in Design and Graphics …………………………………… (112)
Lesson 33　Lathes and Cutting Parameters in Turning Process ……………………………… (115)
Lesson 34　Engine Lathes ………………………………………………………………………… (119)

Lesson 35	Milling Machines and Grinding Machines	(123)
Lesson 36	Drilling Operations and Machines	(127)
Lesson 37	Milling Operations (*Reading Material*)	(132)
Lesson 38	Gear Manufacturing Methods	(136)
Lesson 39	Laser-Assisted Machining and Cryogenic Machining Technique	(140)
Lesson 40	Machine Tool Motors	(144)
Lesson 41	Development of Metal Cutting (*Reading Material*)	(148)
Lesson 42	History of Machine Tools (*Reading Material*)	(151)
Lesson 43	Nontraditional Manufacturing Processes	(154)
Lesson 44	Overview of Nontraditional Manufacturing Processes	(157)
Lesson 45	Machining of Engineering Ceramics	(161)
Lesson 46	Definitions and Terminology of Vibration	(164)
Lesson 47	Mechanical Vibrations	(167)
Lesson 48	Automated Assembly	(170)
Lesson 49	Roles of Engineers in Manufacturing	(173)
Lesson 50	Manufacturing Enterprises	(176)
Lesson 51	Careers in Manufacturing	(179)
Lesson 52	Manufacturing Research Centers at U. S. Universities (*Reading Material*)	(182)
Lesson 53	Developments in Manufacturing Technology (*Reading Material*)	(185)
Lesson 54	Mechanical Engineering and Mechanical Engineers	(188)
Lesson 55	Cost Estimating	(191)
Lesson 56	Quality and Inspection	(194)
Lesson 57	Quality in the Modern Business Environment	(197)
Lesson 58	Coordinate Measuring Machines	(201)
Lesson 59	Reliability Requirements	(205)
Lesson 60	Product Reliability	(208)
Lesson 61	Effect of Reliability on Product Salability	(212)
Lesson 62	Computers in Manufacturing	(215)
Lesson 63	Computer Applications in Design and Manufacturing	(218)
Lesson 64	Computer-Aided Analysis of Mechanical Systems	(221)
Lesson 65	Computer-Aided Process Planning	(224)
Lesson 66	Numerical Control	(227)
Lesson 67	Numerical Control Software	(231)
Lesson 68	Computer Numerical Control	(235)
Lesson 69	Training Programmers	(238)
Lesson 70	History of Numerical Control (*Reading Material*)	(241)
Lesson 71	Industrial Robots	(244)
Lesson 72	Robotics	(247)
Lesson 73	Basic Components of an Industrial Robot	(251)

Lesson 74	Robot Sensors	(254)
Lesson 75	Mechanical Engineering in the Information Age	(258)
Lesson 76	How to Write a Scientific Paper	(261)
Lesson 77	How to Write a Technical Report	(264)

参考译文 (268)

第1课	力学基本概念	(268)
第4课	轴、联轴器和花键	(269)
第7课	紧固件和弹簧	(270)
第8课	滚动轴承	(271)
第13课	机械零件的强度	(272)
第17课	材料选择	(273)
第21课	润滑	(274)
第22课	摩擦学概论	(275)
第23课	产品图样	(276)
第26课	尺寸公差	(278)
第28课	公差与配合	(279)
第29课	机械设计概论	(280)
第31课	几条机械设计准则	(282)
第32课	计算机在设计和制图中的应用	(283)
第33课	车床和车削参数	(284)
第34课	普通车床	(285)
第35课	铣床和磨床	(286)
第38课	齿轮制造方法	(287)
第43课	特种加工工艺	(288)
第45课	工程陶瓷的机械加工	(290)
第46课	振动的定义和术语	(291)
第49课	工程师在制造业中的作用	(292)
第55课	成本估算	(293)
第56课	质量与检验	(294)
第59课	可靠性要求	(295)
第61课	可靠性对产品适销性的影响	(297)
第62课	计算机在制造业中的应用	(298)
第65课	计算机辅助工艺设计	(299)
第66课	数字控制	(300)
第69课	培训编程人员	(302)
第71课	工业机器人	(303)
第73课	工业机器人的基本组成部分	(304)
第75课	信息时代的机械工程	(306)
第76课	如何撰写科技论文	(307)

参考文献 (309)

Lesson 1　Basic Concepts in Mechanics

The branch of scientific analysis which deals with motions, time, and forces is called mechanics and is made up of two parts, statics and dynamics. Statics deals with the analysis of stationary systems, i.e., those in which time is not a factor, and dynamics deals with systems which change with time.

When a number of bodies are connected together to form a group or system, the forces of action and reaction between any two of the connecting bodies are called constraint forces. These forces constrain the bodies to behave in a specific manner. Forces external to this system of bodies are called applied forces.[1]

Forces are transmitted into machine members through mating surfaces, e.g., from a gear to a shaft or from one gear through meshing teeth to another gear (see Fig. 1.1), from a V-belt to a pulley, or from a cam to a follower. It is necessary to know the magnitudes of these forces for a variety of reasons. For example, if the force operating on a journal bearing becomes too high, it will squeeze out the oil film and cause metal to metal contact, overheating, and rapid failure of the bearing. If the forces between gear teeth are too large, the oil film may be squeezed out from between them. This could result in spalling of the metal, noise, and eventual failure. In the study of mechanics, we are principally interested in determining the magnitude, direction, and location of the forces.

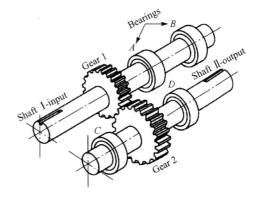

Figure 1.1　Two shafts carrying gears in mesh

Some of the terms used in mechanics are defined below.

Force　Our earliest ideas concerning forces arose because of our desire to push, lift, or pull various objects. So force is the action of one body on another. Our intuitive concept of force includes such ideas as place of application, direction, and magnitude, and these are called the characteristics of a force.

Couple　Two equal and opposite forces acting along two non-coincident parallel straight lines in a body cannot be combined to obtain a single resultant force. Any two such forces acting on a body constitute a couple (see Fig. 1.2). The only effect of a couple is to produce a rotation or tendency of rotation in a specified direction.

Mass　Mass is a measure of the quantity of matter that a body or an object contains. The mass of the body is not dependent on gravity and therefore is different from but proportional to its weight. Thus, a moon rock has a certain constant amount of substance, even though its moon weight is different from its earth weight. This constant amount of substance is called the mass of the rock.

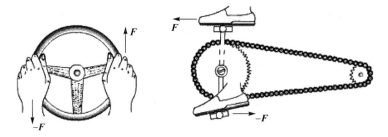

Figure 1.2　Common examples of couples

Inertia　Inertia is the property of a body that causes it to resist any effort to change its motion.

Weight　Weight is the force with which a body is attracted to the earth or another celestial body, equal to the product of the object's mass and the acceleration of gravity.

Particle　A particle[2] is a body whose dimensions are so small that they may be neglected.

Rigid Body　A rigid body does not change size and shape under the action of forces. Actually, all bodies are either elastic or plastic and will be deformed if acted upon by forces. When the deformation of such bodies is small, they are frequently assumed to be rigid, i.e., incapable of deformation, in order to simplify the analysis. A rigid body is an idealization of a real body.

Deformable Body　The rigid body assumption cannot be used when internal stresses and strains due to the applied forces are to be analyzed. Thus we consider the body to be capable of deforming. Such analysis is frequently called elastic body analysis, using the additional assumption that the body remains elastic within the range of the applied forces.

Newton's Laws of Motion　Newton's three laws are:

Law 1　If all the forces acting on a body are balanced, the body will either remain at rest or will continue to move in a straight line at a uniform velocity.

Law 2　If the forces acting on a body are not balanced, the body will experience an acceleration. The acceleration will be in the direction of the resultant force, and the magnitude of the acceleration will be proportional to the magnitude of the resultant force and inversely proportional to the mass of the body.

Law 3　The forces of action and reaction between interacting bodies are equal in magnitude, opposite in direction, and have the same line of action.

Mechanics deals with two kinds of quantities: scalars and vectors. Scalar quantities are those with which a magnitude alone is associated. Examples of scalar quantities in mechanics are time, volume, density, speed, energy, and mass. Vector quantities, on the other hand, possess direction as well as magnitude. Examples of vectors are displacement, velocity, acceleration, force, moment, and momentum.

Words and Expressions

mechanics [mi'kæniks]　*n.* 力学，机械学

statics ['stætiks] *n.* 静力学
dynamics [dai'næmiks] *n.* 动力学
i. e. 也就是，即（that is）
constraint [kən'streint] *n.* 约束，强制
transmit [trænz'mit] *v.* 传递，传送（to cause something to pass on from one place to another）
mating surface 啮合表面，配合表面，接触表面
e. g. 例如（for example）. It can be pronounced as "e. g." or "for example".
gear [giə] *n.* 齿轮（轮缘上有齿、能连续啮合传递运动和动力的机械零件）
shaft [ʃɑːft] *n.* 轴（支承转动件，传递运动或动力的机械零件）
meshing ['meʃiŋ] *n.* 啮合（一对带有齿状部分的零件在传动过程中的相互连接）
bearing ['bɛəriŋ] *n.* 轴承（用于确定旋转轴与其他零件相对运动位置，起支承或导向作用的零部件），支承，承载
pulley ['puli] *n.* 滑轮，带轮（a wheel used to transmit power by means of a belt or belts）
cam [kæm] *n.* 凸轮（具有曲线或曲面轮廓且作为高副元素的构件）
magnitude ['mægnitjuːd] *n.* 大小，尺寸，量度，数值
journal ['dʒəːnl] *n.* 轴颈（轴上被径向轴承支承着的，在其中旋转的部分）
journal bearing 滑动轴承，径向滑动轴承（承受径向载荷的滑动轴承）
squeeze [skwiːz] out 挤出，压出
spalling ['spɔːliŋ] *n.* 剥落（疲劳磨损时从摩擦表面以鳞片形式分离出磨屑的现象）
resultant [ri'zʌltənt] *a.* 合的，组合的，总的；*n.* 合力，合成矢量，组合
couple 力偶（大小相等、方向相反、不作用在同一直线上的两个力所组成的力系）
noncoincident [ˌnɔnkəu'insidənt] *a.* 不重合的，不一致的，不符合的
parallel ['pærəlel] *a.* 平行的（being an equal distance apart everywhere）；*n.* 平行线
inertia [i'nəːʃjə] *n.* 惯性（物体抵抗其运动状态被改变的性质），惯量，惰性
celestial [si'lestʃəl] body 天体（宇宙中各种实体的统称）
incapable [in'keipəbl] *a.* 无能力的，不能的
deformation [ˌdiːfɔː'meiʃən] *n.* 变形（物体在外来因素作用下产生的形状和尺寸的改变）
deformable [di'fɔːməbl] *a.* 可变形的
proportional 成比例的（having a size, number, or amount that is directly related to something）
acceleration [æk͵selə'reiʃən] *n.* 加速度（速度对时间的变化率）
stress [stres] *n.* 应力（内力的集度，即单位面积上的内力）
strain [strein] *n.* 应变（由外力所引起的物体尺寸和形状的单位变化量）
scalar ['skeilə] *n.* ；*a.* 数量（的），标量（的）
vector ['vektə] *n.* 矢量，向量
displacement [dis'pleismənt] *n.* 位移（物体或质点从初位置到末位置的有向线段）
moment ['məumənt] *n.* 力矩（力和力臂的乘积），亦为 moment of force
momentum [məu'mentəm] *n.* 动量（物体的质量和速度的乘积）

Notes

[1] applied force 意为"外力，作用力（能够产生运动或运动趋势的力）"。
[2] particle 意为"质点，动力学中用来代替物体的有质量的点，是一个理想化模型"。

Lesson 2　Forces and Their Effects

A study of any machine or mechanism shows that each is made up of a number of movable parts. These parts transform a given motion to a desired motion. In other words, these machines perform work. Work is done when motion results from the application of force. Thus, a study of mechanics and machines deals with forces and the effects of forces on bodies. [1]

A force is a push or pull. The effect of a force either changes the shape or motion of a body or prevents other forces from making such changes. Every force produces a stress in the part on which it is applied. Forces may be produced by an individual using muscular action or by machines with mechanical motion.

Forces are produced by physical or chemical change, gravity, or changes in motion. When a force is applied which tends to stretch an object, it is called a tensile force. A part experiencing a tensile force is said to be in tension. A force can also be applied which tends to shorten or squeeze the object. Such a force is a compressive force.

A third force is known as a torsional force, or a torque since it tends to twist an object. Still another kind of force, which seems to make the layers or molecules of a material slide or slip on one another, is a shearing force. [2]

Each of these forces may act independently or in combination. For example, if a steel beam is placed in a horizontal position and a force F is applied in the middle (see Fig. 2.1), the bottom of the beam tends to stretch and is in tension. At the same time, the top area is being pushed together and is in compression. [3]

The turning of a part in a lathe is another example of several forces in action (see Fig. 2.2). [4] As the work revolves and the cutting tool moves into the work, the cutting edge produces a shearing force. This force causes the metal to seem to flow off the work in the form of chips.[5] If this work is held between the two centers of the lathe, the centers exert a compressive force against the work.[6] The lathe dog which drives the work tends to produce a shearing force. The pressure of the cutting tool against the work produces tension and compression, as well as a shearing action.

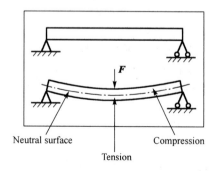

Figure 2.1　Deflection of a simple beam

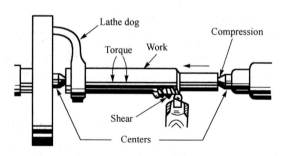

Figure 2.2　Cylindrical turning operation in a lathe

Considerable attention is given to the action of centrifugal force in grinding wheels. That is, the bond that holds the abrasive particles on the wheel (see Figs. 2.3 and 2.4) must be stronger than the forces which tend to make the revolving wheel fly apart at high speeds. For this reason, the speed of a grinding wheel should not exceed the safe surface speed limit specified by the manufacturer. Centrifugal force increases with speed.

Figure 2.3 Grinding wheels

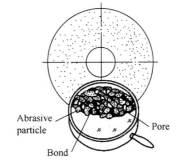

Figure 2.4 Grinding wheel composition

The principles of centrifugal force are used in the design of centrifuges. Some centrifuges are used to separate chemicals; others are used to remove impurities in metals by centrifugal casting processes. One of the most common applications of centrifugal force is the use of washing machine spin dryers in household and commercial laundry. [7]

Centripetal force causes an object to travel in circular path. This action is caused by the continuous application of forces which tend to pull the object to the center. In other words, the inward force which resists the centrifugal force is called the centripetal force. The centripetal force of objects spinning at a constant rate produces an acceleration toward the center which is equal and opposite to the centrifugal force (see Fig. 2.5). [8]

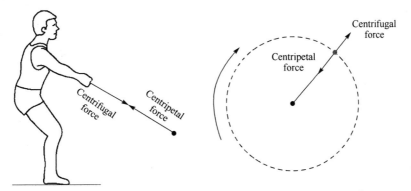

Figure 2.5 Centrifugal force and centripetal force associated with circular motion

The materials used in the construction of rapidly moving machine parts and mechanisms must be structurally strong enough to provide the centripetal force required to hold the parts to a circular path. At the same time, the materials must be able to withstand the centrifugal force which tends to pull the parts apart. [9]

Motion and the basic laws which affect motion are important considerations because of the numerous applications of these principles to produce work through mechanical devices. There are two primary mechanical motions: rotary and rectilinear.

Rotary Motion　Rotary motion is required to drill holes, mill surfaces, or turn parts in a lathe.[10]

Rectilinear Motion　The feed of a cutting tool on a lathe (see Fig. 2.2) and the shaping of materials are situations in which rectilinear motion produces work.[11] In each of these situations a part or mechanism is used to change rotary motion to straight line motion. The screw of a micrometer (Fig. 2.6) and the threads in a nut are still other applications where the direction of motion is changed from rotary to rectilinear.[12]

Figure 2.6　Micrometers

Simple Harmonic and Intermittent Motion　Any simple vibration, such as the regular back-and-forth movement of the end of a pendulum (see Fig. 2.7), is simple harmonic motion.[13] However, many manufacturing processes require intermittent or irregular motion. For example, the fast return stroke of a shaper (see Fig. 2.8) ram is desirable because no cutting is done on the return stroke. Therefore, as more time is saved in returning the cutting tool to the working position, the less expensive is the operation.

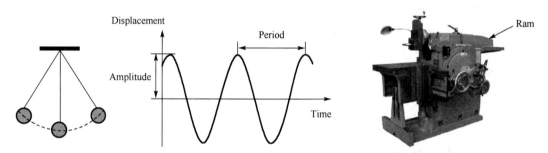

Figure 2.7　Simple harmonic motion of an undamped pendulum　　　Figure 2.8　A shaper

The combinations of rotary and rectilinear motion obtainable are unlimited because of the large variety of parts such as gears, cams, pulleys, screws, links, and belts which can be combined in many arrangements.

Words and Expressions

mechanism ['mekənizəm]　*n.* 机构（两个或两个以上构件通过活动连接形成的构件系统）
given motion　给定运动，某种特定的运动
muscular ['mʌskjulə]　*a.* 肌肉的
compressive [kəm'presiv]　*a.* 有压力的，压缩的
torsional ['tɔːʃənəl] force　扭力，扭转力，亦为 torsion force

torque [tɔːk] *n.* 扭矩，转矩，也可写为 torsional moment
shearing force 剪力，剪切力，亦为 shear force
compression [kəmˈpreʃən] *n.* 压缩（the act or process of pressing something together）
simple beam 简支梁，亦可写为 simply supported beam
neutral surface 中性面，也可写为 neutral plane（中性层）
lathe [leið] *n.* 车床（主要用车刀在工件上加工旋转表面的机床），亦为 turning machine
lathe dog 卡箍，鸡心夹头，亦为 dog 或 lathe carrier
cutting edge 切削刃（刀具前刀面上作切削用的刃）
cutting tool 刀具（能从工件上切除多余材料或切断材料的带刃工具）
cylindrical turning 车外圆（用车削方法加工工件的外圆表面），也可写为 turning
centrifugal [senˈtrifjugəl] *a.* 离心的（moving or directed outward from the center）
grinding [ˈgraindiŋ] *n.* 磨削（磨具以较高的线速度旋转，对工件表面进行加工的方法）
grinding wheel 砂轮（用磨粒和结合剂等制成的中央有通孔的圆形磨具）
bond [bɔnd] *n.* 结合剂（用来固结磨粒形成磨具的黏结材料），亦为 bonding agent
abrasive [əˈbreisiv] *n.* 磨料（在磨削、研磨和抛光中起切削作用的硬质材料）
abrasive particle 磨粒，也可以写为 abrasive grain（以人工方法制成特定粒度，用以制造切除材料余量的磨削、抛光和研磨工具的颗粒材料）
pore [pɔː] *n.* 气孔（磨具所具有的空隙），孔隙（物体内的孔洞）
revolving [riˈvɔlviŋ] *a.* 旋转的（able to be turned around a center point）
fly apart 飞散（to break apart, throwing pieces around），粉碎
grinding wheel composition 砂轮组成，砂轮构造
centrifuge [ˈsentrifjuːdʒ] *n.* 离心机（a machine using centrifugal force for separating substances）
impurity [imˈpjuəriti] *n.* 杂质，混杂物
centrifugal casting 离心铸造（将熔融金属浇入旋转的铸型，在离心力作用下凝固形成铸件）
centripetal [senˈtripitl] *a.* 向心的（acting in a direction toward a center）
spin dryer 衣物甩干机（a dryer using centrifugal force for removing water out of clothes）
rectilinear [ˌrektiˈliniə] *a.* 直线的，由直线组成的
intermittent [ˌintəˈmitənt] *a.* 间歇的，断续的（coming and going at intervals; not continuous）
micrometer [maiˈkrɔmitə] *n.* 千分尺（利用精密螺旋副原理测量长度的量具）
thread [θred] *n.* 螺纹（螺钉、螺母或螺栓上螺旋状的凸棱）
vibration [vaiˈbreiʃən] *n.* 振动（物体经过它的平衡位置所做的往复运动）
back-and-forth 往复的（from one place to another and back again repeatedly）
amplitude [ˈæmplitjuːd] *n.* 振幅（振动物体离开平衡位置的最大距离）
undamped [ˌʌnˈdæmpt] *a.* 无阻尼的，无衰减的（not tending toward a state of rest）
pendulum [ˈpendjuləm] *n.* 摆，单摆（simple pendulum）
simple harmonic motion 简谐运动（随时间按余弦或正弦函数变化规律进行的运动），简称 SHM
irregular motion 不均匀运动，不规则运动
return stroke 返回行程（工作结束后，刀具或工件返回原始位置的行程）
shaper [ˈʃeipə] *n.* 牛头刨床，也可写为 shaping machine
cutting [ˈkʌtiŋ] *n.* 切削（用工具将工件上多余材料切除的过程）
ram 滑枕（装有刀架或主轴等部件，具有导轨可在床身上或其他部件上移动的枕状部件）

obtainable [əbˈteinəbl] *a.* 能获得的，可得到的

link [liŋk] *n.* 构件（机构中的运动单元体），杆（机构中只具有低副元素的构件）

Notes

[1] work 这里指"功"，mechanics 这里指"机械学（the science of designing, constructing, and operating machines）"。这段可译为：研究表明，机器或机构都是由许多可以运动的零件组成的。这些零件将给定运动转换为需要的运动。换句话说，这些机器能做功。在力的作用下产生运动就实现了做功。因此，对机械学和机器的研究都涉及力和力对物体的影响。

[2] 这段可译为：第三种力被称为扭转力或扭矩，这是因为它会扭曲物体。此外还有一种力，它看起来会使材料的层与层之间或分子间产生相对滑移，被称为剪切力。

[3] 这段可译为：这些力可以单独作用或共同作用在物体上。例如，在一根水平放置的钢梁的中间位置有一个作用力 F（见图 2.1），梁的下部区域受到拉伸，承受拉力；同时，梁的上部区域受到挤压，承受压力。

[4] turning 这里指"车削加工"，与外圆车削"cylindrical turning"意思相同。全句可译为：几个力共同作用的另一个例子是在车床上对一个零件进行车削加工（见图 2.2）。

[5] chip 这里指"切屑"，work 这里指"工件，即 workpiece，加工过程中的生产对象"。全句可译为：这个力使得金属看起来像以切屑的形式从工件上流出来一样。

[6] centers of the lathe 意为"车床的顶尖"。全句可译为：如果一个工件被夹持在车床的两个顶尖之间，则顶尖对工件施加一个压力。

[7] 全段可译为：在设计离心机时要应用离心力原理。一些离心机被用来分离化学物质，还可以通过离心铸造工艺去除金属中的杂质。家用和商用洗衣机中的衣物甩干机是离心力最常见的应用之一。

[8] 这段可译为：向心力可使物体做圆周运动。这种作用是由一个持续作用的将物体拉向圆心的力所产生的。换句话说，这个向内的、抵抗离心力的力被称为向心力。以恒定速率做旋转运动的物体的向心力会产生一个向心加速度，向心力与离心力大小相等，方向相反。

[9] 这段可译为：用于制造高速运动的机械零件和机构的材料必须具有足够的结构强度，以提供零件做圆周运动时所需要的向心力。与此同时，这些材料必须能够承受试图将零件拉断的离心力的作用。

[10] rotary motion 意为"旋转运动"，drill, mill, turn 均为动词，分别指"钻削、铣削、车削"。全句可译为：钻孔、铣平面或在车床上车削零件等都需要旋转运动。

[11] feed 意为"进给"，shaping 意为"用牛头刨床（shaper）进行刨削加工"。全句可译为：车床中刀具的进给（见图 2.2）和采用牛头刨床进行刨削加工都属于直线运动做功。

[12] screw 这里指"螺杆，即利用本身的螺纹传递运动的杆状零件"，nut 这里指"螺母"。全句可译为：千分尺（见图 2.6）中的螺杆和螺母上的螺纹是把运动方向从旋转变为直线的另外一些应用实例。

[13] intermittent motion 意为"间歇运动"。这两句可译为：**简谐运动和间歇运动** 任何一种简单的振动，例如，摆的下端有规律的往复运动（见图 2.7）是简谐运动。

Lesson 3 Overview of Engineering Mechanics

As we look around us we see a world full of "things": Machines, devices, and tools; things that we have designed, built, and used; things made of wood, metals, ceramics, and plastics. We know from experience that some things are better than others; they last longer, cost less, look better, or are easier to use. [1]

Ideally, however, every such item has been designed according to some set of "functional requirements" as perceived by the designers—that is, it has been designed so as to answer the question, "Exactly what function should it perform?" From the beams in our homes to the wings of an airplane, there must be an appropriate melding of materials, dimensions, and fastenings to produce structures that will perform their functions reliably for a reasonable cost over a reasonable lifetime. [2]

Engineering mechanics describes the behavior of a body subjected to the action of forces. It is used in many fields of engineering, especially in mechanical engineering and civil engineering. In practice, the engineering mechanics methods are used in two quite different ways:

(1) The development of any new device requires an interactive, iterative consideration of form, size, materials, loads, durability, safety, and cost.

(2) When a device fails (unexpectedly) it is often necessary to carry out a study to pinpoint the cause of failure and to identify potential corrective measures. [3] Our best designs often evolve through a successive elimination of weak points.

To many engineers, both of the above processes can prove to be absolutely fascinating and enjoyable.

In any "real" problem there is never sufficient good, useful information; we seldom know the actual loads and operating conditions with any precision, and the analyses are seldom exact. [4] While our mathematics may be precise, the overall analysis is generally only approximate, and different skilled people can obtain different solutions. In the study of engineering mechanics, most of the problems will be sufficiently "idealized" to permit unique solutions, but it should be clear that the "real world" is far less idealized, and that you usually will have to perform some idealization in order to obtain a solution.

The technical areas we will consider are frequently called "statics", "dynamics", and "strength of materials". Statics is concerned with the equilibrium of a body that is either at rest or moves with constant velocity. In other words, the sum of the forces on the body must equal zero. Dynamics deals with the accelerated motion of a body. Strength of materials, also called mechanics of materials, deals with the effects of forces on the structure (stresses, strains, deformations, etc.).

In engineering mechanics, we will use various types of approximations and simplified models.[5] Primarily, we will consider many structural members to be "weightless"—but they never are. We will deal with forces that act at a "point", but all forces act over an area. We will consider some parts to be "rigid"—but all bodies will deform under load.

We will make many assumptions that clearly are false. But these assumptions should always render the problem easier, more tractable. You will discover that the goal is to make as many simplifying assumptions as possible without seriously degrading the result.

Generally, there is no clear method to determine how completely, or how precisely, to treat a problem: If our analysis is too simple, we may not get a pertinent answer[6]; if our analysis is too detailed, we may not be able to obtain any answer. It is usually preferable to start with a relatively simple analysis and then add more detail as required to obtain a practical solution.

The study of engineering mechanics is based on the understanding of a few basic concepts and on the use of simplified models. This approach makes it possible to develop all the necessary formulas in a rational and logical manner, and to clearly indicate the conditions under which they can be safely applied to the analysis and design of actual engineering structures and machine components.[7]

Figure 3.1 Flange couplings

During the past several decades, there has been a tremendous growth in the availability of computerized methods for solving problems that previously were beyond solution because the time required to solve them would have been prohibitive. At the same time, the cost of computers has decreased by orders of magnitude.[8] The computer programs not only remove the drudgery of computation, they allow fairly complicated problems to be solved with ease. Students gain a greater understanding of the subject by simply changing input values and seeing what happens. For example, suppose a flange coupling (see Fig.3.1) has four M14 bolts. What would happen to the maximum power rating if the design were changed to six M12 bolts? The answer can be seen immediately.

Words and Expressions

overview [ˈəuvəvjuː] *n.* 概述 (a general description or an outline of something), 综述

meld [meld] *v.* 混合, 合并 (to combine with something else; to merge)

dimension [diˈmenʃən] *n.* 尺寸 (用特定长度或角度单位表示的数值); *v.* 标注尺寸

fastening [ˈfɑːsniŋ] *n.* 连接, 紧固件 (something used to attach one thing to another firmly)

mechanical engineering 机械工程 (与机械和动力生产有关的一门工程学科)

civil engineering 土木工程 (与建造房屋、道路、桥梁等各类工程设施有关的一门学科)

iterative [ˈitərətiv] *a.* 反复的, 迭代的, 重复的

durability 耐久性 (ability to exist for a long time without significant deterioration in quality)

pinpoint ['pɪnpɔint] *v.* 准确定位（to locate or identify with precision），正确指出
evolve [i'vɔlv] *v.* 逐渐发展（to change or develop gradually），逐渐完成
strength of materials　材料力学，亦为 mechanics of materials
equilibrium [ˌi:kwi'libriəm] *n.* 力平衡（某一系统的合力与合力矩同时为零）
etc.　等等（前面要有逗号，用于无生命的场合；用于人的场合时为 et al.，意为"等人"）
primarily ['praiˈmərili] *ad.* 首先（at first），起初（originally），主要地（chiefly; mainly）
render ['rendə] *v.* 提供（to give or make available; to provide），使得（to cause to become）
tractable ['træktəbl] *a.* 易处理的，易管理的（easily managed, controlled or handled）
practical solution　实用的解决方案
prohibitive [prə'hibitiv] *a.* 不能被人们所接受的，禁止的，抑制的，昂贵的
drudgery ['drʌdʒəri] *n.* 苦差事，辛苦的工作
greater understanding of the subject　对这个问题（门学科）更深入地理解
flange coupling　凸缘联轴器（用螺栓连接两半联轴器的凸缘以实现两轴连接的联轴器）
maximum power rating　最大功率额定值
M12 bolt　M12 螺栓（M 意为 Metric thread，公制螺纹）

Notes

［1］last longer 意为"持续时间更长"。全句可译为：根据经验我们知道，某些物品要优于其他的物品；它们的使用寿命更长、成本更低、外表更美观或更易于使用。

［2］so as to 意为"使得……如此，以便"。这两句可译为：然而，在理想情况下，每种物品都是设计人员根据其对某些"功能要求"的理解而设计出来的，也就是说，在设计过程中，应该回答这样的问题，即"它应该具有哪种确切的功能？"从我们家中的房梁到飞机的机翼，都必须采用适当的材料、尺寸和连接方式，以形成一个可以用合理的成本制造出来的，在合理的使用寿命期间内能够可靠地实现其功能的结构。

［3］carry out 意为"进行，完成"，corrective measure 意为"改正措施"。全句可译为：当一个装置意外地失效后，通常需要进行研究，以准确地找出失效的原因和确定可能的改正措施。

［4］operating condition 意为"工作状态，工作条件"。全句可译为：对于任何实际的问题，总是缺乏足够完整和有用的信息。我们很少准确地知道实际载荷和工作状态，因此，所做的分析工作也很少是精确的。

［5］approximation 意为"近似方法"，simplified model 意为"简化模型"。全句可译为：在工程力学中，我们将使用各种类型的近似方法和简化模型。

［6］pertinent answer 意为"相应的答案"。全句可译为：如果我们的分析工作太简单，我们可能会得不到相应的答案；如果我们的分析工作考虑太多的细节，我们可能会得不到任何答案。

［7］rational and logical manner 意为"合理的和符合逻辑的方式"。这几句可译为：学习工程力学，要在理解基本概念和应用简化模型的基础上进行。采用这种方法，可以用一种合理的和符合逻辑的方式推导出所有必需的公式，并指明这些公式在什么条件下可以安全地用来分析和设计工程结构和机器零件。

［8］order of magnitude 意为"数量级"。

Lesson 4 Shafts, Couplings, and Splines

Virtually all machines contain shafts. The most common shape for shafts is circular and the cross section can be either solid or hollow (hollow shafts can result in weight savings). Rectangular shafts are sometimes used, as in slotted screwdriver blades (see Fig. 4.1).

A shaft must have adequate torsional strength to transmit torque and not be over stressed. It must also be torsionally stiff enough so that one mounted component does not deviate excessively from its original angular position relative to a second component mounted on the same shaft. Generally speaking, the angle of twist should not exceed one degree in a shaft length equal to 20 diameters.

Figure 4.1 A slotted screwdriver

Shafts are mounted in bearings and transmit power through such devices as gears, pulleys, cams and clutches. These devices introduce forces which attempt to bend the shaft; hence, the shaft must be rigid enough to prevent overloading of the supporting bearings. In general, the bending deflection of a shaft should not exceed 0.5 mm per meter of length between bearing supports.

In addition, the shaft must be able to sustain a combination of bending and torsional loads. Thus, an equivalent load must be considered which takes into account both torsion and bending (see Figs. 4.2 and 4.3). Also, the allowable stress must contain a factor of safety which includes fatigue, since torsional and bending stress reversals occur.

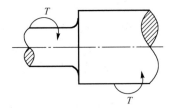   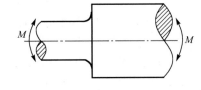

Figure 4.2 Torsion of a stepped shaft   Figure 4.3 Bending of a stepped shaft

For diameters less than 75 mm, the usual shaft material is cold-rolled steel containing about 0.4 percent carbon. Shafts are either cold-rolled or forged in sizes from 75 to 125 mm. For sizes above 125 mm, shafts are forged and machined[1] to the required size. Plastic shafts are widely used for light load applications. One advantage of using plastic is safety in electrical applications, since plastic is a poor conductor of electricity.

Components such as gears and pulleys are mounted on shafts by means of key. The design of the key and the corresponding keyway (see Fig. 4.4) in the shaft must be properly evaluated. For example, stress concentrations occur in shafts due to keyways, and the material removed to form the keyway further weakens the shaft.

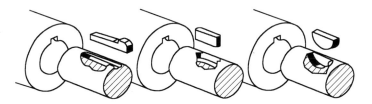

Figure 4.4 Several key and corresponding keyway shapes

If shafts rotate at critical speeds, severe vibrations can occur which can seriously damage a machine. It is important to know the magnitude of these critical speeds so that they can be avoided. Generally, the difference between the operating speed and the critical speed should be at least 20 percent.

Some shafts are supported by three or more bearings, which means that the problem is statically indeterminate. Textbooks on strength of materials give methods of solving such problems. The design effort should be in keeping with the economics of a given situation. For example, if a long shaft supported by three or more bearings is needed, it probably would be cheaper to make conservative assumptions as to moments and design it as though it were statically determinate. The extra cost of an oversize shaft may be less than the extra cost of an elaborate design analysis.

Another important aspect of shaft design is the method of directly connecting one shaft to another. This is accomplished by devices such as rigid or flexible couplings.

A coupling is a device for connecting the ends of adjacent shafts. In machine construction, couplings are used to effect a semipermanent connection between adjacent rotating shafts. The connection is permanent in the sense that it is not meant to be broken during the useful life of the machine, but it can be broken and restored in an emergency or when worn parts are replaced.

A rigid coupling locks two shafts together tightly so that no relative motion can occur between them (see Figs. 3.1 and 4.5). Rigid couplings are desirable for certain kinds of equipment in which precise alignment of two shafts is required and can be provided. If an exceptionally long shaft is required in a manufacturing plant or a propeller shaft on a ship, it is made in sections that are coupled together with rigid couplings.

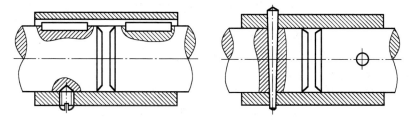

Figure 4.5 Sleeve couplings

In connecting shafts belonging to separate devices (such as an electric motor and a gearbox), precise aligning of the shafts is difficult and a flexible coupling (see Figs. 4.6 and 4.7) is used. This coupling connects the shafts in such a way as to minimize the harmful effects of shaft misalignment.[2] Flexible couplings also permit the shafts to deflect under their separate systems of loads and to move freely (float) in the axial direction without interfering with one another. Flexible couplings can also serve to reduce the intensity of shock loads and vibrations transmitted from one shaft to another.

Figure 4.6　A flexible pin bush coupling　　Figure 4.7　A serpentine spring coupling

When axial movement between the shaft and hub is required, relative rotation is prevented by means of splines machined on the shaft and into the hub (see Fig. 4.8). There are two forms of splines: straight-sided splines and involute splines. The former is relatively simple and employed in some machine tools. The latter has an involute curve in its outline, which is in widespread use on gears. The involute form is preferred because it provides for self-centering of the mating element and because it can be machined with standard hobs (see Fig. 4.9) used to cut gear teeth.

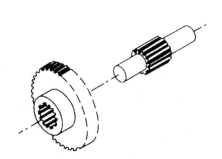 

Figure 4.8　General form of spline connection　　Figure 4.9　Hobs

Words and Expressions

coupling ['kʌpliŋ]　n. 联轴器（连接两轴或轴与回转件，在传递运动和动力过程中一同回转，在正常情况下不脱开的一种装置）

rectangular [rek'tæŋɡjulə]　a. 矩形的

cross section　横截面（a section formed by a plane cutting at right angles to an axis）

slotted screwdriver　一字螺丝刀，平头螺丝刀（flat head screwdriver）

torsional ['tɔːʃənl] strength　抗扭强度

fatigue [fə'tiːɡ]　n. 疲劳（物件在低于其断裂应力的循环加载下，通过一定的循环次数后发生损伤和断裂的现象）

mounted ['mauntid]　a. 安装好的，固定好的，安装在……上的

deviate [ˈdiːvieit] v. 偏离，背离，偏差，偏移
angle of twist 扭转角（两横截面间的相对角位移）
clutch [klʌtʃ] n. 离合器
bending [ˈbendiŋ] n. 弯曲（movement that causes the formation of a curve）
stepped shaft 阶梯轴
deflection [diˈflekʃən] n. 挠度（弯曲变形时，横截面中心在垂直于变形前轴线方向的位移）
reversal [riˈvəːsəl] n. 颠倒，改变方向（a change to an opposite or former state or direction）
cold-roll [kəuldˈrəul] v. 冷轧（to roll metal without applying heat）
forge [fɔːdʒ] n.; v. 锻造，打制，锻工车间
key [kiː] n. 键（置于轴和轴上零件的槽中，使二者周向固定以传递转矩的连接件）
keyway [ˈkiːwei] n. 键槽，也写为 key seat（轴和轮毂孔表面上为安装键而制成的槽）
stress concentration 应力集中（受力零件在形状、尺寸急剧变化时局部应力增大的现象）
critical speed 临界转速（系统产生共振的特征转速），又称共振转速（resonant speed）
statically indeterminate [ˌindiˈtəːminit] 超静定的，静不定的
oversize [ˈəuvəˈsaiz] a. 过大的，加大的（larger than the normal size; too big）
conservative [kənˈsəːvətiv] a. 保守的（likely to be less than the real amount or number）
rigid coupling 刚性联轴器（由刚性传力件组成，不能补偿两轴相对位移的联轴器）
semipermanent [ˌsemiˈpəːmənənt] a. 半永久性的
worn [wɔːn] a. 用旧的，磨损后的；wear 的过去分词（past participle of wear）
alignment [əˈlainmənt] n. 对中（arrangement or position in a straight line）
sleeve coupling 套筒联轴器（利用公用套筒以某种方式连接两轴的联轴器）
propeller [prəˈpelə] n. 螺旋桨（a device with two or more blades that turn quickly and cause a ship or aircraft to move）
gearbox [ˈgiəbɔks] n. 齿轮箱，变速箱（装有变速机构的箱形部件）
flexible coupling 挠性联轴器（能补偿两轴相对位移的联轴器）
misalignment [ˈmisəlainmənt] n. 偏移（the condition of being out of correct position）
flexible pin bush coupling 弹性套柱销联轴器
serpentine [ˈsəːpəntain] spring coupling 蛇形弹簧联轴器
shock [ʃɔk] n. 冲击，冲撞，打击
spline [splain] n. 花键（轴和轮毂上有多个凸起和凹槽构成的周向连接件）
straight-sided spline 矩形花键（分为直齿矩形花键和斜齿或螺旋齿矩形花键）
machine tool 机床（制造机器的机器，亦称工作母机）
involute [ˈinvəluːt] n. 渐开线
involute spline 渐开线花键（齿形是渐开线的花键）
self-centering 自动定心（centering in one's self）
hob [hɔb] n. 滚刀（一种蜗杆状的切齿刀具）
general form 常见形式，通用形式

Notes

[1] machined 这里意为"经过机械加工的"。
[2] shaft misalignment 这里指"被连接两轴轴线的相对偏移，两轴的相对位移"。

Lesson 5 Shafts and Associated Parts

The term shaft usually refers to a relatively long member of round cross section that rotates and transmits power. The diameter of shaft is often varied from point to point (see Fig. 5.1). One or more members such as gears, sprockets (Fig. 5.2), pulleys, and cams are usually attached to the shaft by means of pins, keys, splines, snap rings, and other devices.[1] These members are among the "associated parts" considered in this book, as are couplings and universal joints, which serve to connect the shaft to the source of power or load.[2]

Figure 5.1 A stepped shaft

Figure 5.2 Sprockets

A shaft can have a non-round cross section, and it need not necessarily rotate. It can be stationary and serve to support a rotating member, such as the short shafts that support the nondriving wheels of an automobile.

It is apparent that shafts can be subjected to various combinations of axial, bending, and torsional loads, and that these loads may be static or alternating.[3] Typically, a rotating shaft transmitting power is subjected to a constant torque (producing a mean torsional stress) together with a completely reversed bending load (producing an alternating bending stress).[4]

In addition to satisfying strength requirements, shafts must be designed so that deflections are within acceptable limits. Excessive lateral shaft deflection can hamper gear performance and cause objectionable noise. The associated angular deflection can be very destructive to non-self-aligning bearings. Self-aligning bearings (Fig. 5.3) may eliminate this trouble if the deflection is otherwise acceptable.

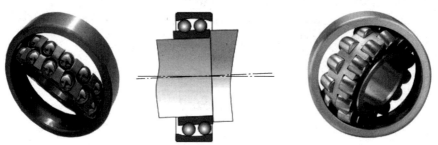

(a) Self-aligning ball bearing (b) Self-aligning roller bearing

Figure 5.3 Self-aligning bearings

Torsional deflection can affect the accuracy of a cam or gear driven mechanism. Furthermore, the greater the flexibility—either lateral or torsional—the lower the corresponding critical speed.[5]

Sometimes members like gears and cams are made integral with the shaft (see Fig. 31.2), but more often such members (which also include pulleys, sprockets, etc.) are made separately and then mounted onto the shaft. The portion of the mounted member in contact with the shaft is the hub. Attachment of the hub to the shaft is made in a variety of ways. The grooves in the shaft and hub into which the key fits are called keyways (see Fig. 4.4).[6]

A simpler attachment for transmitting relatively light loads is provided by pins (see Figs. 4.5 and 24.7). Pins provide a relatively inexpensive means of transmitting both axial and circumferential loads.

An excellent and inexpensive method of axially positioning and retaining hubs and bearings onto shafts is by means of retaining rings (see Fig. 5.4), commonly called snap rings. Shaft snap rings require grooves that weaken the shaft, but this is no disadvantage if they are located where stresses are low.

Figure 5.4   Retaining rings

Perhaps the simplest of all hub-to-shaft attachments is by means of an interference fit[7] (hub bore is slightly smaller than shaft diameter, with assembly being made with a press, or by thermally expanding the hub—sometimes also contracting the shaft, as with dry ice—prior to making a quick assembly before the temperatures equalize). The strongest provision for torque transmission is usually by means of mating splines cut in the shaft and hub (see Fig. 4.8).[8]

Rotating shafts, particularly those that run at high speeds, must be designed to avoid operation at critical speeds. This usually means providing sufficient lateral rigidity to place the lowest critical speed significantly above the operating range.

A few general principles that should be kept in mind are:

1. Keep shafts as short as possible, with bearings close to the applied loads. This reduces deflections and bending moments, and increase critical speeds.

2. Place necessary stress raisers away from highly stressed shaft regions if possible. Consider local surface strengthening processes (as shot peening or rolling).[9]

3. Use inexpensive steels for deflection-critical shafts, as all steels have essentially the same elastic modulus.[10]

4. When weight is critical, consider hollow shafts (see Fig. 5.5).

Figure 5.5   Hollow shafts

17

Allowable shaft deflections for satisfactory gear and bearing performance vary with the gear or bearing application, but the following can be used as a general guide:

1. Deflections should not cause mating gear teeth to separate more than about 0.13 mm, nor should they cause the relative slope of the gear axes to change more than about 0.03 deg.

2. The shaft (journal) deflection across a plain bearing (see Fig. 5.6) must be small compared to the oil film thickness.[11]

3. The shaft angular deflection at a ball or roller bearing should generally not exceed 0.04 deg, unless the bearing is self-aligning.[12]

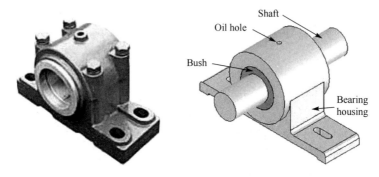

Figure 5.6 Plain bearings

Words and Expressions

associated parts   相关零件
sprocket ['sprɔkit]   n. 链轮（与链条相啮合的带齿的轮形机械零件），亦为 chain wheel
pin [pin]   n. 销（贯穿于两个零件孔中，用于定位、连接或作为安全装置中过载易剪断元件）
snap ring   弹性挡圈（用弹簧钢制成的开口挡圈，起轴向限位作用）
universal joint   万向联轴器，万向节（两轴线呈夹角变化时仍能传递转矩的关节式机械装置）
nondriving wheel   从动轮，非驱动轮
hamper ['hæmpə]   v. 妨碍，阻碍
objectionable [əb'dʒekʃənəbl]   a. 令人不快的，令人讨厌的（unpleasant or offensive）
angular deflection   角偏移，角度偏转
destructive [dis'trʌktiv]   a. 破坏性的
self-aligning bearing   调心轴承（一滚道是球面形的，能对两滚道轴心线间的角偏差及角运动做适应性自调整的滚动轴承）
plain or rolling bearing   滑动或滚动轴承
attachment [ə'tætʃmənt]   n. 连接（a connection by which one thing is joined to another）
self-aligning ball bearing   调心球轴承
self-aligning roller bearing   调心滚子轴承（滚动体是球面滚子的向心轴承）
otherwise acceptable   除此之外是可以接受的
axial and circumferential [sə‚kʌmfə'renʃəl] loads   轴向和周向载荷
retaining ring   挡圈（紧固在轴上的圈形机件，可以防止装在轴上的其他零件窜动）
hub bore   毂孔

press 压力机（能使滑块做往复运动，并按所需方向给模具施加一定压力的机器）
contract ['kɔntrækt] v. 收缩，缩小（to reduce in size）；n. 合同（有约束力的协议）
dry ice 干冰，固态二氧化碳（通常呈块状，在-78.5℃下吸热升华成气态，主要用作冷却剂）
equalize ['i:kwəlaiz] v. 使相等（to make equal），使一致（to make uniform）
bending moment 弯矩（与横截面垂直的分布内力系的合力偶矩）
mating gear 配对齿轮（齿轮副中的任意一个齿轮，均可称为另一齿轮的配对齿轮）
slope [sləup] n. 倾斜，斜率
bush [buʃ] n. （滑动轴承中的）轴套（圆筒形的轴瓦），整体式轴瓦，亦为 bearing bush
bearing housing 轴承座（安装轴承的部件），亦为 bearing block
oil hole 油孔（轴或轴承上的润滑油进出孔）

Notes

[1] attach to 意为"连接到……上，安装在……上"，snap ring 意为"卡环，弹性挡圈"。全句可译为：一个或多个诸如齿轮、链轮（见图 5.2）、带轮和凸轮等类的构件通常通过销、键、花键、弹性挡圈或其他装置与轴相连接。

[2] 全句可译为：这些零件在本书中被称为"相关零件"，此外还有联轴器和万向节，它们被用来实现轴与动力源或与载荷之间的连接。

[3] it is apparent that 意为"很显然，很明显"。全句可译为：很明显，轴可以承受轴向、弯曲、扭转等载荷的各种组合的作用。而且，这些载荷可能是静态的，也可能是交变的。

[4] mean 意为"平均的"，reversed 意为"相反的"，bending load 意为"弯曲载荷"。全句可译为：通常，一个传递动力的转轴要承受一个恒定的扭矩（产生一个平均的扭转应力）和一个方向会完全改变的弯曲载荷（产生一个交变的弯曲应力）。

[5] torsional deflection 意为"扭转变形"。这段可译为：扭转变形会影响凸轮或齿轮传动机构的精度。此外，柔性（横向的或扭转的）越大，则相应的临界转速就越低。

[6] 全句可译为：在轴和轮毂中与键配合的凹槽称为键槽（见图 4.4）。

[7] interference fit 意为"过盈配合（具有过盈的配合，包括最小过盈等于零的配合）"。全句可译为：可能最简单的轴毂连接是采用过盈配合实现的。

[8] mating splines 意为"花键副（相互连接的一对花键）"。全句可译为：最好的传递扭矩的方式是用在轴和轮毂上加工出来的花键副（见图 4.8）实现的。

[9] stress riser 意为"应力集中源"，surface strengthening process 意为"表面强化工艺"，shot peening 意为"喷丸强化"。这两句可译为：如有可能，应该使必要的应力集中源远离轴上承受较高应力的区域。还应该考虑采用局部表面强化工艺（如喷丸强化或滚压）。

[10] deflection-critical shaft 意为"挠度为关键影响因素的轴"。全句可译为：对于挠度是关键影响因素的轴，可以采用廉价钢材制造，因为各种钢材的弹性模量相差很小。

[11] journal 意为"轴颈"，deflection 这里指"偏转量，偏斜量"，plain bearing 意为"滑动轴承"。全句可译为：在滑动轴承中，轴（轴颈）的偏转量必须小于油膜厚度。

[12] ball or roller bearing 意为"球轴承或滚子轴承"。全句可译为：如果没有采用调心轴承，那么轴在球轴承或滚子轴承中的角度偏转通常不能超过 0.04°。

Lesson 6 Belts, Clutches, Brakes, and Chains

Belts are used to transmit power from one shaft to another smoothly, quietly, and inexpensively. Belts are frequently necessary to reduce the higher rotative speeds of electric motors to the lower values required by mechanical equipment. Clutches are required when two shafts having a common axis of rotation must be frequently connected and disconnected. [1] The function of the brake is to turn mechanical energy into heat. Chains provide a convenient and effective means for transferring power between parallel shafts.

There are four main belt types: flat, round, V, and synchronous. A widely used type of belt, particularly in industrial drives and vehicular applications, is the V-belt. The V shape causes the belt to wedge tightly into the groove of the sheave, increasing friction and allowing high torques to be transmitted before slipping occurs. [2] Since the cost of V-belts is relatively low, the power output of a V-belt system may be increased by operating several belts side by side (see Fig. 6.1). Another type of V-belt drive is the poly V-belt drive (see Fig. 6.2). V-belt drive helps protect the machinery from overload, and it damps and isolates vibration.

Figure 6.1 V-belts and V-belt drive   Figure 6.2 Poly V-belt drive

Other belts, such as flat belts, are used for long center distances and high speed applications. Synchronous belts (also known as timing belts) have evenly spaced teeth on the inside circumference (see Fig. 6.3). Synchronous belts do not slip and hence transmits motion and power at a constant velocity ratio. [3]

A clutch is a device for quickly and easily connecting or disconnecting a rotatable shaft and a rotating coaxial shaft (see Fig. 6.4). [4] Clutches are usually placed between the input shaft to a machine and the output shaft from the electric motor, and provide a convenient means for starting and stopping the machine and permitting the motor or engine to be started in an unloaded state.

Figure 6.3 Synchronous belt drive   Figure 6.4 A friction clutch

The rotor (rotating member) in an electric motor has rotational inertia, and a torque is required to bring it up to speed when the motor is started. If the motor shaft is rigidly connected to a load with a large rotational inertia, and the motor is started suddenly by closing a switch, the motor may not have sufficient torque capacity to bring the motor shaft up to speed before the windings in the motor are burned out by the excessive current demands. [5] A clutch between the motor and the load shafts will restrict the starting torque on the motor to that required to accelerate the rotor and parts of the clutch only. On some machine tools it is convenient to let the electric motor run continuously and to start and stop the machine by operating a clutch. Clutches are also used when the transmission of power must be controlled in amount, e.g. electric screwdrivers limit how much torque is transmitted through use of a clutch. [6]

A brake is similar to a clutch except that one of the shafts is replaced by a fixed member. A brake is a mechanical device which inhibits motion (see Fig. 6.5). The basic function of a brake is to absorb energy (i. e., to convert kinetic and potential energy into friction heat)[7] and to dissipate the resulting heat without developing destructively high temperatures. Clutches also absorb energy and dissipate heat, but usually at a lower rate. Where brakes (or clutches) are used more or less continuously for extended periods of time, provision must be made for rapid transfer of heat to the surrounding atmosphere. For intermittent operation, the thermal capacity of the parts may permit much of the heat to be stored, and then dissipated over a longer period of time. Brake and clutch parts must be designed to avoid objectionable thermal stresses and thermal distortion. The rate at which heat is generated on a unit area of friction interface is equal to the product of the normal pressure, coefficient of friction, and rubbing velocity. [8]

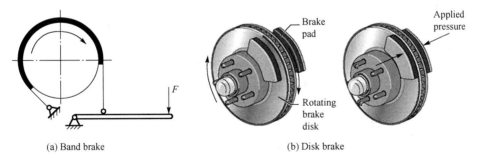

(a) Band brake (b) Disk brake

Figure 6.5 Brakes

Chain drives combine some of the more advantageous features of belt and gear drives. Chain drives provide almost any speed ratio. Their chief advantage over gear drives is that chain drives can be used with arbitrary center distances. Compared with belt drives, chain drives offer the advantage of positive (no slip) drive and therefore greater power capacity. [9] An additional advantage is that not only two but also many shafts can be driven by a single chain at different speeds, yet all have synchronized motions. Primary applications are in conveyor systems, farm machinery, textile machinery, and motorcycles.

In chain drive applications, toothed wheels called sprockets mate with a chain (see

Fig. 6.6) to transmit power from one shaft to another.[10] The most common type of chain is the roller chain (see Fig. 6.7), in which the roller on each bushing provides exceptionally low friction between the chain and the sprockets. Of its diverse applications, the most familiar one is the roller chain drive on a bicycle. A roller chain is generally made of hardened steel. Sprockets are generally made of steel or cast iron, but some applications use aluminum alloy or plastic. Stainless steel and bronze chains are obtainable where corrosion resistance is needed.[11]

Figure 6.6 Chain drives　　　　　　　　Figure 6.7 Portion of a roller chain

Words and Expressions

belt drive 带传动（由柔性带和带轮组成传递运动和动力的机械传动，分为摩擦传动和啮合传动）
clutch [klʌtʃ] n. 离合器（主、从动部分在同轴线上传递动力或运动时，具有接合或分离功能的装置）
brake [breik] n. 制动器（具有使运动部件减速、停止或保持停止状态等功能的装置）
chain [tʃein] n. 链条（由相同或间隔相同的构件以运动副形式串接起来的组合件）
chain drive 链传动（利用链与链轮轮齿的啮合来传递动力和运动的机械传动）
flat belt 平带（横截面为矩形或近似为矩形的传动带，其工作面为宽平面）
round belt 圆带（横截面为圆形或近似为圆形的传动带）
V-belt V带（横截面为等腰梯形或近似为等腰梯形的传动带，其工作面为两侧面）
synchronous ['siŋkrənəs] a. 同步的（happening, moving, or existing at the same time）
synchronous belt 同步带（横截面为矩形或近似为矩形，带面具有等距横向齿的环形传动带）
wedge [wedʒ] n. 楔；v. 楔入（to force something into a very small or narrow space）
sheave [ʃiːv] n. 有沟槽的带轮（a grooved pulley, such as a V-belt pulley）
V-belt drive V带传动（由一条或数条V带和V带轮组成的摩擦传动）
poly V-belt 多楔带（以平带为基体，内表面具有等距纵向楔的环形传动带）
damp [dæmp] v. 阻尼，使衰减，抑止
center distance 中心距（当带处于规定的张紧力时，两带轮轴线间的距离）
friction clutch 摩擦式离合器（靠主、从动件间的摩擦力传递动力的离合器）
starting torque 启动转矩（电动机在静止状态下施加额定电压的瞬间所产生的转矩）
screwdriver ['skruːdraivə] n. 螺丝刀（拧紧或旋松头部带一字或十字槽的螺钉的工具）
dissipate ['disipeit] v. 消散，耗散（to cause something to spread out and disappear）
thermal ['θəːməl] a. 热的，热量的
band brake 带式制动器（用制动带的内侧面作为摩擦副接触面的制动器）

disk brake 盘式制动器（用圆盘的端面作为摩擦副接触面的制动器）
brake pad 刹车片，制动衬块（the part of a brake that presses on the wheel when stopping or slowing down a vehicle）
brake disk 制动盘（以端平面为摩擦工作面的圆盘形运动部件）
normal ['nɔːməl] *n.* 正规，常态，法线（垂直于界面的直线）；*a.* 垂直的，法向的
speed ratio 速比，传动比（在机械传动系统中，其始端主动轮与末端从动轮的角速度或转速的比值），亦为 transmission ratio
conveyor [kən'veiə] *n.* 输送机械（可连续或间断地沿给定线路输送物料或物品的机械设备）
roller chain 滚子链（组成零件中具有回转滚子，且滚子表面在啮合时直接与链轮齿接触的链条）
bushing ['buʃiŋ] *n.* 套筒（与销轴构成铰链副的筒形元件），也可写为 bush
hardened steel 淬硬钢，淬火钢

Notes

［1］axis of rotation 意为"回转轴线，an imaginary straight line around which the object can rotate"。全句可译为：离合器可以实现同一回转轴线上两根轴之间经常性的接合与分离。

［2］slip 这里指"带在带轮上打滑"。全句可译为：V 带的形状可使其紧紧楔入 V 带轮的轮槽内，增大摩擦力，能够在出现打滑现象前传递较大的扭矩。

［3］timing belt 意为"同步齿形带"，velocity ratio 意为"速度比，传动比"。这两句可译为：同步带（也称同步齿形带）的内周表面设有等间距齿（见图 6.3）。同步带不会打滑，因此能够以恒定的传动比传递运动和动力。

［4］rotatable shaft 意为"可以转动的轴，从动轴"，rotating shaft 意为"正在转动的轴，主动轴"。全句可译为：离合器是一个用来使从动轴与位于同一轴线上的主动轴进行快速和顺利的接合或分离的装置（见图 6.4）。

［5］rotational inertia 意为"转动惯量"，也可写为 moment of inertia，winding 意为"绕组"。全句可译为：如果电动机的轴与具有很大转动惯量的负载刚性地连接在一起，则当合上开关使电动机突然启动时，有可能在电动机还没有来得及产生足够的转矩，使电动机的轴在达到应有的转速之前，电动机内的绕组就会因为需要过大的电流而被烧坏。

［6］electric screwdriver 意为"电动螺丝刀（一种装有调节和限制扭矩的机构，用于拧紧和旋松螺钉用的电动工具）"。全句可译为：离合器也用来对所传递的功率的大小进行控制，例如，在电动螺丝刀中采用离合器对扭矩进行控制。

［7］kinetic and potential energy 意为"动能与势能"。全句可译为：将动能与势能转化为摩擦热。

［8］coefficient of friction 意为"摩擦系数"，product 意为"乘积"。全句可译为：在摩擦面之间单位面积上产生热量的速率等于正压力、摩擦系数和摩擦速度三者的乘积。

［9］positive (no slip) drive 意为"强制（无滑动）传动"。全句可译为：与带传动相比，链传动的优点是强制（无滑动）传动，因此具有更强的传递动力的能力。

［10］mate with 意为"与……啮合"。全句可译为：在采用链传动时，被称为链轮的带齿的轮形零件与链条啮合（见图 6.6），将动力从一根轴传送到另一根轴。

［11］obtainable 意为"可以得到的"，corrosion resistance 意为"耐蚀性"。全句可译为：在需要具有耐蚀性的场合，可以使用不锈钢和青铜链条。

Lesson 7 Fasteners and Springs

Fasteners are devices which permit one part to be joined to a second part and, hence, they are involved in almost all designs. The acceptability of any product depends not only on the selected components, but also on the means by which they are fastened together. The principal purposes of fasteners are to provide the following design features:

(1) Disassembly for inspection and repair;

(2) Modular design where a product consists of a number of subassemblies. Modular design aids manufacturing as well as transportation.

There are three main classifications of fasteners, which are described as follows:

(1) Removable. This type permits the parts to be readily disconnected without damaging the fastener. An example is the ordinary nut-and-bolt fastener (see Fig. 7.1).

(2) Semipermanent. For this type, the parts can be disconnected, but some damage usually occurs to the fastener. One such example is a cotter pin (Fig. 7.2).

(3) Permanent. When this type of fastener is used, it is intended that the parts will never be disassembled. Examples are riveted and welded joints.

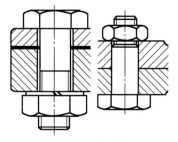

Figure 7.1 Threaded fasteners

Figure 7.2 Cotter pins

The following factors should be taken into account when selecting fasteners for a given application:

(1) Primary function;

(2) Appearance;

(3) A large number of small size fasteners versus a small number of large size fasteners (an example is bolts);

(4) Operating conditions such as vibration, loads and temperature;

(5) Frequency of disassembly;

(6) Adjustability in the location of parts;

(7) Types of materials to be joined;

(8) Consequences of failure or loosening of the fastener.

The importance of fasteners can be realized when referring to any complex product. In the case of the automobile, there are thousands of parts which are fastened together to produce the total product. The failure or loosening of a single fastener could result in a simple nuisance such as a door rattle or in a serious situation such as a wheel coming off. Such possibilities must be considered in the selection of the type of fastener for the specific application.

Springs are mechanical members which are designed to give a relatively large amount of elastic deflection under the action of an externally applied load. Hooke's Law, which states that deflection[1] is proportional to load, is the basis of behavior of springs. However, some springs are designed to produce a nonlinear relationship between load and deflection. The following is a list of the important purposes and applications of springs:

(1) Control of motion in mechanisms. This category represents the majority of spring applications such as operating forces in clutches and brakes. Also, springs are used to maintain contact between two members such as a cam and its follower.

(2) Reduction of transmitted forces as a result of impact or shock loading. Applications here include automotive suspension system springs and rubber springs.

(3) Storage of energy. Applications in this category are found in clocks, watches, and lawn mowers.

(4) Measurement of force. Scales used to weigh people is a very common application for this category.

The three major classifications of springs are compression, extension, and torsion (see Fig. 7.3). Compression and extension springs are the springs most often used. Deflection is linear in these types. Torsion springs are characterized by angular instead of linear deflection. Leaf springs are of the simple beam or cantilever type. Rubber springs and cushioning devices are finding an increasing range of application in industry. Rubber does not follow Hooke's law, but becomes increasingly stiff as the deformation is increased.

(a) Helical compression spring   (b) Helical extension spring   (c) Helical torsion spring

Figure 7.3　Three types of helical springs

Most springs are made of steel, although silicon bronze, brass and beryllium copper are also used. Springs are universally made by companies which specialize in the manufacture of springs. The cylindrical helical spring is the most popular type of spring; torsion bar springs and leaf springs are also widely used. If the wire diameter (assuming a cylindrical helical spring) is less than 8 mm, the spring will normally be cold wound from hard-drawn or oil-tempered wire[2]. For larger diameters, springs are usually hot wound. After forming, the spring is heat treated by quenching and tempering to produce the desired physical properties.

It is good practice to consult with a spring manufacturing company when selecting a spring, especially if high loads or temperatures are to be encountered, or if stress reversals occur or corrosion resistance is required. To properly select a spring, a complete study of the spring requirements, including space limitations, must be undertaken. Many different types of special springs are available to satisfy unusual requirements or applications.

Words and Expressions

fastener ['fɑːsnə]   *n.*   紧固件（用于连接和紧固零部件的元件），连接件
spring [sprɪŋ]   *n.*   弹簧（使变形与载荷之间保持规定关系的一种弹性元件）
acceptability [əkˌseptə'biliti]   *n.*   可接受性
disassembly [ˌdisə'sembli]   *n.*   拆卸（将产品零部件逐个分离的过程），分解
modular ['mɔdjulə]   *a.*   模数的，制成标准组件的，预制的，组合的
subassembly ['sʌbə'sembli]   *n.*   部件（机械的一部分，由若干装配在一起的零件所组成）
nut [nʌt]   *n.*   螺母（具有内螺纹并与螺栓配合使用的紧固件）
threaded fastener   螺纹紧固件（a fastener, a portion of which has some form of screw thread）
versus ['vəːsəs]   *prep.*   ……对……，与……比较，……与……的关系曲线，作为……的函数
cotter ['kɔtə] pin   开口销，也可写为 split pin
riveted joint   铆接，铆钉连接（借助铆钉形成的不可拆卸连接），亦为 riveting
welded joint   焊接接头（两个或两个以上零件用焊接方法连接的接头）
nuisance ['njuːsəns]   *n.*   讨厌的人或事（an annoying or troublesome person, thing, or situation）
rattle ['rætl]   *v.*   发出嘎嘎声；*n.* 嘎嘎声（a series of short, loud sounds）
suspension [səs'penʃən]   *n.*   悬架（汽车车架或车身与车桥或车轮之间弹性连接的装置）
scale [skeil]   *n.*   秤（an instrument for weighing people or things），天平
lawn mower ['məuə]   割草机（一种用于修剪草坪、植被等的机械工具）
torsion ['tɔːʃən] spring   扭转弹簧
leaf spring   板弹簧，由单片或多片板材（簧板）制成的弹簧
cantilever ['kæntiliːvə]   *n.*   悬臂，悬臂梁（a beam that is supported at only one end）
rubber spring   橡胶弹簧（利用橡胶弹性起缓冲、减震作用的弹簧）
cushion ['kuʃən]   *v.*   缓冲，减震（to lessen the shock of; to protect from impacts）
helical spring   螺旋弹簧（呈螺旋状的弹簧）
cylindrical helical spring   圆柱螺旋弹簧（外廓呈圆柱形的螺旋弹簧）
beryllium [bə'riljəm] copper   铍铜，亦为 beryllium bronze（铍青铜）
torsion bar spring   扭杆弹簧（一端固定而另一端与工作部件连接的杆形弹簧，主要作用是靠扭转弹力来吸收振动能量）
quenching and tempering   淬火和回火，调质（淬火及高温回火的复合热处理工艺）
temper ['tempə]   *v.*；*n.*   回火（一种热处理方式）
corrosion [kə'rəuʒən]   *n.*   腐蚀（the process of destroying sth by chemical action slowly）

Notes

[1] deflection 这里指"变形量（弹簧沿载荷作用方向产生的相对位移）"。
[2] oil-tempered wire 意为"油淬火-回火钢丝"，也可写为 oil-hardened and tempered wire。

Lesson 8 Rolling Bearings

Rolling bearings can carry radial, thrust or combination of the two loads. Accordingly, most rolling bearings are categorized in one of the three groups: radial bearings for carrying loads that are primarily radial, thrust bearings for supporting loads that are primarily axial, and angular contact bearings or tapered roller bearings for carrying combined radial and axial loads. The three basic types of rolling bearings are ball bearings, roller bearings, and needle roller bearings. Figure 8.1a shows a common single-row, deep groove ball bearing. The bearing consists of an inner ring, an outer ring, the balls and the separator. To increase the contact area and hence permit larger loads to be carried, the balls run in curvilinear grooves in the rings called raceway. This type of bearing can stand a radial load as well as some axial load. Some other types of rolling bearings are shown in Figs. 8.1b-d, 9.1, and 10.1.

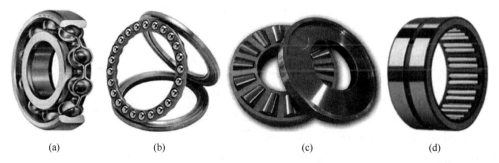

(a)　　　　　　(b)　　　　　　(c)　　　　　　(d)

Figure 8.1 Some types of rolling bearings: (a) deep groove ball bearing; (b) thrust ball bearing; (c) tapered roller thrust bearing; (d) needle roller bearing

The concern of a machine designer with ball and roller bearings is fivefold as follows: (a) life in relation to load; (b) stiffness, i.e. deflections under load; (c) friction; (d) wear; (e) noise. For moderate loads and speeds the correct selection of a standard bearing on the basis of load rating will usually secure satisfactory performance. The deflection of the bearing elements will become important where loads are high, although this is usually of less magnitude than that of the shafts or other components associated with the bearing. Where speeds are high special cooling arrangements become necessary which may increase frictional drag. Wear is primarily associated with the introduction of contaminants, and sealing arrangements must be chosen with regard to the hostility of the environment.

Because the high quality and low price of ball and roller bearings depends on quantity production, the task of the machine designer becomes one of selection rather than design. Rolling bearings are generally made with steel which is through-hardened[1] to about 900 HV. Owing to the cyclic stresses developed at bearing contact surface during operation, a predominant form of failure should be metal fatigue, and a good deal of work is currently in progress intended to improve the reliability of this type of bearing. Design can be based on accepted values of life and it is general practice in the bearing industry to define the load capacity of the bearing as that value below which 90 percent of a batch will exceed a life of one million revolutions.

Notwithstanding the fact that responsibility for the design of ball and roller bearings rests with the bearing manufacturer, the machine designer must form a correct appreciation of the duty to be performed by the bearing and be concerned not only with bearing selection but with the conditions for correct installation.

The fit of the bearing rings onto the shafts or onto the housings[2] is of critical importance because of their combined effect on the internal clearance of the bearing as well as preserving the desired degree of interference fit. The inner ring is frequently located axially by abutting against a shoulder. A fillet radius at this point is essential for the avoidance of stress concentration and the inner ring is provided with a corner radius or chamfer to allow space for this (see Fig. 8.2).

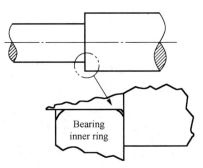

Figure 8.2　The shoulder fillet

Where life is not the determining factor in design, it is usual to determine maximum loading by the amount to which a bearing will deflect under load. Thus the concept of "static load-carrying capacity" is understood to mean the load that can be applied to a bearing, which is either stationary or subject to slight swiveling motions, without impairing its running qualities for subsequent rotational motion. This has been determined by practical experience as the load which when applied to a bearing results in a total deformation of the rolling element and raceway at any point of contact not exceeding 0.01 percent of the rolling-element diameter. This would correspond to a permanent deformation of 0.0025 mm for a ball 25 mm in diameter.

The successful functioning of many bearings depends upon providing them with adequate protection against their environment, and in some circumstances the environment must be protected from lubricants or products of deterioration of the bearing surfaces. Achievement of the correct functioning of seals is an essential part of bearing design. Moreover, seals which are applied to moving parts for any purpose are of interest to tribologists. Bearing seals (see Fig. 10.2) can only be designed satisfactorily on the basis of the appropriate bearing theory, because they are components of bearings. Notwithstanding their importance, the amount of research effort that has been devoted to the understanding of the behavior of seals has been small when compared with that devoted to other aspects of bearing technology.

Words and Expressions

radial bearing　向心轴承（主要用于承受径向载荷的滚动轴承）
thrust bearing　推力轴承（主要用于承受轴向载荷的滚动轴承）
angular contact bearing　角接触轴承

radial and axial loads　径向和轴向载荷（垂直于和平行于旋转轴线的载荷）
tapered roller bearing　圆锥滚子轴承（滚动体是圆锥滚子的滚动轴承）
single row bearing　单列轴承（沿圆周有一列滚动体的滚动轴承）
deep groove ball bearing　深沟球轴承
inner ring　内圈（滚道在外表面的轴承套圈）
outer ring　外圈（滚道在内表面的轴承套圈）
separator ['sepəreitə]　*n.* 保持架，也称为 cage 或 retainer
curvilinear [kə:vi'liniə]　*a.* 曲线的，由曲线而成的
raceway ['reiswei]　*n.* 滚道（滚动轴承承受负荷的表面，用作滚动体的轨道）
thrust ball bearing　推力球轴承（滚动体是球的推力滚动轴承）
tapered roller thrust bearing　推力圆锥滚子轴承（滚动体是圆锥滚子的推力滚动轴承）
needle roller bearing　滚针轴承（滚动体是滚针的向心滚动轴承）
fivefold ['faiv'fəuld]　*a.*; *ad.* 五倍的（地），五重的（地）
stiffness ['stifnis]　*n.* 刚度（作用在弹性元件上的力或力矩的增量与相应的位移或角位移之比）
load rating　额定负荷，额定载荷
frictional ['frikʃənl] drag　摩擦阻力
wear [wɛə]　*v.* 磨损（物体表面因相对运动而出现的材料不断迁移或损失的过程）
contaminant [kən'tæminənt]　*n.* 污染物，异物（a substance that makes something less pure）
HV　维氏硬度（Hardness according to Vickers; Vickers Hardness）
notwithstanding [ˌnɔtwiθ'stændiŋ]　*prep.*; *ad.*; *conj.* 虽然，尽管……（还是）
clearance ['kliərəns]　*n.* 间隙（the amount of space between two objects）
bearing ring　轴承套圈（具有一个或几个滚道的向心滚动轴承的环形零件）
interference [ˌintə'fiərəns]　*n.* 过盈（孔的尺寸减去相配合的轴的尺寸所得的代数差为负）
fit　配合（基本尺寸相同的，相互结合的孔和轴公差带之间的关系）
interference fit　过盈配合（具有过盈的配合，包括最小过盈等于零的配合）
abut [ə'bʌt] against　紧靠着，紧挨着
shoulder ['ʃəuldə]　*n.* 轴肩（阶梯轴上截面尺寸变化的部位）
chamfer ['tʃæmfə]　*n.*; *v.* 在……开槽，倒棱，倒角，圆角
fillet ['filit]　*n.* 内圆角（a rounded interior corner），过渡圆角
swiveling ['swivliŋ]　*a.* 转动的，旋转的
lubricant ['lu:brikənt]　*n.* 润滑剂（为减少表面磨损和摩擦力而施加在摩擦表面上的物质）
deterioration [diˌtiəriə'reiʃən]　*n.* 变质，退化，恶化，变坏
bearing seal　轴承密封圈（由一个或几个零件组成的环形罩，固定在轴承的一个套圈或垫片上并与另一套圈或垫片接触或形成窄的迷宫间隙，防止润滑油漏出及外物侵入）
tribologist [trai'bɔlədʒist]　*n.* 摩擦学家，摩擦学研究人员

Notes

[1] through-hardened 意为"透淬（工件从表面至心部全部硬化）的，整体淬火的"。
[2] housing 这里指"轴承座"，也可写为 bearing housing。

Lesson 9 Turbine Engine Bearings for Ultra-High Temperatures

The advanced gas turbine engines which will be used to power aircraft in the next decade are already at the design stage. These engines will be extremely efficient and in many instances will produce aircraft speeds above Mach 3. The operating conditions for the main shaft bearings of these engines can be considered as extremely demanding. Shaft speeds in excess of 30,000 r/min and bearing temperatures greater than 650°C are predicted.

In applications where relatively long bearing lives are required, the present temperature limit for liquid-lubricated steel bearings is 200°C. [1] For short-life bearing applications it is possible to operate up to 450°C. Even using the most technologically advanced liquid lubricants and metallic alloys, the temperature capability of the bearings operating in very limited life applications is 500°C.

New thinking is required which crosses the boundaries of conventional bearing design and exploits the latest developments of research into high temperature materials and solid lubricants. The predicted extreme operating temperatures (800°C to 900°C) appear impossible to attain, given present temperature limitations. Ceramic bearings (Fig. 9.1) offer the hope of increasing operating temperatures considerably above 650°C. The selection of effective bearing and lubricant materials depends on their thermal, physical, chemical and mechanical properties, as well as the operational environment and engineering constraints of the application.

Figure 9.1 Ceramic bearings

Materials for balls and rings

The most important criteria for evaluating materials to be used for balls and rings of high temperature bearings are: high temperature strength, high temperature hardness and oxidation characteristics. Tool steels are currently the most common materials used for aeroengine bearings with a practical temperature limit of approximately 400°C. [2] At such temperatures normal bearing steels rapidly lose their hardness.

A group of materials which appears promising for ultra-high temperature bearing operation is high performance ceramics. At temperatures above 1,100°C these ceramic materials have a higher hardness than normal bearing steels. For the past decade one ceramic material has been developed to the virtual exclusion of all others for high speed, high temperature rolling bearings: silicon nitride (Si_3N_4). Silicon nitride is desirable because it has good high temperature strength and hardness, an advantageous strength/weight relationship and excellent resistance to rolling contact fatigue when there is adequate lubrication.

However, there are drawbacks with silicon nitride, including low tensile strength, low fracture toughness and an extremely low coefficient of thermal expansion. Because of these properties, considerable development work is needed for the manufacture and application of ceramic bearings. [3]

Other ceramic materials, such as silicon carbide (SiC) and titanium carbide (TiC) are being evaluated for their suitability as ball and ring materials. Although not as popular and not as mature as silicon nitride, they do have certain properties which make them candidates for high temperature rolling bearing materials. [4] For example, silicon carbide has been used as a ball material in a 40,000 r/min bearing test, and although the temperature was not extreme, it was above the liquid lubrication range. The lubrication system consisted solely of a film of solid lubricant.

The positive properties of silicon carbide for use in high temperature bearings are its good thermal conductivity and thermal diffusivity, its oxidation resistance and high purity (properties are little affected by impurities). [5] One of the negative properties of the material is its high modulus of elasticity. This is about 50% higher than silicon nitride and has been cited as a potential problem because of the risk of high Hertzian contact stresses (see Figs. 9.2 and 9.3).

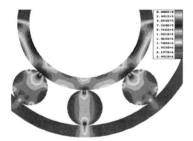 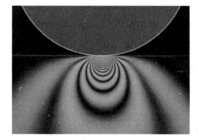

Figure 9.2　Hertzian contact stress analysis using finite element method

Figure 9.3　Stresses in Hertzian contact as measured by photoelasticity

Solid lubricants

It is worth noting that the temperature limits for the most sophisticated synthetic lubricants are almost equal to those of the most advanced bearing steels. The calculated operating temperatures for future turbine engines greatly exceed the capabilities of these materials. The only solution is to utilize unconventional lubricants.

If a rolling bearing is well lubricated and sealed against contaminants, bearing life is normally limited by material fatigue. If liquid lubricants cannot be used, then some form of boundary lubrication is necessary to minimize frictional heat and wear. [6] Oxides formed on the surfaces of the bearing's contacting components can offer reasonably effective lubrication for a limited period.

The difficulty when selecting a solid lubricant is to find a compound which will remain resistant to heat and oxidation over a wide temperature range, e. g. from −50°C to +980°C. It is often the case that solid lubricants which are good at low temperatures break down or become abrasive at high temperatures and vice versa.

The importance of requiring a lubricating film cannot be over emphasized, even when using ceramic materials.[7] Unlubricated silicon nitride or silicon carbide do not inherently have low friction, nor good wear resistance. These properties can be obtained with the aid of solid lubricants which are compatible with the materials. Silicon nitride, lubricated for example with graphite containing high temperature additives, can form a tribo-chemical film which reduces the coefficient of friction and, as a consequence, minimizes heat generation.

For bearing operation at ultra-high temperatures—above 550°C—solid lubricants which are more heat stable than graphite are being considered. It is imperative that the development of a complex tribological system such as a high temperature solid lubricated, ceramic bearing is conducted with full understanding of the individual tribological relationships of the various components.

Words and Expressions

gas turbine ['təːbain] engine　燃气涡轮发动机，或称燃气轮机

Mach [mɑːk]　n. 马赫，马赫数（速度单位，速度与音速之比。1 马赫等于音速）

main shaft　主轴（a principal drive shaft）

demanding [diˈmændiŋ]　a. 要求高的（requiring much effort or attention; hard to satisfy）

shaft speed　轴的转速（轴在单位时间内的转数）

exploit [ˈeksplɔit]　v. 充分利用（to make full use of and derive benefit from something）

operating temperature　工作温度（设备工作时环境温度范围）

engineering constraint　工程技术约束（限制）

aeroengine [ˈɛərəuˈendʒin]　n. 航空发动机

ultra- [ˈʌltrə-]　超过（beyond the range, scope, or limit of），极端（extreme or extremely）

silicon nitride　氮化硅

lubrication [ˌluːbriˈkeiʃən]　n. 润滑（用润滑剂减少两摩擦表面之间的摩擦和磨损或其他形式的表面破坏的措施）

toughness [ˈtʌfnis]　n. 韧性（材料在塑性变形和断裂过程中吸收能量的能力）

fracture toughness　断裂韧性（材料抵抗裂纹启裂和扩展的能力），断裂韧度

silicon carbide　碳化硅

titanium carbide　碳化钛

thermal conductivity　热导率，导热系数（单位时间内单位面积上通过的热量与温度梯度的比例系数）

thermal diffusivity　热扩散系数（反映温度不均匀的物体中温度均匀化速度的物理量），又称热扩散率或导温系数

impurity [imˈpjuəriti]　n. 杂质（an unwanted substance that is found in something else）

modulus of elasticity　弹性模量（材料在弹性变形范围内，正应力与正应变的比值），亦为 elasticity modulus 或 Young's modulus（杨氏模量）

Hertzian contact stress　赫兹接触应力
finite element method (FEM)　有限元法，又称"有限元分析 (finite element analysis, FEA)"
photoelasticity [ˌfəutəuilæs'tisiti] *n.* 光弹性，光弹性法（利用偏振光通过透光的弹性变形模型产生的双折射效应，测定光程差来确定物体弹性应力的实验应力分析方法）
synthetic [sin'θetic] *a.* 合成的（made by combining different substances; not natural）
synthetic lubricant　合成润滑剂（a lubricant produced by chemical synthesis）
oxide ['ɔksaid] *n.* 氧化物（a compound of oxygen and another substance）
compound ['kɔmpaund] *n.* 化合物（a substance that contains atoms of two or more chemical elements held together by chemical bonds）
vice versa ['vaisi'vəːsə] *ad.* 反之亦然（the opposite of a statement is also true）
adherence [əd'hiərəns] *n.* 黏着，黏附，黏着力
unlubricated ['ʌn'luːbrikeitid] *a.* 无润滑的（not lubricated）
inherently [in'hiərəntli] *ad.* 固有地（according to or because of the basic nature of something）
wear resistance　耐磨性（材料在一定摩擦条件下抵抗磨损的能力，以磨损率的倒数来评定）
compatible [kəm'pætəbl] *a.* 相容的，兼容的，可共存的，相适应的
additive ['æditiv] *n.* 添加剂（添加到润滑剂中以提高某些原有特性或获得新特性的物质）
tribo-chemical film　摩擦化学反应膜
coefficient of friction　摩擦系数（the ratio between friction force and normal force）
tribological [ˌtraibəu'lɔdʒikəl] *a.* 摩擦学的

Notes

[1] temperature limit 意为"温度极限"。全句可译为：对于需要有较长的轴承寿命的用途，当采用液体润滑的钢制轴承时，目前的温度极限是200℃。

[2] tool steel 意为"工具钢（适宜制造刃具、模具和量具等各式工具的钢）"。全句可译为：工具钢是目前制造航空发动机轴承的最常用材料，它的实际温度极限大约是400℃。

[3] development work 意为"开发工作"。全句可译为：由于具有了这些性能，在陶瓷轴承的制造和应用方面需要做大量的开发工作。

[4] 全句可译为：尽管不像氮化硅那么普及和成熟，它们确实具有某些能够使其成为高温滚动轴承备选材料的性能。

[5] thermal conductivity and thermal diffusivity 意为"热导率和热扩散率"。全句可译为：碳化硅有利于在高温轴承中应用的性能是其良好的热导率和热扩散率、抗氧化性和高纯度（性能几乎不受杂质的影响）。

[6] boundary lubrication 意为"边界润滑"。全句可译为：如果不能采用液态润滑剂，则有必要采用某种形式的边界润滑来减少摩擦热和磨损。

[7] cannot be over emphasized 意为"怎么强调都不过分"。全句可译为：润滑膜的重要性怎么强调都不过分，即使采用陶瓷材料时也是这样。

Lesson 10　Spindle Bearings
(*Reading Material*)

A machine tool spindle must provide high rotational speed, transfer torque and power to the cutting tool, and have reasonable load carrying capacity and life. The bearing system, one of the most critical components of any high-speed spindle design, must be able to meet these demands, or the spindle will not perform.

High-speed spindle bearings available today include roller, tapered roller, and angular contact ball bearings (Fig. 10.1). Selection criteria depend on the spindle specifications and the speed needed for metal cutting.

(a) Roller bearing　　(b) Tapered roller bearing　　(c) Angular contact ball bearing

Figure 10.1　Some types of spindle bearings

Angular contact bearings are the type most commonly used in very high-speed spindle design. These bearings provide precision, load carrying capacity, and the speed needed for metal cutting. Angular contact ball bearings have a number of precision balls fitted into a precision steel race, and provide both axial and radial load carrying capacity when properly preloaded.

In some cases, tapered roller bearings are used because they offer higher load carrying capacity and greater stiffness than ball bearings. However, tapered roller bearings do not allow the high speeds required by many spindles.

Angular contact ball bearings are available with a choice of preloading magnitude, typically designated as light, medium, and heavy. Light preloaded bearings allow maximum speed and less stiffness. These bearings are often used for very high-speed applications, where cutting loads are light.

Heavy preloading allows less speed, but higher stiffness. To provide the required load carrying capacity for a metal cutting machine tool spindle, several angular contact ball bearings are used together. In this way, the bearings share the loads, and increase overall spindle stiffness.

Hybrid ceramic bearings (Fig. 10.2) are a recent development in bearing technology that uses ceramic (silicon nitride) material to make precision balls (Fig. 10.3). The ceramic balls, when used in an angular contact ball bearing, offer distinct advantages over typical bearing-steel balls.

Ceramic balls have 60% less mass than steel balls. This is significant because as a ball bearing is operating, particularly at high rotational speeds, centrifugal forces push the balls to the outer race, and even begin to deform the shape of the ball. This deformation leads to rapid wear and bearing deterioration. Ceramic balls, with less mass, will not be affected as much at the same speed. In fact, the use of ceramic balls allows up to 30% higher speed for a given ball bearing size, without sacrificing bearing life.

Figure 10.2 A hybrid ceramic bearing

Figure 10.3 Silicon nitride balls

Due to the nearly perfect roundness of the ceramic balls, hybrid ceramic bearings operate at much lower temperatures than steel ball bearings, which results in longer life for the bearing lubricant. Tests show that spindles utilizing hybrid ceramic bearings exhibit higher rigidity and have higher natural frequencies, making them less sensitive to vibration.

Bearing lubrication is necessary for angular contact ball bearings to operate properly. The lubricant provides a microscopic film between the rolling elements to prevent abrasion. In addition, it protects surfaces from corrosion, and protects the contact area from particle contamination.

Grease is the most common and most easily applied type of lubricant. It is injected into the space between the balls and the races, and is permanent. Grease requires minimal maintenance. Generally, high-speed spindles utilizing grease lubrication do not allow replacement of the grease between bearing replacements. During a bearing replacement, clean grease is carefully injected into the bearing.

Often at high rotational speeds, lubrication with grease is not sufficient. Oil is then used as a lubricant, and delivered in a variety of ways. Maintenance of the lubrication system is vital, and must be closely monitored to ensure that proper bearing conditions are maintained. Also, use of the correct type, quantity, and cleanliness of lubricating oil is critical.

How much life can be expected? All bearings will have a useful life, defined as operation time until the bearing specifications are lost, or a complete failure of the bearing occurs. The most common cause of bearing failure is fatigue, in which the races become rough, leading to heating, and eventual mechanical failure. Bearing life, in general, is affected by axial and radial bearing loads, vibration levels, lubrication quality and quantity, maximum speed, and average bearing temperature. Today, computer models are often used to forecast bearing life.

Words and Expressions

spindle ['spindl]   *n.*   主轴（带动工件或加工工具旋转的轴）
spindle bearing   主轴轴承
rotational speed   转速（单位时间内，物体绕某轴线转动的圈数）
cutting tool   刀具（能从工件上切除多余材料或切断材料的带刃工具）
load carrying capacity   承载能力
speed   速度，转速（a rate of rotation, usually expressed in revolutions per unit time）
angular contact ball bearing   角接触球轴承
be designated as   被指定为，被定名为，被分类为，用……表示
specification [ˌspesifiˈkeiʃən]   *n.*   规范，说明书，(*pl.*) 技术参数，技术要求
race   （轴承的）滚道（a groove for the balls in a ball bearing or rollers in a roller bearing）
preload [ˈpriːˈləud]   *n.*   预紧，预载荷（在施加"使用"载荷（外部载荷）前，通过相对于另一轴承的轴向调整而作用在轴承上的力，或由轴承内滚道与滚动体的尺寸形成"负游隙"（内部预载荷）而产生的力）
hybrid [ˈhaibrid]   *a.*   混合的（consisting of diverse components; combining two different things）
hybrid ceramic bearing   混合陶瓷轴承
cutting load   切削载荷
silicon nitride   氮化硅
roundness [ˈraundnis]   *n.*   圆度（指工件的横截面接近理论圆的程度）
deterioration [diˌtiəriəˈreiʃən]   *n.*   恶化（the act or process of becoming worse），劣化
bearing deterioration   轴承性能退化，轴承性能劣化
natural frequency   固有频率（物体或结构系统做自由振动时所具有的振动频率）
microscopic [maikrəˈskɔpik]   *a.*   用显微镜可见的，极小的，微小的
abrasion [əˈbreiʒən]   *n.*   磨损（与 wear 相同），磨料磨损（与 abrasive wear 相同）
grease [griːs]   *n.*   脂（一种稠厚的油脂状半固体，起润滑和密封作用），润滑脂（lubricating grease）
lubrication [ˌluːbriˈkeiʃən]   *n.*   润滑（用润滑剂减少两摩擦表面之间的摩擦和磨损或其他形式的表面破坏的措施）
cleanliness [ˈklenlinis]   *n.*   清洁度（the quality or state of being clean）
axial [ˈæksiəl]   *a.*   轴的，轴向的
radial [ˈreidjəl]   *a.*   径向的
mechanical failure   机械故障

Lesson 11 Machine Tool Frames

The frame is a machine's fundamental element. Most frames are made from cast iron, welded steel, or composites. The following factors govern material choice.

The material must resist deformation and fracture. Hardness must be balanced against elasticity. The frame must withstand impact, yet yield under load without cracking or permanent deformation. [1] The frame material must eliminate or block vibration transmission. It must withstand the hostile shop floor environment, including the newer coolants and lubricants. The material must not build up too much heat, must retain its shape for its lifetime, and must be dense enough to distribute forces throughout the machine.

Pros and Cons

Either castings or welded sections can be used in most applications. The decision on which is best depends on the costs in a given design situation.

Cast iron Almost all machine tool frames were traditionally made of cast iron because features difficult to obtain any other way can be cast in. Castings (see Fig. 11.1) have a good stiffness-to-weight ratio and good damping qualities. Modifying wall thickness and putting the metal where it's needed is fairly easy.

Figure 11.1 Castings

Although cast iron is a fairly cheap material, each casting requires a mold. Larger sizes are a limiting factor because of mold cost. Smaller, high-volume machines usually have cast iron frames. Welded frames may be cheaper for lower volume machines.

Welded steel Machine builders fabricate steel frames from welded steel sections when casting is impractical. Because steel has a higher tensile strength, the welded steel frames are much thinner compared to cast-iron ones. The steel frames are usually ribbed to provide the necessary stiffness. [2] The number of welds is a design tradeoff: with welding, it's easy to make large sections and add features even after the initial design is complete, but the heat can introduce distortion and also adds cost. Welds also help block vibration transmission through the steel frame. Builders sometimes increase damping by circulating coolant through the welded structure or adding lead or sand to frame cavities. [3]

Composites Advanced forms of these materials, including those with polymer, metal, and ceramic matrices, may change machine tool design dramatically. Both matrix and reinforcing material can be tailored to provide strength in specific axes. [4]

Designers must consider the different expansion coefficients between the composite and the metal sections to which it is joined. The most common applications for this material are high accuracy machine tools and grinders.

Ceramics The Japanese introduced experimental machine tools with ceramic frames in the 1980s. Ceramics offer high strength, stiffness, dimensional stability, corrosion resistance, and excellent surface roughness, but they are brittle and expensive. Their lack of conductivity can be an advantage or not.

Foundations

Foundations ensure the machine's stiffness; shock absorption and isolation are secondary considerations. [5] In selecting a foundation, designers must consider the machine's weight, the forces it generates, accuracy requirements, and the loads being transmitted to the ground by adjacent machines. Soil condition can be a problem because long-term changes can influence machine stability.

Frame Design

The major considerations in frame design are loads, damping, heat distortion, and noise.

Loads Understanding the static and dynamic loads a machine generates is essential. The basic load is static; the mass of the machine and its workpiece. The dynamic load adds all that happens once the machine is running. This includes the forces of acceleration, deceleration, tool action, irregular loads caused by an unbalanced condition, or self-exciting loads from load and vibration interaction. Finite element analysis (see Fig. 11.2) gives a good indication of how a frame will react to loads.

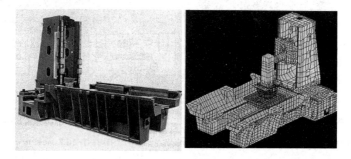

Figure 11.2 A machine tool frame and finite element analysis model

Damping Though frame material and design should handle damping, dampers are sometimes built into frame sections to handle specific problems. They are effective only when the designer has a good understanding of all the loads involved. For example, a damper that works well under static conditions may do more harm than good under dynamic conditions. [6]

Heat distortion Heat from external or internal sources can be a major cause of error if the frame distorts. External sources include ambient shop conditions, cooling and lubricating media, and the sun. [7] The machine tool also has its own heat sources: motors, friction from machine motion, and the cutting action of the tool on the workpiece. Ideally, frame heating should be minimized and kept constant.

Noise Reduction of noise for health and safety reasons is a fairly recent concern. Enclosures prevent sound transfer through the machine, and sound damping materials help reduce objectionable sound.

Words and Expressions

machine tool　机床（制造机器的机器。一般分为金属切削机床、锻压机床和木工机床等），又称"工作母机"或"工具机"

frame [freim] *n.* 机架（构成本体的框架）

weld [weld] *v.* 焊接；*n.* 焊接，焊缝（焊件经焊接后所形成的接合部分）

composite　复合材料（由有机高分子、无机非金属或金属等几类不同材料通过复合工艺组合而成的新型材料。既保留原组成材料的重要特点，又通过复合效应获得原组分所不具备的性能），亦为 composite [kəmˈpɔzət] material

fracture [ˈfræktʃə] *n.* 断裂（the result of breaking sth）；*v.* 断裂（to cause a break in sth）

elasticity [ˌilæsˈtisiti] *n.* 弹性（材料在外力作用下变形，外力卸除后能恢复原状的性能）

block [blɔk] *v.* 阻碍，阻塞（to stop movement through something），妨碍（to obstruct）

coolant [ˈkuːlənt] *n.* 冷却液（起冷却作用的调温流体），切削液

lubricant [ˈluːbrikənt] *n.* 润滑剂（加入两个相对运动表面间，能减少或避免摩擦磨损的物质）

build up　逐渐增加，逐渐积累（to increase or accumulate gradually）

accuracy [ˈækjurəsi] *n.* 准确，精度，精确性

shop floor　工人，工作场所（the ordinary workers in a factory or the area where they work）

circulate [ˈsəːkjuleit] *v.* 循环（to move in or flow through a circle or circuit），流通

slide [slaid] *n.* 滑板（可沿导轨移动的部件），滑块

pros and cons　优缺点（advantages and disadvantages），有利和不利因素

casting [ˈkɑːstiŋ] *n.* 铸件（将熔融金属浇入铸型，凝固后所得到的金属工件或毛坯）

stiffness [ˈstifnis] *n.* 刚度（材料或结构在受力时抵抗变形的能力）

stiffness-to-weight ratio　刚度重量比，比刚度（specific stiffness，材料的弹性模量与其密度的比值）

mold [məuld] *n.* 铸型（用型砂、金属或其他耐火材料制成，包括形成铸件的型腔、芯子和浇冒口系统的组合整体），亦为 mould

rib [rib] *n.* 肋板；*v.* 用肋板支撑或增强（to support or strengthen with a rib or ribs）

tradeoff [ˈtreidɔf] 折中，权衡（a balancing of factors all of which are not attainable at the same time）

distortion [disˈtɔːʃən] *n.* 变形，畸变（the act of twisting out of shape or making inaccurate）

lead [led] *n.* 铅（a heavy soft gray metallic element that is easily bent and shaped）

cavity [ˈkæviti] *n.* 空腔，（铸造）型腔，室穴，腔体，凹处

polymer [ˈpɔlimə] *n.* 聚合物，高分子材料

ceramics [siˈræmiks] *n.* 陶瓷

matrix [ˈmeitriks] (*pl.* matrices 或 matrixes) *n.* 基体，母体，本体

reinforce [ˌriːinˈfɔːs] *v.* 加强，增强（to make a structure or material stronger, especially by adding another material to it）

tailor... to...   使……适合(满足)……(的要求，需要，条件)
grinder ['graində]   n. 磨床(用磨具或磨料加工工件各种表面的机床)，亦为 grinding machine
dimensional [di'menʃənəl]   a. 尺寸的，空间的，维的
stability [stə'biliti]   n. 稳定性(a situation in which something is not likely to move or change)
conductivity [ˌkɔndʌk'tiviti]   n. 传导率(the ability to conduct heat, electricity, or sound)
expansion coefficient   膨胀系数
foundation [faun'deiʃən]   n. 基础(支承机械系统的结构)
heat distortion   热变形(在加工过程中，机架各部分温度不同引起的变形)
deceleration [diːˌseləˈreiʃən]   n. 减速(度)，降速
self-exciting   自激的，自励的
finite element analysis   有限元分析(将连续体离散化为若干有限大小的单元体，对实际物理问题进行模拟求解的分析方法)，又称"有限元法"
damper ['dæmpə]   n. 阻尼器(一种利用阻尼特性来减缓机械振动及消耗动能的装置)
assembly [ə'sembli]   n. 装配(将零件或部件进行配合和连接，使之成为半成品或成品的过程)
maintenance ['meintinəns]   n. 维护，保养，维修
workpiece ['wəːkpiːs]   n. 工件(加工过程中的生产对象)
enclosure [in'kləuʒə]   n. 外壳，机壳，罩
damping material   阻尼材料(将机械振动能转变为热能而耗散的材料，用于振动和噪声控制)
objectionable [əb'dʒekʃənəbl]   a. 引起反对的，讨厌的，令人不愉快的

Notes

[1] cracking or permanent deformation 意为"产生裂纹和永久变形"。此句与后面两句可译为：机架必须能够承受冲击，在载荷作用下不产生裂纹或永久变形。机架材料必须能够消除或阻碍振动的传递。它必须能够承受加工车间的恶劣环境，其中包括新型的冷却液和润滑剂。

[2] 这两句可译为：由于钢具有较高的抗拉强度，相对铸铁机架而言，焊接钢机架的尺寸要单薄很多。通常采用设置肋板的方式使钢机架具有所需要的刚度。

[3] 这两句可译为：焊缝还可以阻碍振动在钢制机架中的传播。制造商有时通过在焊接结构中循环冷却液或在机架的空腔中添加铅或沙子来增加阻尼。

[4] reinforcing material 意为"增强材料"。全句可译为：可以使基体和增强材料在某些指定的轴线方向上具有所需要的强度。

[5] shock absorption and isolation 意为"减震和隔震"。全句可译为：基础应该保证机器的刚度，减震和隔震是次要的考虑事项。

[6] do more harm than good 意为"弊大于利"。全句可译为：例如，一个在静载荷作用下效果很好的阻尼器，在动载荷作用下有可能会效果很差。

[7] external source 这里指"外部热源"。全句可译为：外部热源包括车间的环境条件、冷却剂、润滑剂和阳光。

Lesson 12 Spur Gears

Gears are used to transmit torque, rotary motion, and power from one shaft to another (see also Fig. 1.1). They have a long history. In about 2600 B.C., the Chinese used primitive gears like those illustrated in Fig. 12.1. The gears were made of wood, their teeth merely pegs inserted in wheels. Early examples of metal gears date from the 4th century B.C. in China (Warring States period—late Eastern Zhou dynasty), which have been preserved at the Luoyang Museum of Henan Province, China. In the 15th century A.D., Leonardo da Vinci designed a multitude of devices incorporating many kinds of gears.

Among the various means of mechanical power transmission (including primarily gears, belts, and chains) gears are generally the most durable. Their power transmission efficiency is as high as 98 percent. On the other hand, gears are usually more costly than chains and belts. As would be expected, gear manufacturing costs increase sharply with increased precision—as required for the combination of high speeds and heavy loads, and for low noise levels. [1] Standard tolerances for various degrees of manufacturing precision have been established by the AGMA, American Gear Manufacturers Association.

Spur gears (Fig. 12.2) are the simplest and most common type of gears. They are used to transfer motion between parallel shafts, and have teeth that are parallel to the shaft axes. [2]

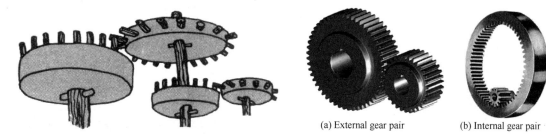

(a) External gear pair (b) Internal gear pair

Figure 12.1 Primitive gears   Figure 12.2 Spur gears

The basic requirement of gear-tooth geometry is the provision of angular velocity ratios that are exactly constant. [3] For example, the angular velocity ratio between a 20-tooth and a 40-tooth gear must be precisely 2 in every position. It must not be, for example, 1.99 as a given pair of teeth comes into mesh, and then 2.01 as they go out of mesh. Of course, manufacturing inaccuracies and tooth deflections will cause slight deviations in velocity ratio; but acceptable tooth profiles are based on theoretical curves that meet this criterion.

As gears mesh, both rolling and sliding occur, causing pitting and wear; pitting is primarily a consequence of rolling contact and wear is the result of sliding contact. [4] Although sliding causes wear, it also creates hydrodynamic action that counteracts wear. This relative motion, minute as it may be, is sufficient to provide conditions necessary for hydrodynamic lubrication. [5] Each time a gear goes through mesh, surface and subsurface material is subjected to compressive, shear, and tensile stresses. Pitting is initiated when contact stresses are high; it is hastened by sliding action.

Pitting first occurs at the pitch line (see Fig. 12.3), where the absence of sliding favors early breakdown of protective oil films. Excessive contact stress causing surface fatigue is the real cause of pitting. After what is usually a very large number of stress repetitions, surface failure may occur. Minute cracks form in and below the surface, then grow and join. Eventually, small bits of metal are separated and forced out, leaving pits. [6]

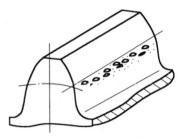

Figure 12.3 Gear pitting

Tooth breakage may be caused by an unexpectedly heavy load being imposed on the teeth. A more common type of failure is due to bending fatigue, which results from the large number of repetitions of load imposed on the tooth as the gear rotates.

The least expensive gear material is usually ordinary cast iron, ASTM (or AGMA) grade 20. Grades 30, 40, 50, and 60 are progressively stronger and more expensive. [7] Cast iron gears typically have greater surface fatigue strength than bending fatigue strength. Their internal damping tends to make them quieter than steel gears. Nodular cast iron gears have substantially greater bending strength, together with good surface durability. A good combination is often a steel pinion mated to a cast iron gear.

Unheat-treated steel gears are relatively inexpensive, but have low surface endurance capacity. Heat-treated steel gears must be designed to resist warping; hence, alloy steels and oil hardening are usually preferred. For hardnesses over 250–350 Bhn, machining must usually be done before hardening. [8] Greater profile accuracy is obtained if the surfaces are finished after heat-treating, as by grinding (see Figs. 38.5 and 38.7). But if grinding is done, care must be taken to avoid residual tensile stresses at the surface.

Of the non-ferrous metals, bronzes are most often used for making gears. Nonmetallic gears (nylon and other plastics, see Fig. 12.4) are generally quiet, durable, reasonably priced, and often can operate under light loads without lubrication. Their teeth deflect more easily than those of corresponding metal gears. This promotes effective load sharing among teeth in simultaneous contact. Since nonmetallic

Figure 12.4 Nylon gears

materials have low thermal conductivity, special cooling provisions may be required. Furthermore, these materials have relatively high coefficients of thermal expansion, and this may require installation with greater backlash than metal gears.

Nonmetallic gears are usually mated with cast iron or steel pinions. For best wear resistance, the hardness of the mating metal pinion should be at least 300 Bhn. Design procedures for gears made of plastics are similar to those for gears made of metals, but are not yet as reliable. Hence, prototype testing is even more important than for metal gears.

Words and Expressions

spur gear 直齿轮（齿线为分度圆柱面直母线的圆柱齿轮）
primitive ['primitiv] a. 原始的，早期的（of or relating to an earliest or original stage）
peg [peg] n. 木销，圆销，销钉（a cylindrical or tapered pin of wood, metal, etc.）
date from 起源于，始于（to have existed since a particular time in the past）
Warring States period 战国时代，战国时期（公元前475年—公元前221年）
late Eastern Zhou dynasty 东周（公元前770年—公元前256年）晚期
Leonardo da Vinci 莱昂纳多·达·芬奇（1452—1519年，意大利画家、工程师、科学家）
a multitude of 许多的，大量的
transmission [trænz'miʃən] n. 传动，传动装置（传递运动和动力的装置）
efficiency [i'fiʃənsi] n. 效率（有用功率对驱动功率之比值）
standard tolerance 标准公差（标准规定的，用以确定公差带大小的任意一个公差）
external gear 外齿轮（齿顶曲面位于齿根曲面之外的齿轮）
internal gear 内齿轮（齿顶曲面位于齿根曲面之内的齿轮）
gear pair 齿轮副（由两个相啮合的齿轮组成的基本机构）
external gear pair 外齿轮副（两齿轮均为外齿轮的齿轮副）
internal gear pair 内齿轮副（有一个齿轮是内齿轮的齿轮副）
axis ['æksis] n. (pl. axes ['æksi:z]) 轴（机床部件直线运动或旋转运动的方向），轴线
mesh [meʃ] n. 啮合（一对带有齿状部分的零件在传动过程中的相互连接）
deviation [,di:vi'eiʃən] n. 偏离，偏差（a difference from what is usual or expected）
tooth profile 齿廓（齿面被一个与齿线相交的既定平面或曲面所截的截线）
hydrodynamic lubrication 流体动力润滑（用借助相对速度而产生的流体膜将两摩擦表面完全隔开的润滑），液体动力润滑
hasten ['heisn] v. 促进，加速
counteract [,kauntə'rækt] v. 抵消（to reduce or remove the effect of something unwanted by producing an opposite effect），对抗，减少
subsurface ['sʌb'sə:fis] n. 亚表面（固体表面下紧靠表面的部分，无明确尺寸界定）
pitch line 节线
breakdown ['breikdaun] n. 故障，损坏，破裂
tooth breakage 轮齿折断
ASTM (American Society of Testing Materials) 美国材料实验学会
AGMA (American Gear Manufacturers Association) 美国齿轮制造商协会
nodular ['nɔdjulə] cast iron 球墨铸铁（在其中石墨主要以球状存在的高强度铸铁）
unheat-treated 没有经过热处理的
warping ['wɔ:piŋ] n. 变形，翘曲，扭曲
oil hardening 油冷淬火（将合金加热到相变点上某一温度，保温适当时间后在油中急冷）
Bhn 布氏硬度值（Brinell hardness number），亦为BHN
residual tensile stress 残余拉应力
non-ferrous metal 有色金属（除铁、铬、锰三种金属以外的所有金属元素的统称）
nonmetallic [,nɔnmi'tælik] a. 非金属的；n. 非金属物质

thermal conductivity　导热性，热导率，导热系数
coefficient of thermal expansion　热膨胀系数
backlash ['bæklæʃ]　n. 齿侧间隙（齿轮啮合传动时，为了在啮合齿廓之间形成润滑油膜，避免因轮齿摩擦发热膨胀而卡死，齿廓之间留有的间隙，简称侧隙）
pinion ['pinjən]　n. 小齿轮（齿轮副中齿数较少的那个齿轮，齿数较多的那个齿轮称为 gear）
wear resistance　耐磨性（材料在一定摩擦条件下抵抗磨损的能力，以磨损率的倒数来评定）
mating metal pinion　配对的金属小齿轮
prototype testing　原型试验，样机试验

Notes

[1] 全句可译为：正如预期的那样，随着齿轮精度的提高，其制造成本也会大幅度增加。在高速、重载和低噪声的工作环境中，必须使用高精度的齿轮。

[2] shaft axes 意为"轴线"。全句可译为：它们被用来在平行轴之间传递运动，其轮齿与轴线平行。

[3] gear-tooth geometry 意为"轮齿几何形状"，the provision of 意为"提供"。此句和后面两句可译为：对轮齿几何形状的基本要求是提供精确恒定的角速度比。例如，一个齿数为 20 的齿轮和一个齿数为 40 的齿轮的角速度比在任何位置都应该精确地为 2。举例来说，不能有一对轮齿的角速度比在其开始啮合时为 1.99，在其终止啮合时为 2.01。

[4] pitting 意为"点蚀（因表面疲劳作用导致材料流失，在摩擦表面留下小而浅的锥形凹坑的损伤形式）"。全句可译为：当齿轮啮合时，会同时产生滚动和滑动，引起点蚀和磨损。点蚀主要是由滚动接触造成的，而磨损是由滑动接触造成的。

[5] minute 是一个多音词，这里指"微小的（very small, tiny）"，其发音为 [mai'nju:t]，conditions 这里指"条件"。全句可译为：这种相对运动，尽管可能很微小，但足以提供产生流体动力润滑的必要条件。

[6] 这两句可译为：若干微小的裂纹在齿面和齿面以下生成，然后扩展并连接到一起。最终，很多小金属颗粒被分离出来，留下凹坑。

[7] grade 这里指"牌号"，与 designation 意思相同。全句可译为：牌号为 30、40、50 和 60 的这些铸铁的强度逐渐提高，价格也随之增加。

[8] Bhn 为 Brinell hardness number，指"布氏硬度值"，machining 意为"机械加工"；在下一句中，profile accuracy 意为"齿形精度"，heat-treating 意为"热处理"。这两句可译为：当硬度超过 250～350Bhn 时，通常在淬火前完成机械加工。如果在热处理后，采用磨削方式对齿面进行精加工，则可以获得更高的齿形精度。

Lesson 13　Strength of Mechanical Elements

One of the primary considerations in designing any machine or structure is that the strength must be sufficiently greater than the stress to assure both safety and reliability. To assure that mechanical parts do not fail in service, it is necessary to learn why they sometimes do fail. Then we shall be able to relate the stresses with the strengths to achieve safety.

Ideally, in designing any machine element, the engineer should have at his disposal[1] the results of a great many strength tests of the particular material chosen. These tests should have been made on specimens having the same heat treatment, surface roughness, and size as the element he proposes to design; and the tests should be made under exactly the same loading conditions as the part will experience in service. This means that, if the part is to experience a bending load, it should be tested with a bending load. If it is to be subjected to combined bending and torsion, it should be tested under combined bending and torsion. Such tests will provide very useful and precise information. They tell the engineer what factor of safety to use and what the reliability is for a given service life. Whenever such data are available for design purposes, the engineer can be assured that he is doing the best possible job of engineering. The cost of gathering such extensive data prior to design is justified if failure of the part may endanger human life, or if the part is manufactured in sufficiently large quantities. Automobiles and refrigerators, for example, have very good reliabilities because the parts are made in such large quantities that they can be thoroughly tested in advance of manufacture. The cost of making these tests is very low when it is divided by the total number of parts manufactured.

You can now appreciate the following four design categories:

(1) Failure of the part would endanger human life, or the part is made in extremely large quantities; consequently, an elaborate testing program is justified during design.

(2) The part is made in large enough quantities so that a moderate series of tests is feasible.

(3) The part is made in such small quantities that testing is not justified at all; or the design must be completed so rapidly that there is not enough time for testing.

(4) The part has already been designed, manufactured, and tested and found to be unsatisfactory. Analysis is required to understand why the part is unsatisfactory and what to do to improve it.

It is with the last three categories that we shall be mostly concerned. This means that the designer will usually have only published values of yield strength, tensile strength,[2] and percentage elongation. With this meager information the engineer is expected to design against static and dynamic loads, biaxial and triaxial stress states, high and low temperatures, and large and small parts! The data usually available for design have been obtained from the simple tension test, where the load was applied gradually and the strain given time to develop.

Yet these same data must be used in designing parts with complicated dynamic loads applied thousands of times per minute. No wonder machine parts sometimes fail.

To sum up, the fundamental problem of the designer is to use the simple tension test data and relate them to the strength of the part, regardless of the stress state or the loading situation.

It is possible for two metals to have exactly the same strength and hardness, yet one of these metals may have a superior ability to absorb overloads, because of the property called ductility. One way to measure the ductility of a material is to calculate its percentage elongation after fracture. Figure 13.1 shows a specimen before and after a standard tensile test. The elongation of a material is usually measured over 50 mm gauge length. For determination of percentage elongation, the two halves of the broken specimen are carefully fitted back together so that their axis lie in a straight line. The percentage elongation after fracture A is expressed as:

$$A = \frac{L_u - L_o}{L_o} \times 100\% \tag{13.1}$$

where L_u is the final gauge length after fracture and L_o is the original gauge length.

The usual dividing line between ductility and brittleness is 5 percent elongation. A material having less than 5 percent elongation at fracture is said to be brittle, while one having more is said to be ductile.

The percentage of decrease in cross-sectional area of a tension test specimen after fracture is another way to measure ductility. It is defined within the region of necking as follows:

$$Z = \frac{S_o - S_u}{S_o} \times 100\% \tag{13.2}$$

where Z is the percentage reduction of area, S_o is the original cross-sectional area of a tension test specimen and S_u is the area of its smallest cross section after fracture.

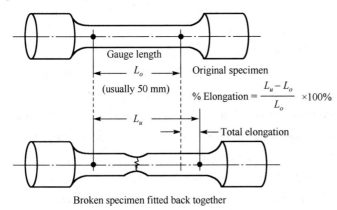

Figure 13.1 Measurement of percentage elongation

Materials that exhibit little or no yielding before failure are referred to as brittle materials (see Fig. 13.2). Grey cast iron is an example. Any material that can be subjected to large strains before its ruptures (see Fig. 13.3) is called a ductile material. Low carbon steel is a typical example. Engineers often choose ductile materials for design because these materials are capable of absorbing shock or energy, and if they become overloaded, they will usually exhibit large deformation before failing.

When a material is to be selected to resist wear or plastic deformation, hardness is generally the most important property. Several methods of hardness testing are available, depending upon which particular property is most desired. The four hardness numbers in greatest use are the Brinell, Rockwell, Vickers, and Knoop.

Figure 13.2　Tensile fracture of a brittle material

Figure 13.3　Necking of a ductile material

Most hardness-testing systems employ a standard load which is applied to an indenter with the shape of a ball, pyramid, or cone in contact with the material to be tested. The hardness is then expressed as a function of the size of the resulting indentation (see Fig. 13.4). This means that hardness is an easy property to measure, because the test is nondestructive and test specimens are not required. Usually the test can be conducted directly on an actual machine element.

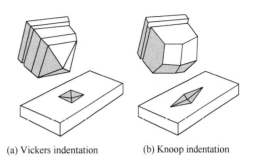

(a) Vickers indentation　　(b) Knoop indentation

Figure 13.4　Indentations

Words and Expressions

reliability　可靠性（产品在规定的条件下和规定的时间区间内完成规定功能的能力）
specimen ['spesimən]　n. 试样（从被测材料或零件上获取的符合相应规范的试验样品）
justify ['dʒʌstifai]　v. 证明……是对的（to show that something is right or reasonable）
elongation [ˌiːlɔŋ'geiʃən]　n. 伸长（试验期间任一时刻原始标距的增量）
necking ['nekiŋ]　n. 颈缩（在拉伸应力下，材料发生的局部截面缩减的现象）
biaxial [bai'æksiəl]　a. 二轴的，二维的
triaxial [trai'æksiəl]　a. 三轴的，三维的，空间的
brittle ['britl] fracture　脆性断裂（断裂前宏观塑性变形很小甚至为零的断裂方式）
ductility [dʌk'tiliti]　n. 延性（材料在断裂前塑性变形的能力）
ductile ['dʌktail] fracture　延性断裂，又称"韧性断裂"
gauge length　标距（在所测定的应变或长度变化范围内，标出的试样原始长度）
grey cast iron　灰口铸铁，简称"灰铸铁"，亦为 gray cast iron 或 gray iron
low carbon steel　低碳钢（碳含量低于 0.25% 的碳素钢），亦为 mild steel
indenter [in'dentə]　n. 压头（硬度计上压入试样而无永久变形的零部件）
pyramid ['pirəmid]　n. 棱锥，四面体；v. 成角锥形
indentation [ˌinden'teiʃən]　n. 压痕（在试验力作用下，压头压入后试件表面产生的变形）
nondestructive [ˌnɔndis'trʌktiv]　a. 非破坏性的，无损的

Notes

[1] at one's disposal 意为"供……使用（available for use as someone prefers）"。
[2] yield strength, tensile strength 意为"屈服强度，抗拉强度"。

Lesson 14　Physical Properties of Materials

One of the important considerations in material selection is their physical properties (that is, density, melting point, specific heat, thermal conductivity, thermal expansion, and corrosion resistance). Physical properties can have several important influences on manufacturing and the service life of components. For example, high-speed machine tools require lightweight components to reduce inertial forces and, thus, keep machines from excessive vibration. [1]

1. Density　The density of a material is its mass per unit volume. Another term is specific gravity, which expresses a material's density in relation to that of water, and thus, it has no units. Weight saving is important particularly for aircraft and aerospace structures, for automotive bodies and components, and for other products where energy consumption and power limitations are major concerns. Substitution of materials for the sake of weight saving and economy is a major factor in the design both of advanced equipment and machinery and of consumer products, such as automobiles. [2]

2. Melting Point　The melting point of a metal is the temperature at which it changes state from solid to liquid at atmospheric pressure. At the melting point the solid and liquid phase exist in equilibrium. [3] The melting temperature of a metal alloy can have a wide range (depending on its composition) and is unlike that of a pure metal, which has a definite melting point.

3. Specific Heat　A material's specific heat is the energy required to raise the temperature of a unit mass by one degree. Alloying elements have a relatively minor effect on the specific heat of metals. The temperature rise in a workpiece resulting from machining operations is a function of the work done and of the specific heat of the workpiece material. [4] Temperature rise in a workpiece, if excessive, can decrease product quality by adversely affecting its surface roughness and dimensional accuracy, and can cause excessive tool wear.

4. Thermal Conductivity　Thermal conductivity indicates the rate at which heat flows within and through a material. Metals generally have high thermal conductivity, while ceramics and plastics have poor conductivity.

When heat is generated by plastic deformation or due to friction, the heat should be conducted away at a rate high enough to prevent a severe rise in temperature. The main difficulty experienced in machining titanium, for example, is caused by its very low thermal conductivity. Low thermal conductivity can also result in high temperature gradients and, in this way, cause inhomogeneous deformation of workpieces in metalworking processes. [5]

5. Thermal Expansion　The density of a material is a function of temperature. The general relationship is that density decreases with increasing temperature. Thermal expansion is the name given to this effect that temperature has on density.

In manufacturing operations, thermal expansion is put to good use in shrink fitting and expansion fitting. To assemble by shrink fitting, the external part is heated to enlarge it by thermal expansion, and the internal part remains at room temperature. The parts are then assembled and brought back to room temperature, so that the external part shrinks to form an interference fit. To assemble by expansion fitting, the internal part is cooled to decrease its size to permit insertion into the mating component. When the part returns to room temperature, an interference fit assembly is obtained. These assembly methods are used to fit gears, pulleys, and other components onto solid and hollow shafts.

6. Corrosion Resistance Corrosion resistance is an important aspect of material selection for applications in the chemical, food, and petroleum industries, as well as in manufacturing operations. In addition to various possible chemical reactions, environmental corrosion of components and structures is a major concern, particularly at elevated temperatures.

Resistance to corrosion depends on the composition of the material and on the particular environment. Corrosive media may be chemicals (acids, alkalis, and salts), the environment (oxygen, moisture, pollution, and acid rain), and water (fresh or salt water).[6] Nonferrous metals, stainless steels, and nonmetallic materials generally have high corrosion resistance. Steels and cast irons usually have poor resistance and must be protected by various coatings and surface treatments.

The usefulness of some level of oxidation is exhibited in the corrosion resistance of aluminum, titanium, and stainless steel. Pure aluminum naturally forms a thin (a few atomic layers) surface layer of aluminum oxide (Al_2O_3) on contact with oxygen in the atmosphere, that protects the surface from further environmental corrosion.[7] Titanium develops a film of titanium oxide (TiO_2). A similar phenomenon occurs in stainless steels, which develop an extremely thin (1–5 nm), strongly adherent protective film of chromium oxide in the presence of oxygen on their surfaces.[8] When the protective film is scratched and exposes the metal underneath, a new oxide film begins to form.

Words and Expressions

physical property 物理性质，物理性能
density ['densiti] *n.* 密度（单位体积中所含物质的质量）
machinery [mə'ʃiːnəri] *n.* 机械（机器与机构的总称）
melting point 熔点（在标准压强之下固体熔化成液体的温度）
specific gravity 比重（物质的密度和某一标准物质的密度之比）
phase [feiz] *n.* 相（体系内物理和化学性质均一的部分）
specific heat 比热（单位质量物质的温度升高 1℃ 所需的热量）
alloying element 合金元素（组成合金的化学元素）
a function of ……的函数（something that is related to and changes with something else）
thermal conductivity 热导率，导热系数（coefficient of thermal conductivity）
thermal expansion 热膨胀
corrosion resistance 耐（腐）蚀性（在给定的腐蚀体系中金属所具有的抗腐蚀能力）

surface roughness　表面粗糙度（加工表面上由较小间距和峰谷所组成的微观几何形状特征）
dimensional accuracy　尺寸精度（实际尺寸变化所达到的标准公差的等级范围）
tool wear　刀具磨损（刀具与工件材料作用引起的刀具损伤）
temperature gradient　温度梯度（在温度降低的方向上，单位距离内温度降低的数值）
inhomogeneous [ˌinhəuməˈdʒi:niəs]　*a.* 不均匀的
metalworking [ˈmetlwə:kiŋ]　*n.* 金属加工
put to good use　充分利用
shrink fitting　热装（具有过盈量配合的两个零件，装配时先将包容件加热胀大，再将被包容件装入配合位置的过程）
expansion fitting　冷装（具有过盈量配合的两个零件，装配时先将被包容件用冷却剂冷却，使其尺寸收缩，再装入包容件使其达到配合位置的过程）
external part and internal part　包容件和被包容件
mating component　相配零件，配合的零件
assemble [əˈsembl]　*v.* 装配（to connect or put together the parts of something）
assembly　装配，装配件（由零件、组合件通过多种形式的装配而连接在一起形成的单元）
scratch [skrætʃ]　*v.* 擦伤，划伤，刮伤

Notes

［1］lightweight 意为"重量轻的"，inertial force 意为"惯性力"。全句可译为：例如，高速机床需要采用重量轻的部件以减少惯性力，使机床不会产生过大的振动。

［2］consumer product 意为"用于消费的产品，消费品"。这两句可译为：对于飞机和航空航天器结构、汽车车身和部件，以及其他主要考虑能耗和功率限制的产品而言，减轻重量是特别重要的。在设计先进设备和机械及汽车等消费品时，采用新的替代材料来减轻重量和降低成本是一个应该着重考虑的问题。

［3］atmospheric pressure 意为"大气压力"。这两句可译为：金属的熔点是其在大气压力下，从固体转换为液体的温度。在熔点时，固相和液相呈平衡状态。

［4］work done 意为"所做的功（the force multiplied with the distance moved by the force）"。全句可译为：在机械加工过程中，工件温度的升高是所做的功和工件材料比热的函数。

［5］in this way 意为"这样，因此"。全段可译为：当塑性变形或摩擦产生热量时，这些热量应该以足够高的速率被传导出去，以防止温度急剧升高。例如，在加工钛金属时遇到的主要困难就是由于这种材料的导热系数非常低所造成的。导热系数低可以产生高的温度梯度，从而导致在金属加工过程中出现工件不均匀变形。

［6］fresh water 意为"淡水（water that is not salty）"。全句可译为：腐蚀性介质可以是化学品（酸、碱和盐）、环境（氧气、湿气、污染和酸雨）和水（淡水或盐水）。

［7］naturally 意为"自然地，轻而易举地"。全句可译为：纯铝与大气中的氧气接触后，其表面会自然地形成氧化铝薄膜（几个原子层厚），它会保护表面避免遭受进一步的环境腐蚀。

［8］protective film 意为"保护膜，防护膜"。全句可译为：在不锈钢中也会出现类似的现象，在有氧环境中，会生成非常薄（1～5nm）、牢固附着在其表面的氧化铬保护膜。

Lesson 15 Kinematics and Dynamics

One principal aim of kinematics is to design the desired motions of the mechanical parts and then mathematically compute the positions, velocities, and accelerations, which those motions will create on the parts. [1] Since engineering design is charged with creating systems which will not fail during their expected service life, the goal is to keep stresses within acceptable limits for the materials chosen and the environmental conditions encountered. [2] This obviously requires that all forces be kept within desired limits. Kinematics is the study of motion without regard to forces that cause it. Very basic and early decisions in the design process involving kinematic principles, which can be crucial to the success of any mechanical design. A design which has poor kinematics will prove troublesome and perform badly. Dynamics is the study of motions that result from forces. In machinery, the largest forces encountered are often those due to the dynamics of the machine itself. These dynamic forces are proportional to acceleration, which brings us back to kinematics, the foundation of mechanical design.

Any mechanical system can be classified according to the number of degrees of freedom (DOF) which it possesses. The system's DOF is equal to the number of independent parameters which are needed to uniquely define its position in space at any instant of time.

A rigid body free to move within a reference frame will, in the general case, have complex motion, which is a simultaneous combination of rotation and translation. [3] In three-dimensional space, there may be rotation about any axis and also simultaneous translation which can be resolved into components along three axes. In a plane, or two-dimensional space, complex motion becomes a combination of simultaneous rotation about one axis (perpendicular to the plane) and translation resolved into components along two axes in the plane. [4] For simplicity, we will limit our present discussions to the case of planar (2D) kinematic systems. We will define these terms as follows for our purposes, in planar motion:

Pure rotation The body possesses one point (center of rotation) which has no motion with respect to the "stationary" frame of reference. All other points on the body describe arcs about that center. A reference line drawn on the body through the center changes only its angular orientation (see Fig.15.1).

Pure translation All points on the body describe parallel (rectilinear or curvilinear) paths. A reference line drawn on the body changes its linear position but does not change its angular orientation (see Fig.15.2). [5]

Complex motion A simultaneous combination of rotation and translation. Any reference line drawn on the body will change both its linear position and its angular orientation. Points on the body will travel non-parallel paths, and there will be, at every instant, a center of rotation, which will continuously change location. [6]

Translation and rotation represent independent motions of the body. Each can exist without the other. If we define a 2D coordinate system as shown in Fig. 15.3, the x and y terms represent the translation components of motion, and the θ term represents the rotation component.[7]

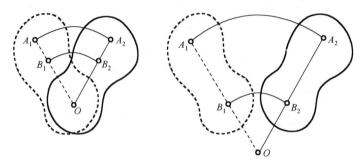

Figure 15.1 Pure rotation

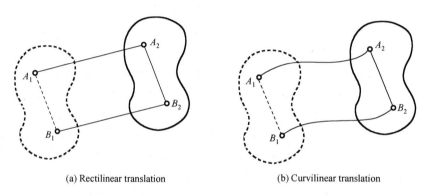

(a) Rectilinear translation   (b) Curvilinear translation

Figure 15.2 Pure translation

The pencil in this example represents a rigid body, or link, which for purposes of kinematic analysis we will assume to be incapable of deformation. This is merely an assumption to allow us to more easily define the motion of the body. We can later superpose any deformation due to external or internal loads on the body to obtain a more complete and accurate solution. But remember, we are typically facing a blank sheet of paper at the beginning stage of the design process. We cannot define deformations of a body until we define its size, shape, material properties, and loadings. Thus, at this stage we will assume that our kinematic bodies are rigid and massless.[8]

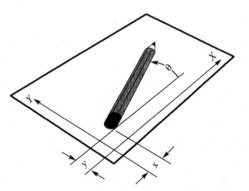

Figure 15.3 A rigid body in the plane

When designing machinery, we must first do a complete kinematic analysis of our design, in order to obtain information about the accelerations of the moving parts. We next want to use Newton's second law to calculate the dynamic forces. But to do so we need to know the masses of all the moving parts which have these known accelerations. These parts do not exit yet! As with any design problem, we lack sufficient information at this stage of the design to accurately determine the best sizes and shapes of the parts. We must estimate the masses of the links and other parts of the design in order to make a first pass at the calculation. We will then have to iterate to better and better solutions as we generate more information.

A first estimate of your parts' masses can be obtained by assuming some reasonable shapes and size for all the parts and choosing appropriate materials. Then calculate the volume of each part and multiply its volume by the material's density to obtain a first approximation of its mass. These mass values can then be used in Newton's equation.

How will we know whether our chosen sizes and shapes of links are even acceptable, let alone optimal? Unfortunately, we will not know until we have carried the computations all the way through a complete stress and deflection analysis of the parts. It is often the case, especially with long, thin elements such as shafts or slender links, that the deflections of the parts under their dynamic loads will limit the design even at low stress level.

We will probably discover that the parts fail under the dynamic forces. Then we will have to go back to our original assumptions about the shapes, sizes, and materials of these parts, redesign them, and repeat the force, stress, and deflection analysis.

It is also worth noting that, unlike a static force situation in which a failed design might be fixed by adding more mass to the part to strengthen it, to do so in a dynamic force situation can have a deleterious effect.[9] More mass with the same acceleration will generate even higher forces and thus higher stresses! The machine designer often need to remove mass (in the right places) from parts in order to reduce the stresses and deflections due to $\boldsymbol{F} = m\boldsymbol{a}$. Thus the designer needs to have a good understanding of material properties, stress and deflection analysis to properly shape and size parts for minimum mass while maximizing the strength and stiffness needed to withstand the dynamic forces.[10]

Words and Expressions

kinematics [ˌkainiˈmætiks] *n.* 运动学
service life 使用寿命 (the time period over which a product is expected to be used)
dynamic force 动力，动态力（随时间变化的力）
without regard to 不考虑 (not taking something into account)，无须顾及
crucial [ˈkruːʃiəl] *a.* 至关紧要的，非常重要的 (extremely important)
troublesome [ˈtrʌblsəm] *a.* 棘手的，令人烦恼的 (causing difficulty or annoyance)
degree of freedom 自由度（在任意时刻完全确定机械系统位置所需要的独立的广义坐标数）
parameter [pəˈræmitə] *n.* 参数（在给定系统内描述变量关系的量），也叫参变量
uniquely [juːˈniːkli] *ad.* 独特地，唯一地
in the general case 通常，总地来说，概括地说
axis 轴 (a reference line from which distances or angles are measured in a coordinate system)
translation 平移（在平面内构件上各点均向同一方向移动相同距离的运动）
rectilinear translation or curvilinear translation 直线平移或曲线平移
orientation [ˌɔːrienˈteiʃən] *n.* 方向，方位 (the state or fact of facing a particular direction)
link 构件（机构中的运动单元体），杆（机构中只具有低副元素的构件，亦为 bar）

multiply by 乘以，使相乘
let alone 更不用说
all the way through 一直到，自始至终
dynamic load 动载荷（any load that moves, changing magnitude or direction over time）
deleterious [ˌdeliˈtiəriəs] a. 有害的

Notes

[1] principal 意为"最重要的，主要的（most important; main）"，desired motion 意为"需要的运动"。全句可译为：运动学的一个主要目的是设计机械零件应该具有的运动，然后用数学方法计算在实现这些运动时零件的位置、速度和加速度。

[2] charge with 意为"承担（to impose a duty or task on）"。全句可译为：由于工程设计所承担的任务是建立一些不会在其预期使用寿命内失效的系统，因此目标是对于所选用的材料和使用环境，应力必须始终处于安全的范围内。

[3] reference frame 或 frame of reference 意为"参考系"。全句可译为：在一个参考系内自由运动的刚体，通常会做复杂运动，即同时进行转动和平移。

[4] resolve into 意为"分解为（to separate into its parts）"，component 意为"分量"。全句可译为：在平面上或二维空间中，复杂运动变成了同时发生的绕一个（垂直于这个平面的）轴的转动和可以被分解为在这个平面上沿两个坐标轴的平移分量。

[5] describe 意为"沿……运动（to make a movement that has a particular shape）"，reference line 意为"基准线"。这两句可译为：物体上所有的点都沿平行（直线或曲线）的轨迹运动。在物体上画的一条基准线，只能发生位置变化，不会改变其角度方向（见图 15.2）。

[6] linear position 意为"线性位置，位置"。这几句可译为：同时产生转动和平移的合成运动称为复杂运动。此时物体上任意一条基准线的位置和角度方向都会发生变化，物体上各点的运动轨迹是非平行的。在每一个瞬间都会有一个转动中心，转动中心的位置是不断变化的。

[7] x and y terms 意为"x 项和 y 项"，translation component 意为"平移分量"，rotation component 意为"转动分量"。全句可译为：如果我们定义一个如图 15.3 所示的平面坐标系，x 项和 y 项代表运动的平移分量，θ 项代表转动分量。

[8] superpose [sjuːpəˈpəuz] 意为"叠加"，a blank sheet of paper 意为"一张白纸"。这四句可译为：随后，我们可以将由外部载荷或内部载荷使物体产生的任何变形与之叠加，以获得更完整和更精确的答案。但是要记住，在设计的初始阶段，我们通常面对一张白纸。只有在确定了物体的尺寸、形状、材料和载荷之后，我们才能确定它们的变形。因此，在这个阶段，我们假定运动物体是刚体并且没有质量。

[9] it is also worth noting 意为"同样值得注意的是（it is also important to note）"，fix 这里指"解决（问题），修复，处理"。全句可译为：同样值得注意的是，在静态力作用下，可以通过增加零件的质量来提高其强度，将不合格的设计变为合格；而在动态力作用的情况下，这样做可能会产生有害的后果。

[10] shape 和 size 在这里均为动词，意为"确定形状和尺寸"。全句可译为：因此，设计人员需要对材料的性能及应力和变形分析有透彻的理解，才能确定合适的零件形状和尺寸，使其在质量最小的情况下同时具有承受动态力所需的最大强度和刚度。

Lesson 16 Basic Concepts of Mechanisms

A mechanism is a component of a machine consisting of two or more bodies arranged so that the motion of one compels the motion of the others. Kinematics is the study of motion in mechanisms without reference to the forces that act on the mechanism. [1] Dynamics is the study of motion of individual bodies and mechanisms under the influence of forces and torques. The study of forces and torques in stationary systems is called statics.

The definitions of some terms and concepts fundamental to the study of kinematics and dynamics of mechanisms are presented below.

Link A link is one of the rigid bodies joined together to form a kinematic chain. [2] The term rigid link, or sometimes simply link, is an idealization used in the study of mechanisms that does not consider small deflections due to strains in machine parts. A perfectly rigid or inextensible link can exist only as a textbook type of model of a real machine part. For typical machine parts, the maximum changes in dimension are on the order of only a one-thousandth of the part length. We are justified in neglecting this small motion when considering the much greater motion characteristic of most mechanisms. The word link is used in a general sense to include cams, gears, and other machine parts in addition to cranks and couplers. [3]

Frame The fixed or stationary link in a mechanism is called the frame. When there is no link that is actually fixed, we may consider one as being fixed and determine the motion of the other links relative to it. In an automotive engine, for example, the engine block (Fig. 16.1) is considered as the frame, even though the automobile may be moving.

Kinematic Pair The connections between two or more links that permit constrained relative motion are called kinematic pairs. [4]

Lower and Higher Pairs Theoretically, the two elements of a lower pair [5] are in surface contact with one another, while the two elements of a higher pair have theoretical point or line contact (if we disregard deflections).

Figure 16.1 An engine block

Kinematic Chain A kinematic chain is an assembly of links and kinematic pairs. Kinematic chains may be either open or closed (see Fig. 16.2). Each link in a closed kinematic chain is connected to two or more other links. A kinematic chain failing to meet this criterion is an open kinematic chain. [6] In this case, one (or more) of the links is connected to one other link. The industrial robot shown in Fig. 71.1 is an open kinematic chain.

Degrees of Freedom The number of degrees of freedom (DOF) of a mechanism is the number of independent parameters required to specify the position of every link relative to the frame or fixed link.

An unconstrained rigid body has six degrees of freedom: translation in three coordinate directions and rotation about three coordinate axes.[7] If the body is restricted to motion in a plane, there are three degrees of freedom: translation in two coordinate directions and rotation within the plane (see Fig. 15. 3).

Linkage A linkage is a mechanism in which all connections between links are lower pairs. The slider-crank mechanism and four-bar linkage (see Figs. 16. 3 and 16. 4) are typical examples.[8]

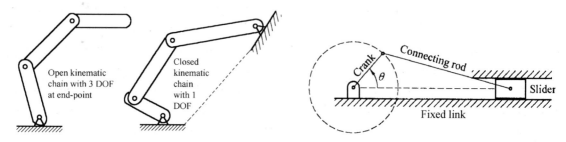

Figure 16. 2 Examples of kinematic chains   Figure 16. 3 A slider-crank mechanism

Planar Linkage If all points in a linkage move in parallel planes, the system undergoes planar motion and the linkage may be described as a planar linkage. Most of the mechanisms in common use may be treated as planar linkages.

Spatial Linkage The more general case in which motion cannot be described as taking place in parallel planes is called spatial motion, and the linkage may be described as a spatial linkage.

Kinematic Inversion The absolute motion of a linkage depends on which of the links is fixed, that is, which link is selected as the frame. An inversion is created by selecting a different fixed link in the linkage. Thus there are as many inversions of a given linkage as it has links.[9] Figure 16. 4 shows the four possible inversions of a four-bar linkage: two crank-rocker linkages (Figs. 16. 4a and 16. 4b), a double-rocker linkage (Fig. 16. 4c), and a double-crank linkage (Fig. 16. 4d). In Fig. 16. 4, the link that has no connection to the fixed link is called a coupler, both of its ends are able to move; the link which can rotate 360° is referred to as a crank.

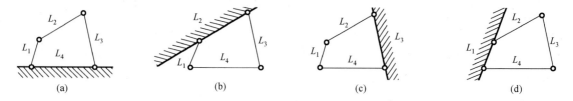

Figure 16. 4 Kinematic inversions of a four-bar linkage

Words and Expressions

mechanism [ˈmekənizəm] *n.* 机构（两个或两个以上构件通过活动连接形成的构件系统）

kinematic chain 运动链（用运动副连接而成的相对可动的构件系统）

rigid link　刚性构件（受力变形可忽略不计的构件）
idealization [aiˌdiəlaiˈzeiʃən]　n. 理想化，理想化的事物
crank [kræŋk]　n. 曲柄（与机架用转动副相连并能绕该转动副轴线整圈旋转的构件）
coupler [ˈkʌplə]　n. 连杆（机构中不与机架相连的杆件），亦为 floating link
frame [freim]　n. 机架（机构中固结于定参考系的构件），固定构件（fixed link）
engine block　发动机缸体，机体（曲轴箱及与其连成一体的汽缸或汽缸水套）
slider-crank mechanism　曲柄滑块机构（通常指具有一个曲柄和一个滑块的平面四杆机构）
connecting rod　连杆（机构中与滑块、活塞等做往复运动的构件用转动副相连的运动构件）
slider　滑块（机构中与机架用移动副相连又与其他运动构件用转动副相连的构件）
kinematic inversion [inˈvəːʃən]　机架变换，运动倒置
rocker [ˈrɔkə]　n. 摇杆（与机架用转动副相连但只能绕该转动副轴线摆动的构件）
crank-rocker linkage　曲柄摇杆机构（有一个曲柄和一个摇杆的铰链四杆机构）
double-rocker linkage　双摇杆机构（两连架杆均为摇杆的铰链四杆机构）
double-crank linkage　双曲柄机构（两连架杆均为曲柄的铰链四杆机构）

Notes

[1] a component of 意为"……的组成部分"，without reference to 意为"与……无关"。这两句可译为：机构是机器的组成部分，它由两个或多个物体组成，其中某一个物体的运动能使其他物体产生运动。运动学只研究机构的运动，而不考虑作用在机构上的力。

[2] link 意为"构件"。全句可译为：连接在一起组成运动链的每个刚体称为构件。

[3] on the order of 意为"近似（close to），大约（approximately）"。这四句可译为：一个绝对刚性或不可伸长的构件只能作为真实机器零件在教科书上的模型而存在。对于典型的机器零件，它最大的尺寸变动量大约仅是该零件长度的千分之一。因此，在大多数机构中，在考虑比其大得多的运动特性时，忽略这个很小的运动是完全可以的。一般来说，构件这个词除曲柄和连杆外，也包括凸轮、齿轮和其他机器零件。

[4] constrained 意为"受到限制的"，kinematic pair 意为"运动副"。全句可译为：运动副是两个或多个构件之间的可动连接，它限制了构件之间的某些相对运动。

[5] element 这里指"运动副元素"，lower pair 意为"低副"。全句可译为：在理论上，低副的两个元素之间为面接触，而高副的两个元素之间为点或线接触（如果我们忽略变形的话）。

[6] closed kinematic chain 意为"闭式运动链"。这四句可译为：运动链是由若干个构件和运动副连接而成的。运动链可以是开式的或闭式的（见图 16.2）。在闭式运动链中，每个构件至少与其他两个构件相连接。不符合这个标准的运动链是开式运动链。

[7] unconstrained 意为"自由的"，coordinate 意为"坐标"。全句可译为：一个自由刚体有六个自由度：沿三个坐标轴方向的移动自由度和绕这三个坐标轴的转动自由度。

[8] linkage 意为"连杆机构"。这两句可译为：连杆机构是一个构件间只用低副连接的机构。曲柄滑块机构和四杆机构（见图 16.3 和图 16.4）都是典型的连杆机构。

[9] inversion 这里指"机架变换"，given 意为"给定的"。这三句可译为：连杆机构的绝对运动取决于选择哪个构件作为固定构件，也就是说，选择哪个构件作为机架。如果在一个连杆机构中选择不同的构件作为固定构件，就产生了机架变换。因此，在一个给定的连杆机构中，有多少个构件就有多少种机架变换形式。

Lesson 17　Material Selection

During recent years the selection of engineering materials has assumed great importance. Moreover, the process should be one of continual reevaluation. New materials often become available and there may be a decreasing availability of others. Concerns regarding environmental pollution, recycling and worker health and safety often impose new constraints. The desire for weight reduction or energy savings may dictate the use of different materials. Pressures from domestic and foreign competition, increased maintainability[1] requirements, and customer feedback may all promote materials reevaluation. The extent of product liability actions, often the result of improper material use, has had a marked impact. In addition, the interdependence between materials and their processing has become better recognized. The development of new processes often forces reevaluation of the materials being processed. Therefore, it is imperative that design and manufacturing engineers exercise considerable care in selecting, specifying, and utilizing materials if they are to achieve satisfactory results at reasonable cost and still assure quality.

The first step in the manufacture of any product is design, which usually takes place in several distinct stages: (a) conceptual; (b) functional; (c) production design. During the conceptual-design[2] stage, the designer is concerned primarily with the functions the product is to fulfill. Usually several concepts are visualized and considered, and a decision is made either that the idea is not practical or that the idea is sound and one or more of the conceptual designs should be developed further. Here, the only concern for materials is that materials exist that can provide the desired properties. If no such materials are available, consideration is given as to whether there is a reasonable prospect that new one could be developed within cost and time limitations.

At the functional- or engineering-design stage, a practical, workable design is developed. Fairly complete drawings are made, and materials are selected and specified for the various components. Often a prototype or physical model[3] is made that can be tested to permit evaluation of the product as to function, reliability, appearance, maintainability, and so on. Although it is expected that such testing might show that some changes may have to be made in materials before the product is advanced to the production-design stage, this should not be taken as an excuse for not doing a thorough job of material selection. Appearance, cost, and reliability factors should be considered in detail, together with the functional factors. There is much merit to the practice of one very successful company which requires that all prototypes be built with the same materials that will be used in production and, insofar as possible, with the same manufacturing techniques. It is of little value to have a perfectly functioning prototype that cannot be manufactured economically in the expected sales volume, or one that is substantially different from what the products will be in regard to quality and reliability. Also, it is much better for design engineers to do a complete job of material analysis, selection, and specification at the development stage of design rather than to leave it to the production-design stage, where changes may be made by others, possibly less knowledgeable about all of the functional aspects of the product.

At the production-design stage, the primary concern relative to materials should be that they are specified fully, that they are compatible with, and can be processed economically by, existing equipment, and that they are readily available in the needed quantities.

As manufacturing progresses, it is inevitable that situations will arise that may require modifications of the materials being used. Experience may reveal that substitution of cheaper materials can be made. In most cases, however, changes are much more costly to make after manufacturing is in progress than before it starts. Good selection during the production-design phase will eliminate the necessity for most of this type of change. The more common type of change that occurs after manufacturing starts is the result of the availability of new materials. These, of course, present possibilities for cost reduction and improved performance. However, new materials must be evaluated very carefully to make sure that all their characteristics are well established[4]. One should always remember that it is indeed rare that as much is known about the properties and reliability of a new material as about those of an existing one. A large proportion of product failure and product liability[5] cases have resulted from new materials being substituted before their long-term properties were really known.

Product liability actions have made it imperative that designers and companies employ the very best procedures in selecting materials. The five most common faults in material selection have been: (a) failure to know and use the latest and most useful information available about the materials utilized; (b) failure to foresee, and take into account the reasonable uses for the product (where possible, the designer is further advised to foresee and account for misuse of the product, as there have been many product liability cases in recent years where the claimant, injured during misuse of the product, has sued the manufacturer and won); (c) the use of materials about which there was insufficient or uncertain data, particularly as to its long-term properties; (d) inadequate, and unverified, quality control procedures; and (e) material selection made by people who are completely unqualified to do so.

An examination of the faults above will lead one to conclude that there is no good reason why they should exist. Consideration of them provides guidance as to how they can be eliminated. While following the very best methods in material selection may not eliminate all product-liability claims, the use of proper procedures by designers and industries can greatly reduce their numbers. From the previous discussion, it is apparent that those who select materials should have a broad, basic understanding of the nature and properties of materials and their processing.

Words and Expressions

assume [əˈsjuːm]　*v.*　设想，承担责任（to take or begin to have responsibility）

reevaluation [riːˌivæljuˈeiʃən]　*n.*　重新评价

become available　可供使用的，在市场销售（to come on the market; to be offered for sale）

availability [əˌveiləˈbiliti]　*n.*　可用性（the fact or possibility that you can buy, get, or have sth）

recycling [ˌriːˈsaikliŋ] n. 回收再利用（a process to change waste materials into new products）
dictate [dikˈteit] v. 规定，限定，确定，指挥
maintainability [menˌteinəˈbiliti] n. 可维修性，可维护性，维修性
feedback [ˈfiːdbæk] n. 反馈（the return of information about the result of a process or activity）
promote [prəˈməut] v.; n. 促进，发扬，引起
liability [ˌlaiəˈbiliti] n. 责任，法律责任（the state of being responsible for something, especially by law），不利条件（something disadvantageous）
conceptual [kənˈseptjuəl] a. 概念上的
interdependence [ˌintədiˈpendəns] n. 互相依存，相关性
imperative [imˈperətiv] a. 非常重要的（extremely important），必要的（necessary）
visualize [ˈvizjuəlaiz] v. 想象，设想（to form a mental image of），显现（to make visible）
sound 合理的（based on valid reasoning），正确的（free from mistakes），坚固的（firm）
prospect [ˈprɔspekt] n. 期待（something expected），可能性（a possibility），展望
workable [ˈwəːkəbl] a. 切实可行的（capable of being put into effective operation），可使用的
substantially [səbˈstænʃəli] ad. 大幅地（to a great extent or degree），基本上（essentially）
compatible [kəmˈpætəbl] a. 相容的，可共存的，相适应的
insofar as [ˌinsəuˈfɑːrəz] conj. 到这样的程度，在……情况下，既然，因为
inevitable [inˈevitəbl] a. 不可避免的，必然的
reveal [riˈviːl] v. 显示（to show clearly），揭示（to make known）
account for 是（某事）的原因（to be the cause of）
claim 索赔（a right to seek a judicial remedy arising from a wrong or injury suffered）
misuse [ˈmisˈjuːz] n.; v. 错用，误用（the wrong or improper use of sth; to use incorrectly）
claimant [ˈkleimənt] n. 提出要求者，原告（the party who initiates a lawsuit）
sue [sjuː] v. 控告，提起诉讼（to seek justice or right by bringing legal action）
unverified [ʌnˈverifaid] a. 未经证实的（not having been confirmed or proven to be true）
unqualified [ˈʌnˈkwɔlifaid] a. 无资格的，不能胜任的（lacking the necessary qualifications）

Notes

[1] maintainability 意为"可维护性，可维修性，维修性（在规定使用条件下使用的产品，在规定条件下并按规定的程序和手段实施维修时，保持或恢复能执行规定功能状态的能力）"。
[2] conceptual-design 意为"概念设计，方案设计"。
[3] physical mode 意为"物理模型，实体模型（a physical construct whose characteristics resemble the physical characteristics of the modeled system）"。
[4] well established 意为"经过验证的（validated），众所周知的（well-known）"。
[5] product liability 意为"产品责任（又称产品侵权损害赔偿责任，是指产品存在可能危及人身、财产安全的不合理危险，造成消费者人身或除缺陷产品外的其他财产损失后，缺陷产品的生产者、销售者应当承担的特殊的侵权法律责任）"。

Lesson 18 Selection of Materials

An ever-increasing variety of materials is now available, each having its own characteristics, applications, advantages, and limitations. The following are the general types of materials used in manufacturing, either individually or in combination with other materials. [1]

1. Ferrous metals: carbon, alloy, stainless, tool, and die steels.
2. Nonferrous metals: aluminum, copper, nickel, titanium, and precious metals.
3. Plastics: thermoplastics, thermoset plastics, and elastomers.
4. Ceramics, glasses, graphite, and diamond.
5. Composite materials: reinforced plastics, metal-matrix and ceramic-matrix composites.
6. Nanomaterials, shape memory alloys, superconductors, and various other materials with unique properties.

As new materials are developed, the selection of appropriate materials becomes even more challenging. Aerospace structures, as well as products such as sporting goods, have been at the forefront of new material usage. The trend has been to use more titanium and composites for the airframes of commercial aircraft, with a gradual decline of the use of aluminum and steel. There are continously shifting trends in the usage of materials in all products, driven principally by economic considerations as well as other considerations. [2]

For instance, plastics are now widely used in numerous applications for such items as children's toys, automotive and electrical parts, and computer equipment, because of their durability and lower manufacturing costs. In the past few decades, an entirely new family of ceramics of oxides, nitrides, and carbides, with superior properties, have been produced. The new generation of ceramic materials called engineering ceramics, structural ceramics, or advanced ceramics has higher strength, better wear and corrosion resistance (even at higher temperatures), and lower thermal conductivity. They are used to make such things as space shuttle heat shield tiles, ceramic bearings (see Fig. 9.1), and ceramic inserts (see Fig. 39.5). [3]

When selecting materials for products, we first consider their mechanical properties: strength, ductility, hardness, elasticity, and fatigue; then consider their physical properties: density, specific heat, thermal expansion and conductivity, melting point, and electrical and magnetic properties. Optimum designs often require a consideration of a combination of mechanical and physical properties. A typical example is the strength-to-weight and stiffness-to-weight ratios of materials for minimizing the weight of structural members. Aluminum, titanium, and reinforced plastics, for example, have higher ratios than steels and cast irons. Weight minimization is particularly important for aerospace and automotive applications, in order to improve performance and fuel economy. [4]

Chemical properties also play a significant role in hostile (such as corrosive) as well as normal environments. Oxidation, corrosion, and flammability of materials are among the important factors to be considered. [5]

Manufacturing properties of materials determine whether they can be cast, formed, machined, welded, and heat treated with relative ease. The method(s) used to process materials to the desired shapes may adversely affect the product's final properties, service life, and its cost. [6]

Cost and availability of raw materials are major concerns in manufacturing. If raw materials are not available in the desired shapes, dimensions and quantities, substitutes and additional processing will be required, which can contribute significantly to product cost. For example, if we need a round bar of a certain diameter and it is not available in standard form [7], then we have to purchase a larger rod and reduce its diameter by some means, such as turning or grinding. It should be noted, however, that a product design can often be modified to take advantage of standard dimensions of raw materials, thus avoiding extra manufacturing costs.

Reliability of supply, as well as demand, affects cost. Most countries import numerous raw materials that are essential for production. The United States, for example, imports most of the cobalt, titanium, aluminum, nickel, natural rubber, and diamond that it needs. [8]

Different costs are involved in processing materials by different methods. Some methods require expensive machinery, others require extensive labor, and still others require personnel with special skills, a high level of education, or specialized training.

The appearance of materials after they have been manufactured into products influences their appeal to the consumer. Color and surface texture are characteristics that we all consider when making a decision about purchasing a product. [9]

Time- and service-dependent phenomena such as wear, fatigue, and dimensional stability are important. These phenomena can significantly affect a product's performance and, if not controlled, can lead to total failure of the product. Wear, corrosion, and other phenomena can shorten a product's life or cause it to fail prematurely. Recycling or proper disposal of materials at the end of their useful service lives has become increasingly important in an age when we are more conscious of preserving resources and maintaining a clean and healthy environment. The proper treatment and disposal of toxic wastes and materials are also a crucial consideration. [10]

Words and Expressions

ferrous ['ferəs] metal 黑色金属（对铁、铬和锰的统称，亦包括这三种金属的合金）
carbon, alloy, tool, and die steels 碳钢、合金钢、工具钢和模具钢
nonferrous metal 有色金属（除铁、铬、锰三种金属以外的所有金属元素的统称）
precious metal 贵金属（指金、银和铂等），也可写为 noble metal
thermoplastics [ˌθɜːməˈplæstɪks] n. 热塑性塑料（具有加热软化、冷却硬化特性的塑料）
thermoset plastics 热固性塑料（具有不溶、不熔特性的塑料，如酚醛塑料等）
elastomer [ɪˈlæstəmə] n. 弹性体（类似橡胶的弹性材料，any of various elastic materials that resemble rubber），弹性高分子材料（elastic polymer）

composite [kəm'pɑːzət] material　复合材料（a combination of two or more different materials that results in a superior, often stronger product）
reinforced plastics　增强塑料（含有增强材料的塑料，是一种重要的高分子复合材料）
ceramic matrix composite　陶瓷基复合材料
nanomaterial ['nænəumə,tiəriəl]　n. 纳米材料
shape memory alloy　形状记忆合金（能在一定条件下恢复原来的形状的功能性金属材料）
superconductor [,sjuːpəkəndkʌktə]　n. 超导体（在足够低的温度和足够弱的磁场下，其电阻为零的物质）
even more challenging　更具有挑战性
aerospace ['ɛərəuspeis]　n. 航空航天（飞行器在地球大气层中和太空航行活动的总称）
sporting goods　体育用品（equipment and clothes that are used in sport）
forefront　最前部（the foremost part or area），最重要位置（the position of most importance）
airframe ['ɛəfreim]　n. 机体结构（飞机的机身、机翼、起落架等，不包括动力装置）
aircraft ['ɛəkrɑːft]　n.（单复同形的名词）航空器，飞机
durability　耐久性（产品在规定的使用与维修条件下，直到极限状态前完成规定功能的能力）
shield [ʃiːld]　n. 防护物（a device or part that serves as a protective cover），罩，屏
insert [in'sət]　n. 刀片（装夹在刀体上的片状物体，并由它形成刀具的切削部分）
ceramic insert　陶瓷刀片（an insert made from engineering ceramics）
strength-to-weight ratio　强度重量比，比强度（specific strength，材料的强度与其密度之比）
stiffness-to-weight ratio　刚度重量比，比刚度（材料的弹性模量与其密度的比值）
flammability [,flæmə'biləti]　n. 易燃性（在规定的试验条件下材料易发生持续有焰燃烧的特性）
raw material　原材料（投入生产过程以制造新产品的物质）
bar　棒材（一种截面均匀的轧材，其截面通常为圆形、矩形或六边形）
appeal to　对……有吸引力（to be interesting or attractive to someone），呼吁
surface texture　表面结构（是表面粗糙度、表面波纹度、表面纹理和表面缺陷等的总称）
service-dependent　与使用有关的
prematurely [,premə'tjuəli]　ad. 过早地
disposal [dis'pəuzəl]　n. 处理，处置
useful service lives　正常使用寿命
preserving resources　保护资源

Notes

[1] ever-increasing 意为"不断增加的，越来越多的"。这段可译为：目前，可以供人们使用的材料的种类越来越多，每一种材料都有其自身的特点、用途、优点和局限性。下面介绍的是一些在当今制造业中可以单独使用或与其他材料组合使用的常见材料。

[2] shifting trend 意为"变化趋势"，consideration 这里指"因素（a factor to be considered in forming a judgment or decision）"。这两句可译为：当前的趋势是在商用飞机的机体结构材料中越来越多地采用钛和复合材料，而铝和钢的用量则逐渐减少。在各种产品中，所使用的材料的趋势是不断变化的，这主要是由经济因素或其他因素决定的。

[3] widely used in numerous applications 意为"广泛应用于众多的领域中"，an entirely new family of 意为"一类完全新型的"，superior property 意为"性能优异的"，heat shield tile

意为"防热瓦"，也可写为 thermal protection tile。这段可译为：例如，由于其耐久性和低制造成本，塑料现在被广泛应用于诸如儿童玩具、汽车部件、电器部件和计算机设备等众多的领域中。在过去的几十年中，一类性能优异的，完全新型的氧化物陶瓷、氮化物陶瓷和碳化物陶瓷被生产出来。这些被称为工程陶瓷、结构陶瓷或先进陶瓷的新一代陶瓷材料具有较高的强度、较好的耐磨性和耐蚀性（即使在较高的温度时也具有这些性能），以及较低的导热系数。它们被用来制造诸如航天飞机防热瓦、陶瓷轴承（见图 9.1）和陶瓷刀片（见图 39.5）等产品。

[4] optimum design 这里指"最佳设计方案"，structural member 意为"结构构件"，fuel economy 意为"燃油经济性（以最小的燃油消耗量完成单位运输工作的能力）"。这四句可译为：最佳设计方案需要综合考虑机械性能和物理性能。一个典型的例子是利用材料的强度重量比和刚度重量比来实现结构构件的重量最小化。例如，铝、钛和增强塑料的上述两项比值高于钢和铸铁。重量最小化可以提高产品性能并节省燃料，因此在航空航天工业和汽车工业领域是非常重要的。

[5] oxidation 意为"氧化（the combination of a substance with oxygen）"，hostile environment 意为"恶劣环境"。这两句可译为：化学性能在正常环境和恶劣环境（如腐蚀环境）中都起着重要的作用。材料的氧化、腐蚀和易燃性都是值得考虑的重要因素。

[6] form 这里指"压力加工（使毛坯材料产生塑性变形的无切屑加工）"，machined 意为"经过机械加工的"。这两句可译为：材料的制造性能决定了是否可以相对容易地对它们进行铸造、压力加工、机械加工、焊接和热处理。在将材料加工成预期形状的过程中，所采用的方法可能会对产品的最终性能、使用寿命和成本产生不利影响。

[7] availability 意为"可得性（在需要时能够提供足够产品的能力）"，standard form 意为"标准形式，标准尺寸规格"。这段可译为：原材料的成本和可得性是制造业重点关注的问题。如果无法获得具有你想要的形状、尺寸和数量的原材料，那么就需要采用替代品，并需要对其做进一步的加工，这会显著提高产品的成本。例如，如果我们需要某一直径的圆棒料，但它不属于标准尺寸规格，那么我们就必须购买尺寸大一些的棒料，然后通过车削或磨削等方法减小其直径。然而，值得注意的是，通常可以对产品设计进行修改，采用标准尺寸的原材料，从而避免产生额外的制造成本。

[8] essential for 意为"对……是必要的"。这两句可译为：材料供应的可靠性和需求都会影响其成本。大部分国家都会进口许多种类的生产所必需的原材料。例如，美国所需的大部分钴、钛、铝、镍、天然橡胶和金刚石都是进口的。

[9] 这段可译为：材料制成产品后的外观会影响其对消费者的吸引力。颜色和表面结构都是我们在决定购买产品时会考虑的特征。

[10] dimensional stability 意为"尺寸稳定性"，total failure 意为"失效，产品丧失规定的功能"，toxic wastes 意为"有毒废物"。这段可译为：诸如磨损、疲劳和尺寸稳定性这些与时间和使用相关的现象是非常重要的。这些现象能显著地影响产品的性能，如果不加控制，则可导致产品失效。磨损、腐蚀和其他现象可以缩短产品的寿命或使其过早失效。在当今这个更加注重保护资源与保持清洁和健康的环境的时代，材料达到使用年限后如何对其回收利用和妥善处理这个问题变得越来越重要。对有毒废物和有毒材料进行妥善处理和处置也是一个值得着重考虑的问题。

Lesson 19　Gear Materials (*Reading Material*)

Gears (see Fig. 19.1) are manufactured from a wide variety of materials, both metallic as well as nonmetallic. As is the case with all materials used in design, the material chosen for a particular gear should be the cheapest available that will ensure satisfactory performance. Before a choice is made, the designer must decide which of several criteria is most important to the problem at hand. If high strength is the prime consideration, a steel should usually be chosen rather than cast iron. If wear resistance is the most important consideration, a nonferrous metal is preferable to a ferrous one. As still another example of how a choice can be made, for problems involving noise reduction, nonmetallic materials perform better than metallic ones. However, as is true in most design problems, the final choice of a material is usually a compromise. In other words, the material chosen will conform reasonably well to all the requirements mentioned previously, although it will not necessarily be the best in any one area. To conclude this discussion, we will consider the characteristics of various metallic and nonmetallic gear materials according to their general classifications.

(a) Helical gear　　(b) Herringbone gear　　(c) Elliptical gears

Figure 19.1　Examples of gears

Cast Irons　Cast iron is one of the most commonly used gear materials. Its low cost, ease of casting, good machinability, high wear resistance, and good noise abatement property make it a logical choice. The primary disadvantage of cast iron as a gear material is its low tensile strength, which makes the gear tooth weak in bending and necessitates rather large teeth.

Another type of cast iron is nodular iron, which is made of cast iron to which a material such as magnesium or cerium has been added. The result of this alloying is a material having a much higher tensile strength while retaining the good wear and machining characteristics of ordinary cast iron.

Very often the combination of cast iron gear and a steel pinion will give a well balanced design with regard to cost, strength, and wear.

Steels　Steel gears are usually made of plain carbon steels or alloy steels. They have the advantage, over cast iron gears, of higher strength without undue increase in cost. However, they usually require heat treatment to produce a surface hard enough to give satisfactory resistance to wear. Unfortunately, the heat treatment process usually produces distortion of the gear, with the result that the gear load is not uniformly distributed across the gear tooth face. Since alloy steels are subject to less distortion due to heat treatment than carbon steels, they are often chosen in preference to the carbon steels.

Gears are often through-hardened by water or oil quenching in order to increase their resistance to wear. Tempering is usually performed immediately after quenching and involves reheating the gear to a temperature of 200℃ to 700℃, soaking at this temperature, and then slowly cooling it in air back to room temperature. If a low degree of hardness is satisfactory, through-hardening is probably the most desirable heat treatment process to be used because of its inexpensiveness.

In many cases, the bulk of the part requires only moderate strength although the surface must have a very high hardness. In gear teeth, for example, high surface hardness is necessary to resist wear as the mating teeth come into contact several million times during the expected life of the gears. At each contact, a high stress develops at the surface of teeth. For applications such as this, case hardening is used, the surface (or case) of the part is given a high hardness to a depth of perhaps 0.25–1.00 mm, although the interior of the part (the core) is affected only slightly, if at all. The advantage of surface hardening is that as the surface receives the required wear-resisting hardness, the core of the part remains in a more ductile form, resistant to impact and fatigue. The processes used most often for case hardening are flame hardening, induction hardening, laser hardening, carburizing, and nitriding. Figure 19.2 shows a drawing of a typical case-hardened gear-tooth section, clearly showing the hard case surrounding the softer, more ductile core.

Nonferrous Metals Copper, zinc, aluminum, and titanium are materials used to obtain alloys that are useful gear materials. The copper alloys, known as bronzes, are perhaps the most widely used. They are useful in situations where corrosion resistance is important. Owing to their ability to reduce friction and wear, bronzes are generally employed for making the worm wheel in a worm gear pair(see Fig. 19.3). Aluminum and zinc alloys are used to manufacture gears by the die casting process.

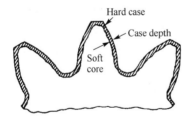

Figure 19.2   Typical case-hardened gear-tooth section

Figure 19.3   A worm gear pair

Nonmetallic Materials Gears have been manufactured of nonmetallic materials for many years. Nylon and various types of plastics have been used (see Fig. 19.4). The advantages obtained by using these materials are quiet operation, internal lubrication, dampening of shock and vibration, and manufacturing economy. Their primary disadvantages are lower load carrying capacity and low heat conductivity, which results in heat distortion of the teeth and may result in a serious weakening of the gear teeth.

Recently thermoplastic resins, with glass-fiber reinforcement and a lubricant as additives, have been used as gear materials. The composite material has resulted in greater load carrying capacity, a reduced thermal expansion, and greater wear resistance.

Figure 19.4　Plastic gears

Words and Expressions

nonmetallic [ˌnɔnmiˈtælik]　*a*. 非金属的；*n*. 非金属物质
metallic [miˈtælik]　*a*. 金属的，金属制的
as is the case with　就像……一样
criterion [kraiˈtiəriən]　（*pl*. criteria [kraiˈtiəriə]）*n*. 标准，规范，准则，依据
wear resistance　耐磨性（材料在一定摩擦条件下抵抗磨损的能力，以磨损率的倒数来评定）
compromise [ˈkɔmprəmaiz]　*n*.；*v*. 妥协（方案），折中，兼顾，综合考虑
nonferrous [ˌnɔnˈferəs]　*a*. 不含铁的，非铁的（not composed of or containing iron）
nonferrous metal　有色金属（除铁、铬、锰三种金属以外的所有金属元素的统称）
ferrous [ˈferəs]　*a*. 铁的，含铁的（of, relating to, or containing iron）
classification [ˌklæsifiˈkeiʃən]　*n*. 分类，归类，类别
conclude [kənˈkluːd]　*v*. 结束（to come to an end），得出结论（to come to a conclusion）
helical gear　斜齿轮（齿线为螺旋线的圆柱齿轮）
herringbone [ˈheriŋbəun] gear　人字齿轮，亦为 double helical gear（双斜齿轮）
elliptical [iˈliptikəl]　*a*. 椭圆的
elliptical gear　椭圆齿轮（分度曲面为椭圆柱面的非圆齿轮）
cast iron　铸铁（铸造法生产的碳含量大于2%的铁碳硅合金）
machinability [məˌʃiːnəˈbiliti]　*n*. 可加工性（在一定生产条件下，材料加工的难易程度）
abatement [əˈbeitmənt]　*n*. 减少，减轻，降低，抑制，削弱
necessitate [niˈsesiteit]　*v*. 需要，使成为必要，以……为条件，迫使
nodular [ˈnɔdjulə]　*a*. 节状的，球状的，团状的
nodular cast iron　球墨铸铁（石墨主要以球状存在的高强度铸铁），亦为 nodular iron
magnesium [mægˈniːzjəm]　*n*. 镁（a light, silvery-white metallic element）
cerium [ˈsiəriəm]　*n*. 铈（an iron-gray, metallic rare-earth element）
pinion [ˈpinjən]　*n*. 小齿轮（the smaller of two gears in mesh）
well balanced　各方面协调的，匀称的，平衡的
carbon steel　碳（素）钢（碳的质量分数小于2.11%而不含有特意加入的合金元素的钢）

plain carbon steel　普通碳素钢（an alloy of iron and carbon, contains carbon from 0.06 to 1.5%）
heat treatment　热处理（采用适当的方式对材料或工件进行加热、保温和冷却，以获得预期的组织结构与性能的工艺）
undue [ʌnˈdjuː]　a. 过度的（exceeding what is appropriate or normal），不适当的（not proper）
in preference to M　优先于 M，（宁取……）而不取 M，比 M 好
the bulk of something　主体部分（the main or largest part of something）
moderate strength　中等强度
case hardening　表面淬火（仅对工件表层进行淬火的工艺），也可写为 surface hardening
hard case　淬硬层（钢件从奥氏体状态急冷的硬化层），也可写为 quench-hardened case
core　心部（热处理工件内部的组织和/或成分未发生变化的部分）
if at all　如果有的话（It indicates that something is unlikely to happen, or rarely happens）
case depth　淬硬深度，也可写为 depth of hardening
quenching [ˈkwentʃiŋ]　n. 淬火（将钢件加热到某一温度，保持一定时间，然后以适当的速度冷却，获得马氏体和/或贝氏体组织的热处理工艺）
water or oil quenching　水冷淬火或油冷淬火
tempering [ˈtempəriŋ]　n. 回火（钢件淬硬后，再加热到形成奥氏体以下的某一温度，保温一定时间，然后冷却到室温的热处理工艺）
soaking [ˈsəukiŋ]　n. 保温（工件在规定温度下，恒温保持一定时间的操作），亦为 holding
inexpensiveness [ˌiniksˈpensivnis]　n. 廉价的，便宜的（low in price; not expensive）
flame hardening　火焰淬火（用可燃气体火焰对工件表面进行加热，随之淬火冷却的工艺）
induction [inˈdʌkʃən]　n. 引导，感应，电感，归纳
induction hardening　感应加热淬火，感应淬火（利用感应电流通过工件所产生的热效应，使工件表面、局部或整体加热并进行快速冷却的淬火工艺）
laser hardening　激光淬火（以高密度能量激光作为能源，迅速加热工件并使其自冷硬化的淬火工艺）
carburizing [ˈkɑːbjuraiziŋ]　n. 渗碳（为增加钢件表层的含碳量，将钢件在渗碳介质中加热并保温使碳原子渗入表层的化学热处理工艺）
nitriding [naitraidiŋ]　n. 渗氮（在一定温度下使活性氮原子渗入工件表面的化学热处理工艺）
worm [wəːm]　n. 蜗杆（只具有一个或几个螺旋齿，并且与蜗轮啮合而组成交错轴齿轮副的齿轮。其分度曲面可以是圆柱面、圆锥面或圆环面）
worm wheel　蜗轮（作为交错轴齿轮副中的大齿轮，与配对蜗杆相啮合的齿轮）
worm gear pair　蜗杆副（由蜗杆及其配对蜗轮组成的交错轴齿轮副）
die casting　压力铸造（熔融金属在高压下高速充型，并在压力下凝固的铸造方法），压铸
nylon [ˈnailən]　n. 尼龙（a strong man-made material used in the making of fabrics and plastics）
dampen [ˈdæmpən]　v. 阻尼，减振，缓冲，抑制，衰减
thermoplastic resin　热塑性树脂（a polymer compound that becomes soft or fluid when heated and then returns to its original solid state when cooled）
glass-fiber reinforcement　玻璃纤维增强

Lesson 20 Friction, Wear, and Lubrication

Our common experience is that objects in motion do not usually stay in motion. In practice we move, or try to move one object against another, and there are interactions that resist the motion. When surfaces in contact move relative to each other, the friction between the two surfaces converts kinetic energy into thermal energy (that is, it converts work into heat).[1]

Friction is harmful or valuable depending upon where it occurs. Friction is necessary for fastening devices such as bolts and nuts (see Fig. 7.1) which depend upon friction to hold the fastener and the parts together. Belt drives, brakes, and tires are additional applications where friction is necessary.

The friction of moving parts in a machine is harmful because it reduces the mechanical advantage of the device. The heat produced by friction is lost energy because no work takes place.[2]

Wear can be defined as the progressive loss of material from the operating surface of a body occurring as a result of relative motion at the surface.[3] The problem of wear arises wherever there are load and motion between surfaces, and is therefore important in engineering practice, often being the major factor governing the life and performance of machine elements. The major types of wear are described next:

Adhesive Wear When two surfaces are loaded against each other, the whole of the contact load is carried on very small area of the asperity contacts. The real contact pressure at these asperities is very high, and adhesion takes place between them. If a tangential force is applied to the model shown in Fig. 20.1, shearing can take place either (a) at the interface or (b) below or above the interface, causing adhesive wear. During sliding, fracture usually occurs in the weaker or softer component, and a wear fragment is generated. Although this fragment is attached to the harder component (upper surface in Fig. 20.1c), it eventually becomes detached during further rubbing at the interface and develops into a loose wear debris.[4]

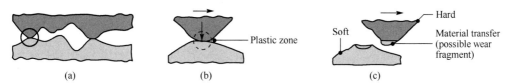

Figure 20.1 Schematic illustration of (a) two contacting asperities, (b) adhesion between two asperities, and (c) the formation of a wear debris

Abrasive Wear This type of wear is caused by a hard, rough surface (or a surface containing hard, protruding particles) sliding across another surface.[5] As a result, microchips (see Fig. 20.2) are produced, thereby leaving grooves or scratches on the softer surface. In fact, processes such as filing (see Fig. 20.3) and grinding act in this manner. The difference is that, in these operations, the process parameters are controlled to produce the desired shapes and surfaces through wear; whereas, abrasive wear generally is unintended and unwanted.

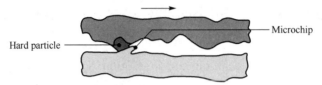

Figure 20.2 Schematic illustration of abrasive wear in sliding

Corrosive Wear Also known as oxidation or chemical wear, this type of wear is caused by chemical or electrochemical reactions between the surfaces and the environment. When the corrosive layer is destroyed or removed through sliding, another layer begins to form, and the process of removal and corrosive layer formation is repeated.

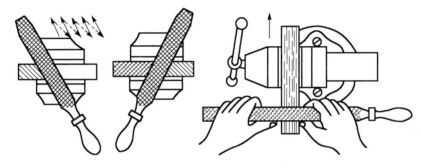

Figure 20.3 Filing operations

Fatigue Wear Fatigue wear is caused when the surface of a material is subjected to cyclic loading, one example of this is the rolling contact in bearings. Another type of fatigue wear is by thermal fatigue. Cracks on the surface are generated by thermal stresses from thermal cycling, such as when a cool forging die repeatedly contacts hot workpieces. These cracks then join, and the surface begins to spall, producing fatigue wear.[6]

Although wear generally alters a part's surface topography and may result in severe surface damage, it also can have a beneficial effect. The running-in period for various machines and engines produces this type of wear by removing the peaks from asperities (see Fig. 20.4). Thus, under controlled conditions, wear may be regarded as a type of smoothing or polishing process.

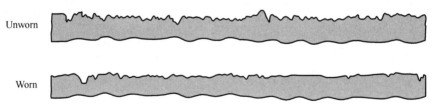

Figure 20.4 Changes in original ground surface profiles after wear

Moving parts are lubricated to reduce friction, wear, and heat. The most commonly used lubricants are oils, greases, and solid lubricants. Each lubricant serves a different purpose. On slow moving parts, an oil groove is usually sufficient to distribute the required quantity of lubricant to the surfaces moving on each other.[7]

A second common method of lubrication is the splash system in which parts moving in a reservoir of lubricant pick up sufficient oil which is then distributed to all moving parts. This system is used in the crankcases of lawn mower to lubricate the crankshaft, connecting rods, and pistons (see Fig. 20.5).[8]

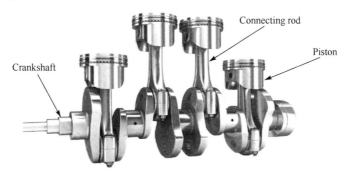

Figure 20.5 Crankshaft with connecting rods and pistons

There are numerous other systems of lubrication and a considerable number of lubricants available for various operating conditions. Modern industry pays greater attention to the use of the proper lubricants than at any other previous time because of the increased speeds, pressures, and operating demands placed on equipment and devices.[9]

Words and Expressions

friction ['frikʃən] n. 摩擦（阻止两物体接触表面发生切向相互滑动或滚动的现象），摩擦力
adhesive wear 黏着磨损，黏附磨损（因黏附作用使两摩擦表面的材料迁移而引起的磨损）
asperity [æs'periti] n. 微凸体（固体表面上微小的不规则凸起）
tangential [tæn'dʒenʃ(ə)l] a. 切线的，切向的
adhesion [əd'hi:ʒən] n. 粘着，黏着（两固体摩擦接触时，由于接触表面间分子力的作用使其产生局部固态连接的现象）
wear debris 磨屑（在磨损过程中从参与摩擦的固体表面上脱落下来的微细颗粒）
abrasive wear 磨料磨损（在摩擦过程中磨粒或凸出物使零件表面材料耗失的一种磨损）
microchip ['maikrəutʃip] n. 细小的切屑，微切屑
scratch 刮伤，划伤（微凸体的滑动作用造成固体摩擦表面上出现划痕的磨损）
filing ['failiŋ] n. 锉削（用锉刀对工件表面进行切削加工）
grinding ['graindiŋ] n. 磨削
unintended [ˌʌnin'tendid] a. 非计划中的，非故意的，无意识的
corrosive wear 腐蚀磨损（以化学或电化学反应为主的磨损）
electrochemical reaction 电化学反应
fatigue wear 疲劳磨损（因循环交变应力引起疲劳使材料表面剥落的一种磨损）
cyclic loading 交变载荷，循环载荷（周期性或非周期性经一定时间后重复出现的动载荷）
thermal fatigue 热疲劳（由于反复加热与冷却产生的循环热应力导致的疲劳现象）
forging die 锻模（模锻时使坯料成型而获得锻件的模具）

surface topography 表面形貌（固体表面与微观峰谷的形态及分布有关的几何形状）
running in 磨合（摩擦初期改变摩擦表面几何形状和表面层物理机械性能的过程）
controlled condition 受控条件，一定条件
polishing ['pɔliʃiŋ] n. 抛光（降低表面粗糙度，使工件获得光亮、平整表面的加工方法）
unworn [ˌʌn'wɔːn] a. 没有磨损的，磨损前的
worn [wɔːn] a. 用旧的，磨损后的；wear 的过去分词
ground surface 磨削表面，经过磨削加工后的表面，这里 ground 是 grind 的过去分词
lubricant ['luːbrikənt] n. 润滑剂（两个相对运动表面之间能减少或避免摩擦磨损的物质）
method of lubrication 润滑方式（向摩擦表面供给润滑剂的方法）
crankshaft ['kræŋkʃɑːft] n. 曲轴（带有若干个曲柄的发动机旋转轴）

Notes

[1] move one object against another 意为"使两个接触的物体做相对运动"，kinetic energy 意为"动能，the energy an object has owing to its motion"。全段可译为：我们日常经验是运动中的物体通常不会持续保持运动状态。在实践中，我们使两个相互接触的物体做相对运动或试图使其做相对运动时，它们之间的相互作用会对运动产生阻抗。当接触表面做相对运动时，两个表面之间的摩擦会把动能转化为热能（也就是说，它把功转化为热）。

[2] mechanical advantage 意为"机械效益"，lost energy 意为"损失的能量"。这两句可译为：在一台机器中，运动零件之间的摩擦是有害的，因为它会降低这个装置的机械效益。由于摩擦所产生的热量不做功，所以它是被损失的能量。

[3] operating surface 意为"工作表面"，be defined as 意为"被定义为，被称为"。全句可译为：物体工作表面材料在相对运动中逐渐损耗的现象被称为磨损。

[4] fragment 意为"碎片，碎屑"。全句可译为：尽管这个碎片黏附在较硬的材料上（图 20.1c 中的上表面），在随后的摩擦过程中，它最终会脱落，变成离散的磨屑。

[5] protruding particle 意为"凸出物，凸出的颗粒"。这句和后面两句可译为：这种磨损是由一个粗糙硬表面（或者一个含有硬质凸出颗粒的表面）与另一表面做相对运动时产生的，其结果是产生了微小的磨屑（见图 20.2），使较软的表面上出现沟槽或划痕。事实上，诸如锉削（见图 20.3）和磨削等类加工过程就是以这种方式工作的。

[6] thermal cycling 意为"冷热交替，温度循环"。全段可译为：材料表面承受交变载荷就会引起疲劳磨损，轴承中的滚动接触就是一个例子。另一种疲劳磨损是热疲劳。当一个凉的锻模反复地与红热的工件相接触时，冷热交替所产生的温度应力就会在锻模表面产生若干裂纹。然后，这些裂纹连接到一起，表面就会有材料剥落，产生疲劳磨损。

[7] oil groove 意为"油槽（滑动表面上供给和分布润滑油的沟槽）"。全句可译为：对于低速运动的零件，一个油槽就足以将所需数量的润滑剂送到相互运动的表面。

[8] splash system 意为"飞溅润滑系统"，crankcase 意为"曲轴箱"，piston 意为"活塞"。这段可译为：第二种常见的润滑方式是飞溅润滑系统，这个系统内的一些零件经过润滑剂储存装置时，带起足够的润滑油，然后将其散布到所有的运动零件上。这种系统被用在割草机的曲轴箱中，对曲轴、连杆和活塞等（见图 20.5）进行润滑。

[9] 全句可译为：由于设备或装置的速度、压力和工作要求的提高，在现代工业界中，人们比以往任何时候都更加注重选用适当的润滑剂。

Lesson 21 Lubrication

One of the main purposes of lubrication is to reduce friction, any substance —liquid, solid, or gaseous —capable of controlling friction and wear between sliding surfaces can be classed as a lubricant.

Varieties of lubrication

Unlubricated sliding Metals that have been carefully treated to remove all foreign materials[1] seize and cold-weld to one another when slid together. In the absence of such a high degree of cleanliness, adsorbed gases, water vapor, oxides, and contaminants reduce friction and the tendency to seize but usually result in severe wear; this is called "unlubricated" or dry sliding.

Fluid-film lubrication Interposing a fluid film that completely separates the sliding surfaces results in fluid-film lubrication (see Fig. 21.1). The fluid may be introduced intentionally as the oil in the main bearings of an automobile, or unintentionally, as in the case of water between a smooth rubber tire and a wet pavement. Although the fluid is usually a liquid such as oil, water, and a wide range of[2] other materials, it may also be a gas. The gas most commonly employed is air.

To keep the parts separated, it is necessary that the pressure within the lubricating film balance the load on the sliding surfaces. If the lubricating film's pressure is supplied by an external source, the system is said to be lubricated hydrostatically. If the pressure between the surfaces is generated as a result of the shape and motion of the surfaces themselves, however, the system is hydrodynamically lubricated.

Boundary lubrication A condition that lies between unlubricated sliding and fluid-film lubrication is referred to as boundary lubrication (see Fig. 21.2), also defined as that condition of lubrication in which the friction between surfaces is determined by the properties of the surfaces and properties of the lubricant other than viscosity. Boundary lubrication encompasses a significant portion of lubrication phenomena and commonly occurs during the starting and stopping of machines.

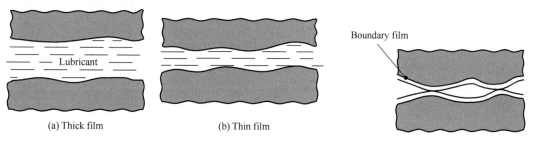

Figure 21.1 Fluid-film lubrication   Figure 21.2 Boundary lubrication

Solid lubrication Solids such as graphite and molybdenum disulfide are widely used when normal lubricants do not possess sufficient resistance to load or temperature extremes. But lubricants need not take only such familiar forms as greases (see Fig. 21.3) and powders; even some metals commonly serve as sliding surfaces in some sophisticated machines.

 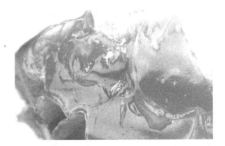

Figure 21.3 Lubricating greases

Functions of lubricants

Although a lubricant primarily controls friction and wear, it can and ordinarily does perform numerous other functions, which vary with the application and usually are interrelated.

Friction control The amount and character of the lubricant made available to sliding surfaces have a profound effect upon the friction that is encountered. For example, disregarding such related factors as heat and wear but considering friction alone between two oil-film lubricated surfaces, the friction can be 200 times less than that between the same surfaces with no lubricant. Under fluid-film conditions, friction is directly proportional to the viscosity of the fluid. Some lubricants, such as petroleum derivatives, are available in a great range of viscosities and thus can satisfy a broad spectrum of functional requirements. Under boundary lubrication conditions, the effect of viscosity on friction becomes less significant than the chemical nature of the lubricant.

Wear control Wear occurs on lubricated surfaces by abrasive, corrosion, and solid-to-solid contact. Proper lubricants will help combat each type. They reduce abrasive and solid-to-solid contact wear by providing a film that increases the distance between the sliding surfaces, thereby lessening the damage by abrasive contaminants and surface asperities (see Fig. 21.4).

Temperature control Lubricants assist in controlling temperature by reducing friction and carrying off the heat that is generated. Effectiveness depends upon the amount of lubricant supplied, the ambient temperature, and the provision for external cooling. To a lesser extent, the type of lubricant also affects surface temperature.

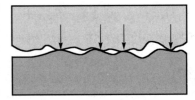

Figure 21.4 Surface asperities

Corrosion control The role of a lubricant in controlling corrosion of the surfaces is twofold. When machinery is idle, the lubricant acts as a preservative. When machinery is in use, the lubricant controls corrosion by coating lubricated parts with a protective film that may contain anti-corrosion additives. The ability of a lubricant to control corrosion is directly related to the thickness of the lubricant film remaining on the metal surfaces and the chemical composition of the lubricant.

Other functions

Lubricants are frequently used for purposes other than the reduction of friction. Some of these applications are described below.

Power transmission Lubricants are widely employed as hydraulic fluids in fluid transmission devices.

Insulation In specialized applications such as transformers (see Fig. 21.5), lubricants with high dielectric constants act as electrical insulators. For maximum insulating properties, a lubricant must be kept free of contaminants and water.

Shock dampening Lubricants act as shock-dampening fluids in energy transfer devices such as shock absorbers[3] and around machine parts such as gears that are subjected to high intermittent loads.

Sealing Lubricating grease frequently performs the special function of forming a seal to retain lubricants or to exclude contaminants.

Figure 21.5 A transformer

Words and Expressions

sliding ['slaidiŋ] n. 滑动（摩擦副公接面上两表面速度的大小和/或方向不同的相对运动）
lubricant ['lju:brikənt] n. 润滑剂（加入两个相对运动表面间，能减少或避免摩擦磨损的物质）
unlubricated [ʌn'lu:brikeitid] a. 无润滑的
dry sliding 干滑动（两相对运动表面间无润滑剂的滑动）
seize [si:z] v. 咬死（摩擦表面产生严重黏着或转移，使相对运动停止）
cold-weld [kəuld-weld] v. 冷焊（摩擦学中，两直接接触表面在常温、低温下形成黏着）
cleanliness ['klenlinis] n. 清洁度（the quality or state of being clean）
adsorb [æd'sɔ:b] v. 吸附，吸取
fluid-film lubrication 流体膜润滑
interpose [ˌintə'pəuz] v. 插入，加入……之间（to come between; to introduce between parts）
intentionally [in'tenʃənli] ad. 故意地
main bearing 主轴承（发动机内支承曲轴在其中旋转的轴承）
unintentionally [ˌʌnin'tenʃənli] ad. 不是存心地，非故意地
hydrostatical [ˌhaidrəu'stætikəl] a. 流体静力（学）的，液压静力的
hydrodynamical ['haidrəudai'næmikəl] a. 流体动力（学）的
viscous ['viskəs] a. 黏的，黏性的，黏稠的
boundary lubrication 边界润滑（摩擦表面被一层极薄的、呈非流动状态的润滑膜隔开时的润滑）
viscosity [vis'kɔsiti] n. 黏度（液体受外力作用而流动时，分子间所呈现的内摩擦或流动内阻力）
encompass [in'kʌmpəs] v. 环绕，包围，包括，包含
graphite ['græfait] n. 石墨（a soft black carbon used in making pencils and as a lubricant）

molybdenum [mə'libdinəm] *n.* 钼（a silver-white metallic element）
disulfide [dai'sʌlfaid] *n.* 二硫化物，亦为 disulphide 或 bisulfide
molybdenum disulfide 二硫化钼（分子式为 MoS_2，有金属光泽的黑稍带银灰色的固体粉末）
temperature extremes 温度极限，极高/极低温度，极限温度，极端温度
profound [prə'faund] *a.* 深奥的，深刻的，极度的
derivative [di'rivətiv] *n.* 衍生物（一种简单化合物中的氢原子或原子团被其他原子或原子团取代而衍生的较复杂的产物）
a broad spectrum of 大范围的，用途广泛的
lessen ['lesn] *v.* 减少，缩小（to reduce in size, extent, or degree）
effectiveness [i'fektivnis] *n.* 有效性（the fact or quality of producing the desired result）
ambient ['æmbiənt] *a.* 环境的，周围的（of the surrounding area or environment）
provision [prə'viʒən] *n.* （预防）措施，保证，保障
to a lesser extent 在较小程度上
preservative [pri'zə:vətiv] *a.* 保存的，防腐的；*n.* 防腐剂，保存剂
additive ['æditiv] *n.* 添加剂（添加到润滑剂中以提高某些原有特性或获得新特性的物质）
anti-corrosion [,æntikə'rəuʒən] *n.* 防腐蚀，耐腐蚀（measures that are used to combat the occurrence and progression of corrosion）；*a.* 抗腐蚀的
chemical composition 化学组成，化学成分
insulation [,insju'leiʃən] *n.* 绝缘，绝缘体（a nonconductor of electricity, heat, or sound）
specialized ['speʃəlaizd] *a.* 专门的，专用的（having developed in order to perform a particular function or suit a particular environment）
transformer [træns'fɔ:mə] *n.* 变压器（利用电磁感应的原理来改变交流电压的装置）
dielectric [daii'lektrik] *a.* 不导电的，绝缘的，介电的
dielectric constant 介电常数，又称电容率（permittivity）
dampen ['dæmpən] *v.* 抑制，使衰减，阻尼，减振，缓冲
intermittent [,intə'mitənt] *a.* 间歇的，断断续续的（not happening regularly or continuously）
sealing ['si:liŋ] *n.* 密封，封接，封口
grease [gri:s] *n.* 脂（一种稠厚的油脂状半固体，起润滑和密封作用）
retain [ri'tein] *v.* 保留，保持（to continue to have or use something）
exclude [iks'klu:d] *v.* 拒绝（to prevent from entering; keep out），排除，排斥（to reject）

Notes

[1] foreign material 意为"杂质，外来物质"。
[2] a wide range of 意为"许多，广泛，各种各样的"。
[3] shock absorber 意为"减振器，冲击吸收器（通过能量的耗散来减弱机械系统冲击响应的装置）"。

Lesson 22 Introduction to Tribology

Tribology is defined as the science and technology of interacting surfaces in relative motion, having its origin in the Greek word tribos meaning rubbing. It is a study of the friction, lubrication, and wear of operating surfaces with a view to[1] understanding surface interactions in detail and then prescribing improvements in given applications. The work of the tribologist is truly interdisciplinary, embodying physics, chemistry, mechanics, thermodynamics, and materials science, and encompassing a large, complex, and interwinded area of machine design, reliability, and performance where relative motion between surfaces is involved.

It is estimated that approximately 23% of the world's energy resources in present use appear as friction and wear in one form or another. Of that, 20% is to overcome friction and 3% to remanufacture worn parts. The purpose of research in tribology is understandably the minimization and elimination of unnecessary waste at all levels of technology where the rubbing of surfaces is involved.

One of the main objectives in tribology is the regulation of the magnitude of frictional forces according to whether we require a minimum (as in machinery) or a maximum (as in the case of anti-skid surfaces). It must be emphasized, however, that this objective can be realized only after a fundamental understanding of the frictional process is obtained for all conditions of temperature, sliding velocity, lubrication, surface roughness, and material properties.

The most important criterion from a design viewpoint in a given application is whether dry or lubricated conditions are to prevail at the sliding interface. In many applications such as machinery, it is known that only one condition shall prevail (usually fluid lubrication), although several regimes of lubrication may exist. There are a few cases, however, where it cannot be known in advance whether the interface is dry or wet, and it is obviously more difficult to proceed with any design. The commonest example of this phenomenon is the pneumatic tire. Under dry conditions it is desirable to maximize the adhesion component of friction by ensuring a maximum contact area between tire and road and this is achieved by having a smooth tread and a smooth road surface. Such a combination, however, would produce a disastrously low coefficient of friction under wet conditions. In the latter case, an adequate tread pattern[2] (see Fig. 22.1) and a suitably textured road surface offer the best conditions, although this combination gives a lower coefficient of friction in dry weather.

The several lubrication regimes which exist may be classified as hydrodynamic, boundary, and elastohydrodynamic. The different types of bearing used today are the best examples of fully hydrodynamic behavior, where the sliding surfaces are completely separated by an interfacial lubricant film. Boundary or mixed lubrication (see Fig. 22.2) is a combination of hydrodynamic and solid contact between moving surfaces, and this regime is normally assumed to prevail when hydrodynamic lubrication fails in a given product design. For example, a journal bearing is designed to operate at a given load and speed in the fully hydrodynamic region, but a fall in speed or an increase in load may cause part solid and part hydrodynamic lubrication conditions to occur between the journal and bearing surfaces. This boundary lubrication condition is unstable, and normally recovers to the fully hydrodynamic behavior or degenerates into complete seizure of the surfaces. The pressures developed in thin lubricant films may reach proportions capable of elastically deforming the boundary surfaces of the lubricant, and conditions at the sliding interface are then classified as elastohydrodynamic.

Figure 22.1   Three types of tire tread patterns   Figure 22.2   Mixed lubrication

Solid lubricants exhibit a compromise between dry and lubricated conditions in the sense that although the contact interface is normally dry, the solid lubricant material behaves as though initially wetted. This is a consequence of a physicochemical interaction occurring at the surface of a solid lubricant lining under particular loading and sliding conditions, and these produce the equivalent of a lubricating effect. Graphite and molybdenum disulfide (MoS_2) are the predominant materials used as solid lubricant. Their structures are shown in Fig. 22.3. The lubricity of these solids is attributable to a lamellar structure.

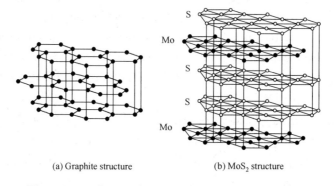

(a) Graphite structure   (b) MoS_2 structure

Figure 22.3   Commonly used solid lubricants' structures

By taking advantage of the new technologies for friction reduction and wear protection, energy losses due to friction and wear in vehicles, machinery and other equipment worldwide could be reduced significantly.

Modern means and technologies to reduce friction and wear are:

1. New lubricant solutions, such as nanotechnology based anti-friction and anti-wear additives, low viscosity oils and vapor phase lubrication;

2. New material solutions, such as new materials and surface modification.

Words and Expressions

tribology [trai'bɔlədʒi] *n.* 摩擦学（研究做相对运动物体的相互作用表面、类型及其机理、中间介质及环境所构成的系统的行为与摩擦及损伤控制的科学与技术）
rubbing ['rʌbiŋ] *n.* 摩擦（moving along a surface with friction and pressure）
interdisciplinary [ˌintə'disiplinəri] *a.* 各学科间的，跨学科的，多种学科的
embody [im'bɔdi] *v.* 体现，包括，包含，合并，补充
thermodynamics [ˌθəːmoudai'næmiks] *n.* 热力学
interwind [ˌintə'waind] *v.* 互相盘绕，互卷
remanufacture [riːˌmænju'fæktʃə] *v.* 再制造（对报废产品进行翻新修理，使其恢复到或接近于新产品的性能标准）
anti-skid [ˌænti'skid] *a.* 防滑的（designed to prevent skidding）
criterion [krai'tiəriən] *n.* 标准，规范，准则，依据
interfacial [ˌintə'feiʃəl] *a.* 分界面的，两表面间的，界面的
pneumatic [njuː'mætik] *a.* 气体的，充气的（filled with air, especially compressed air）
fluid lubrication 流体润滑（相对运动的摩擦表面被气体或液体完全隔开的润滑状态）
lubrication regime 润滑状态（润滑剂在两摩擦表面间存在的条件和状态）
disastrously [di'zɑːstrəsli] *ad.* 灾害性地，造成巨大损失地
elastohydrodynamic [i'læstouˌhaidrəudai'næmik] lubrication 弹性流体动力润滑
mixed lubrication 混合润滑（同时存在流体润滑和边界润滑的润滑状态）
degenerate [di'dʒenəreit] *v.* 退化，恶化（to become worse, weaker, less useful, etc.）
seizure ['siːʒə] *n.* 咬死（摩擦表面产生严重黏着或转移，使相对运动停止的现象）
compromise ['kɔmprəmaiz] *n.; v.* 妥协，折中，兼顾，综合考虑
behave as 性能（作用，表现）像……一样
physicochemical [ˌfizikou'kemikəl] *n.* 物理化学的
lining ['lainiŋ] *n.* 衬层，涂层，覆盖
lubricity [ljuː'brisiti] *n.* 润滑性（润滑剂减少摩擦和磨损的能力）
lamellar [lə'melə] *a.* 薄片状的（having the form of a thin plate），薄层状的
nanotechnology ['nænouˌtek'nɔlədʒi] *n.* 纳米技术
vapor phase lubrication 气相润滑
surface modification 表面改性（改善工件表面层的机械、物理或化学性能的处理方法）

Notes

[1] with a view to 意为"为了（with the aim or intention of），目的是（with the purpose of）"。
[2] tread pattern 意为"胎面花纹（主要由凸起部分和沟槽部分组成的胎面式样）"。

Lesson 23　Product Drawings

Manufacturing companies are established to produce one or more products. These products are completely defined through documents referred to as product drawings.[1] The dimensions and specifications found on the product drawings will ensure part interchangeability and a reliable level of designed performance. Product drawings normally include detail drawings and assembly drawings.

A detail drawing is a dimensioned, multiview drawing of a single part, describing the part's shape, size, material, and surface roughness, in sufficient detail for the part to be manufactured based on the drawing alone. Most parts require three views for complete shape description.

Assembly drawings are used to describe how parts are put together,[2] as well as the function of the entire unit; therefore, complete shape description is not important. The views chosen should describe the relationships of parts, and the number of views chosen should be the minimum necessary to describe the assembly.

All parts must interact with other parts to some degree to yield the desired function from a design. Before detail drawings of individual parts are made the designer must thoroughly analyze the assembly drawing to ensure that the parts fit property with mating parts, that the correct tolerances are applied, that the contact surfaces are properly machined, and that the proper motion is possible between the parts.

The inch is the basic unit of the English system, and virtually all shop drawings in the U. S. are dimensioned in inches.

The millimeter is the basic unit of the metric system. Metric abbreviation (mm) after the numerals is omitted from dimensions because the SI symbol near the title block indicates that all units are metric.

In the U. S., some drawings carry both inch and millimeter dimensions, usually the dimensions in parentheses or brackets are millimeters. The units may also appear as millimeters first and then be converted and shown in brackets as inches. Converting from one unit to the other results in fractional round-off errors. And explanation of the primary unit system for each drawing should be noted in the title block.

Title Blocks　In practice, title blocks usually contain the title of the drawing or part name, drafter, date, scale, name of the company, and drawing number. Other information, such as checkers, and materials, also may be given. Any changes or modifications added after the first version to improve the design is shown in the revision blocks. The title block is generally located in the lower right-hand corner of the drawing sheet.

The scale is the ratio between the linear dimensions of the drawing of an object[3] and the actual dimensions. Whatever scale is used, the dimensions on the drawing indicate the true size of the object, not of the view. Various scales may be used for different drawings in a set of product drawings.

Depending on the complexity of the project, a set of product drawings may contain from one to more than a hundred sheets. Therefore, giving the number of each sheet and the total number of sheets in the set on each sheet is important (for example, sheet 2 of 6, sheet 3 of 6, and so on).

Part Names and Numbers Give each part a name and number, usually using letters and numbers 1/8in (3mm) high. Place part numbers near the views to which they apply, so their association will be clear.

Parts Lists The part numbers and part names in the parts list correspond to those given to each part depicted on the product drawings. In addition, the number of identical parts required is given along with the material used to make each part.

Up to 1900, drawings everywhere were generally made in what is called first-angle projection. In the first-angle projection, the top view is placed under the front view, the left-side view is placed at the right of the front view, and so on. Today, third-angle projection is in common use in the USA, Canada, and the UK, but first-angle projection is still used throughout much of the world. In the third-angle projection, the top view is placed above the front view, the right-side view is placed at the right of the front view, the left-side view is placed at the left of the front view.

Actually, the only difference between first-angle and third-angle projection is the arrangement of views. Confusion and manufacturing errors may result when the user reading a first-angle drawing thinks it is a third-angle drawing, or vice versa. To avoid misunderstanding, projection symbols, shown in Fig. 23.1, have been developed to distinguish between first-angle and third-angle projections on drawings. On drawings where the possibility of confusion is anticipated, these symbols may appear in or near the title block.

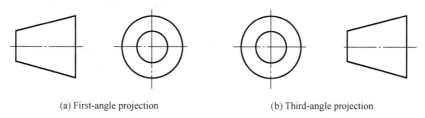

(a) First-angle projection   (b) Third-angle projection

Figure 23.1 Projection symbols

Product drawings are legal contracts that document the design details and specifications as directed by the design engineer. Therefore drawings must be as clear, precise, and thorough as possible. Revisions and modifications of a project at the time of production are much more expensive than when done in the preliminary design stages. To be economically competitive drawings must be as error-free as possible.

People who check drawings must have special qualifications that enable them to identify errors and to suggest revisions and modifications that result in a better product at a lower cost. Checkers inspect assembly or detail drawings for correctness and soundness of design. In addition, they are responsible for the drawing's completeness, quality, and clarity.

In addition to the individual revision records, drafters should keep a log of all changes made during a project. As the project progresses, the drafter should record the changes, dates, and people involved. Such a log allows anyone reviewing the project in the future to understand easily and clearly the process used to arrive at the final design.

Words and Expressions

drawing　图样（根据投影原理、标准或有关规定，表示工程对象，并有必要的技术说明的图）
product drawing　产品图样，产品图纸
interchangeability ['intətʃeindʒə'biliti]　*n.* 互换性（在同一规格的一批零件或部件中，任取其一，不需任何挑选或附加修配，如钳工修理，就能装在机器上，达到规定的性能要求）
assembly [ə'sembli]　*n.* 装配（按规定的技术要求，将零件或部件进行配合和连接，使之成为半成品或成品的工艺过程），装配件
assembly drawing　装配图（表示产品及其组成部分的连接、装配关系的图样）
detail drawing　零件图（表示零件结构、大小及技术要求的图样）
part [pɑːt]　*n.* 零件（组成机械的不可拆分的单个制件，它是机械的基本单元）
dimension [di'menʃən]　*n.* 尺寸（用特定长度或角度单位表示的数值，并在技术图样上用图线、符号和技术要求表示出来）；*v.* 标注尺寸
multiview drawing　有多个视图的图样（如 three-view drawing，三视图）
surface roughness　表面粗糙度（加工表面上由较小间距和峰谷所组成的微观几何形状特征）
view [vjuː]　*n.* 视图（根据有关标准和规定，用正投影法将机件向投影面投影所得到的图形）
specification [ˌspesifi'keiʃən]　*n.* 规格，说明书，规范，技术要求
consolidate [kən'sɔlideit]　*v.* 巩固，统一
fit　配合（基本尺寸相同的，相互结合的孔和轴公差带之间的关系），适合
mating part　配合件，相配零件
tolerance ['tɔlərəns]　*n.* 公差（实际参数值的允许变动量）；*v.* 给（机器部件等）规定公差
machine [mə'ʃiːn]　*n.* 机器（由零件组成的执行机械运动的装置。用来完成所赋予的功能，如变换或传递能量、变换和传递运动和力及传递物料与信息）；*v.* 机械加工（to make or finish with a machine or machine tool）
thorough ['θʌrə]　*a.* 完整的，彻底的（absolute），全面的（exhaustively complete）
English system　英制（以英尺、磅、秒为基本单位的单位制，the foot-pound-second system of units）
shop drawing　车间加工图，制造图样（制造行业中，工厂车间里用来直接指导加工的图样）
metric system　公制，米制（以米、千克、秒等为基本单位的单位制）
SI　*abbr.* 国际单位制（International System of Units）
abbreviation [əˌbriːvi'eiʃən]　*n.* 缩写，缩写词
parenthesis [pə'renθisis]　*n.* (*pl.* parentheses) 圆括号（a round bracket）
bracket ['brækit]　*n.* 方括号，尖括号（one of a pair of marks [ ] or 〈 〉, called also square bracket or angle bracket）
fractional ['frækʃən]　*a.* 分数的，小数的

round-off error　舍入误差，四舍五入产生的误差，又称 rounding error
title block　标题栏（由名称及代号区、签字区、更改区和其他区组成的栏目）
scale [skeil]　*n.*　比例（图中图形与实物相应要素的线性尺寸之比）
drawing number　图号，图样代号
lower right-hand corner　右下角，亦可写为 lower right corner 或 bottom right corner
drawing sheet　图纸，绘图纸（供绘制工程图、机械图、地形图等用的纸）
checker ['tʃekə]　*n.*　审核人员（a person that verifies or examines something）
parts list　明细栏（由序号、代号、名称、数量、材料、重量、备注等内容组成的栏目），亦为 item block 或 item list
soundness ['saundnis]　*n.*　完善，无缺陷
part number　零件序号
review [ri'vju:]　*v.*；*n.*　检查，评审，评阅，审核
projection [prə'dʒekʃən]　*n.*　投影（根据投影法所得到的图形）
first-angle projection　第一角投影（将物体置于第一分角内，并使其处于观察者与投影面之间而得到正投影的方法），第一角画法（first angle method）
third-angle projection　第三角投影（将物体置于第三分角内，并使投影面处于观察者与物体之间而得到正投影的方法），第三角画法（third angle method）
top view　俯视图（由上向下投影得到的视图。在第一角投影中，俯视图配置在主视图的下方；在第三角投影中，俯视图配置在主视图的上方）
front view　主视图（由前向后投影得到的视图）
left-side view　左视图（由左向右投影所得的视图），也可写为 left view
right-side view　右视图（由右向左投影所得的视图），也可写为 right view
projection symbol　投影识别符号，投影符号
log [lɔg]　*n.*　日志，（工程、试验等的）工作记录
keep a log of　将……记录下来

Notes

[1] drawing 的本意为"图"，即用点、线、符号、文字和数字等描绘事物几何特性、形态、位置及大小的一种形式。与图相关，但未在本课中出现的常用词有：diagram（简图），由规定的符号、文字和图线组成示意性的图；sketch（草图），以目测估计图形与实物的比例，按一定画法要求徒手或部分使用绘图仪器绘制的图；figuration drawing（外形图），表示产品外形轮廓的图样；tabular drawing（表格图），用图形和表格，表示结构相同而参数、尺寸、技术要求不尽相同的产品的图样；schematic diagram（原理图，示意图），(1) 表示系统、设备的工作原理及其组成部分的相互关系的简图，(2) 表达产品的工作程序、功能及其组成部分的结构、动作原理的一种简图；block diagram（框图），表示某一系统工作原理的一种简图，其中，整个系统或部分系统连同其功能关系均用称为功能框的符号或图形及连线和字符表示。

[2] put together 意为"装配（to construct or create something out of pieces or parts）"。

[3] object 在与机械制图有关的资料中通常译为"机件"。

Lesson 24　Sectional Views

Sectional views, also called section views or simply sections, are used to show internal construction of an object that is too complicated to be shown clearly by regular views containing many hidden lines. [1] To produce a sectional view, a cutting plane is assumed to be passed through the part, and then removed.

Section Lining　Section lining can serve a double purpose. [2] It indicates the surface that has been cut and makes it stand out clearly, thus helping the observer to understand the shape of the object. Section lining may also indicate the material from which the object is made. The symbols used to distinguish between different materials in sections are shown in Fig. 24.1. However, there are so many different materials used in design that the general symbol (i. e., the one used for cast iron) may be used for most purposes on engineering drawings. The actual type of material required is then noted in the title block or parts list, or entered as a note on the drawing.

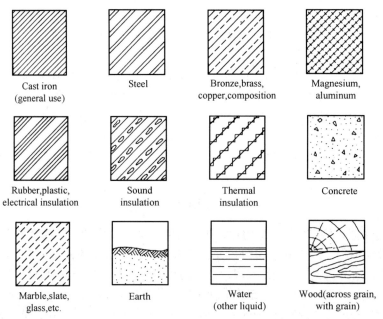

Figure 24.1　ANSI standard section lining symbols for various materials

There are a number of different types of sectional views. A few of the common ones are:

Full Sections　When the cutting plane extends entirely through the object in a straight line and the front half of the object is imagined to be removed, a full section is obtained. [3] This type of section is used for both detail and assembly drawings. When the section is on an axis of symmetry, it is not necessary to indicate its location. Figure 24.2 shows a full section with cutting plane omitted.

Half Sections　If a cutting plane passes halfway through an object, the result is a half section. This type of section is most often used when the object is symmetric or nearly so. [4] A half section has the advantage of showing both the interior and exterior of the object on one view. One half of the object illustrates internal construction, and the other half shows an external view. Figure 24.3 shows the position of the cutting plane in the top view and the resulting half section in the front view. [5] The sectional view and the external view are separated by a centerline. Usually hidden lines are omitted on both sides of a half section. [6]

The half section is not widely used in detail drawings because of difficulties in dimensioning internal shapes that are shown in part only in the sectioned half (see Fig.24.3). The greatest usefulness of the half section is in assembly drawing, in which it is often necessary to show both internal and external constructions on the same view.

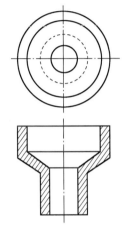
Figure 24.2　Full section

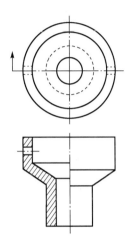
Figure 24.3　Half section

Broken-Out Sections　Where a sectional view of only a portion of the object is needed, broken-out sections may be used (see Fig. 24.4). A break line separates the sectioned portion from the unsectioned portion of the view. [7]

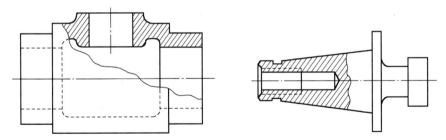
Figure 24.4　Broken-out sections

Offset Sections　In order to show features that are not in a straight line, the cutting plane may be bent at 90 degree angles, so as to include several parallel planes. Such a section is called an offset section (see Fig. 24.5). Note that the change of plane which occurs where the cutting plane is bent 90 degrees is not represented with lines in the section view. [8]

Aligned Sections　To include in a section certain angled features, the cutting plane may be bent to pass through those features. [9] The features cut by the cutting plane are then imagined to be revolved into a plane. Figure 24.6 is an example of aligned section. The aligned section view gives a clearer, more complete description of the geometry of the part.

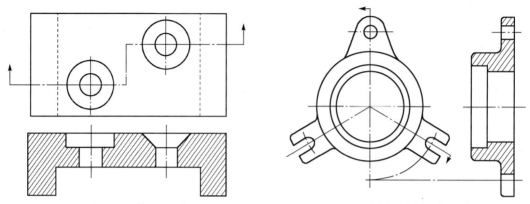

Figure 24.5 Offset section

Figure 24.6 Aligned section

Assembly Sections As the name implies, an assembly section is made up of a combination of parts. All the previously mentioned types of sections may be used to increase the clarity and readability of assembly drawings. The cast iron symbol (evenly spaced section lines) is recommended for most assembly drawings. The section line should be drawn at an angle of 45° with the main outline of the view. On adjacent parts, the section lines should be drawn in the opposite direction, as shown in Fig. 24.7. [10]

Shafts, bolts, nuts, washers, screws, rivets, pins, keys, and similar solid parts, the axes or the symmetry planes of which lie in the cutting plane, should not be sectioned even though the cutting plane passes through them, except that a broken-out section of the shaft may be used to describe more clearly the key, key seat, or pin (see Figs. 4.5 and 24.7). [11] If the shafts, bolts, nuts, rivets, pins, and keys were sectioned, the drawing would be confusing and difficult to read.

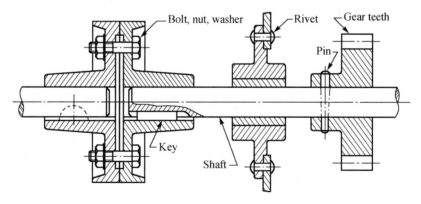

Figure 24.7 Conventional representation of mechanical elements in sectional views

Words and Expressions

sectional view 剖视图（假想用剖切面剖开机件，将处在观察者和剖切面之间的部分移去，将其余部分向投影面投影所得的图形），也可写为 section view

section *n.*; *v.* 剖视，剖面，剖面图，剖视图

internal construction　内部结构
hidden line　虚线（a line used to represent a feature that cannot be seen in the current view）
section lining　剖面线，也可写为 section line
stand out clearly　清晰地显示出来
distinguish between　区别，分辨
section lining symbol　剖面符号，也可写为 section line symbol
general symbol　通用符号
for most purposes　在大多数情况下，在大多数场合
general use　通用，一般用途
engineering drawing　工程图样（根据投影原理和国家标准规定表示工程对象的形状、大小及技术要求的图），简称"工程图"
insulation [ˌinsjuˈleiʃən] *n.* 绝缘（或隔热、隔音）材料，阻滞电、热或声通过的材料
concrete [ˈkɔnkriːt] *n.* 混凝土（a hard, strong construction material）
marble [ˈmɑːbl] *n.* 大理石（a kind of stone that is often polished and used in buildings）
slate [sleit] *n.* 石板，石片（a type of hard rock that splits easily into thin layers）
across grain　（木材）横断面
with grain　（木材）纵断面
ANSI (American National Standards Institute)　美国国家标准协会
full section　全剖视，全剖视图
axis of symmetry　对称轴，对称轴线
half section　半剖视，半剖视图
symmetric [siˈmetrik] *a.* 对称的（having sides or halves that are the same）
centerline [ˈsentəlain] *n.* 中心线
external view　外观图，外形视图
dimensioning [diˈmenʃəniŋ] *n.* 标注尺寸（writing dimensions on a drawing）
broken-out section　局部剖视，也可以写为 partial section
offset [ˈɔːfset] *a.* 偏移的，横向移动的
offset section　阶梯剖视图（用几个相互平行的剖切平面剖开机件所得的剖视图）
aligned section　旋转剖视图（用一个平行于选定的基本投影面和与其相交的剖切平面剖开机件，然后将剖开的结构旋转到与选定的基本投影面平行再投影所得的视图）
assembly section　装配体剖视图，装配件剖视图
a combination of parts　许多零件的组合
readability [ˌriːdəˈbiliti] *n.* 易读性，可读性
evenly spaced　间隔相等的，等间隔的，也可写为 equally spaced
section line　*n.* 剖面线；*v.* 画剖面线
washer [ˈwɔʃə] *n.* 垫圈（放在螺母或螺钉头与被连接件之间的薄金属垫）
screw [skruː] *n.* 螺钉（具有各种结构形状头部的螺纹紧固件）
rivet [ˈrivit] *n.* 铆钉（一种金属制一端有帽的杆状零件，穿入被连接的构件后，在杆的外端打、压出另一头，将构件压紧、固定）
pin　销（贯穿于两个零件孔中，主要用于定位，也可用作连接或安全装置中过载易剪断元件）
gear teeth　轮齿（齿轮上的每一个呈辐射状排列并用于持续啮合的凸起部分）
key seat　键槽（a groove in a mechanical part to receive a key），亦为 keyway

confusing [kənˈfjuːziŋ] *a.* 混乱的，混淆的，令人困惑的
conventional representation 规定画法

Notes

[1] section 这里指"剖视图（the plane figure resulting from the cutting of a solid by a plane）"。全句可译为：剖视图，也可以称为 section views 或简称 sections（剖视）。如果一个机件的内部结构太复杂，采用普通视图会因为虚线太多而不能清楚地表示时，则可以采用剖视图。

[2] a double purpose 意为"双重作用，双重用途"。全句可译为：剖面线能起到双重作用。如果将这句话中的 Section lining 用 Section lines 替代，则此句话可改写为 Section lines serve two purposes.

[3] cutting plane 意为"剖切面（剖切被表达物体的假想平面或曲面）"。全句可译为：用一个剖切面将物体完全剖开后，假想将其前半部分移去而得到的剖视图称为全剖视图。

[4] halfway through 意为"进行到一半"。这两句可译为：当剖切面只通过机件的一半时，所得到的结果为半剖视图。这种类型的剖视图通常用于对称或基本对称的机件。

[5] 全句可译为：图 24.3 在俯视图中标示出了剖切面的位置，其主视图为所产生的半剖视图。（注：在此篇课文中，采用第三角投影视图，俯视图位于主视图上面。）

[6] omitted 这里指"省略不画"。这两句可译为：以中心线为界，一半为剖视图，另一半为外部视图。在半剖视图的这两部分中通常均将虚线省略不画。

[7] break line 意为"波浪线（a freehand line used to show where an object is broken to reveal interior features of a part）"。这两句可译为：当只需要对机件的一部分进行剖视时，可采用局部剖视图（见图 24.4）。视图的剖视部分与未剖部分用波浪线分界。

[8] 全段可译为：为了表示不在同一直线上的一些特征，可以将剖切面以直角的方式进行转折，使其通过几个相互平行的平面。这种剖视图称为阶梯剖视图（见图 24.5）。注意，在剖视图中不应画出剖切平面直角转折处的界线。

[9] angled 意为"成一定角度的（placed or inclined at an angle to something else）"。这两句可译为：为了在一个剖视图上表示一些相互之间成一定角度的特征，可将剖切平面弯折成相交平面，使其通过这些特征，然后将剖开的这些特征假想地旋转到同一平面上。

[10] main outline 意为"主要轮廓线"。这三句可译为：在大多数装配图中，建议采用铸铁的剖面符号（间隔相等的剖面线）。剖面线应该与视图的主要轮廓线成 45°角。相邻零件的剖面线的倾斜方向应该相反（见图 24.7）。

[11] sectioned 意为"按剖切绘制，画出剖面线"，亦为 section-lined。这两句可译为：轴、螺栓、螺母、垫圈、螺钉、铆钉、销、键及类似的实心零件，当剖切面通过其轴线或对称面时，按未剖切绘制，如需对键、键槽或销做更清楚的表达，则可以对轴进行局部剖视（见图 4.5 和图 24.7）。如果给轴、螺栓、螺母、铆钉、销和键都画上剖面线，那么图样将会变得混乱，而且很难识读。

Lesson 25　Computer Graphics (*Reading Material*)

The term computer graphics refers to the entire spectrum of drawing with the aid of a computer, from straight lines to color animation. An immense range of artistic capabilities resides under the heading of computer graphics and drafting is just one of them.

Computer graphics was originally associated with the field of electronics. Companies directly involved in electronics design and manufacturing were the first to experiment and work with computer graphics. As they tested equipment and software, new demands arose that required changes and advances in both hardware and software. Equipment and programs got better, faster, more powerful, less expensive, colorful, and fun to operate.

The potential for this new computer development was endless. And as engineers and drafters learned to work with the new tools in a technically demanding atmosphere, artists were using the new tools in an atmosphere of unlimited structure. We see the structure and freedom brought by engineers and artists to computer graphics as we watch animated television network. Complex mathematical calculations are combined with the creative eye of the artist to achieve effects unseen just a few years ago.

The world of computer graphics is a dynamic and, at times, volatile filed. The developments in technique and equipment have advanced rapidly, and the areas of application continue to spread. The drafting applications of CAD are unlimited.

The speed-of-light capabilities of the computer have shaped the gathering, processing, storage, and retrieval of information into a major industry. Every day our lives are touched by the effects of computers. Contrary to science-fiction works of years past, computers have not grown to occupy the core of the earth, but have shrunk to microscopic proportions, and now occupy tiny spaces inside our machines and tools.

Computer technology is now widely used in drafting trade. The computer chip has found its way into drafting tools. The heart of modern drafting tools is now the computer. The traditional tools of graphics, such as the T-square, drawing compass, and drafting machines (see Figs. 25.1, 25.2, and 25.3) are rapidly becoming obsolete.

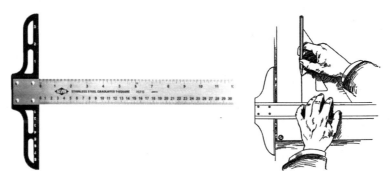

Figure 25.1　T-squares

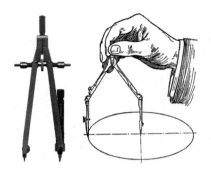

Figure 25.2 Drawing compasses   Figure 25.3 Drafting machines

Now the drafter, or operator, uses electronic hardware to construct drawings, not on paper, but on what looks like a television screen. The screen is called a monitor, video display screen (see Fig. 25.4), or video display terminal (VDT). The drafting process is now called computer-aided design drafting, or CADD. Most people say "computer-aided drafting" and use the acronym of CAD. Actually, CAD means computer-aided design.

Figure 25.4 Video display screens

The first commercially produced computer drafting system was introduced by IBM in 1964. Computer drafting systems began to mature and prove themselves in the late 1970s. Sales of computer graphics terminals reached approximately twenty thousand in 1979. Since that time there has been an exponential increase in the sales of computer graphics systems.

A visit to your local industries and engineering firms will make you aware of the popularity of computer-aided drafting. But why is it so popular? Why were computers introduced as the tool to replace the drafting machine and pencil? Why are companies scrambling to convert to computer drafting?

Foremost in your mind must be the fact that we live in the global market economy environment which thrives on innovation and competition. We are always trying to outdo each other and discover new ways to make money. New technology, such as using a computer, often gives the competitive edge to one company over another. The new electronic technology has given some company an advantage because of a couple of characteristics of the computer that aren't possessed by most humans.

Computers are extremely fast. The contents of an entire encyclopedia can be transmitted in a fraction of a second. Computers are accurate.

Each of these characteristics helps people increase productivity, which is the key to the success of using computer graphics. Without an increase in productivity it is doubtful that computer graphics would have ever gained as much popularity as it has.

But a fast machine cannot account for the productivity gain by itself; it needs competent human operators. The computer may be fast, but it cannot think, make judgments, nor reason. Human, on the other hand, may be slow, but they are intelligent and can use their power of reasoning to solve problems and overcome obstacles. Together, the speed of the computer and the intelligence of the human create a good combination that is very productive.

As you continue your study of drafting and design, keep in mind that creativity has not been replace by the computer. Only the tools have changed, allowing entirely new horizons to be opened for exploration. With the new tools at your fingertips, consider the limitations of your brain as the only obstacles to exploration. Enjoy your computer-aided design and drafting.

Words and Expressions

computer graphics 计算机制图，计算机图形学
spectrum 范围 (a range of different positions, opinions, etc. between two extreme points)
animation [ˌæniˈmeiʃən] n. 动画
unseen [ˈʌnˈsiːn] a. 未被看见的，没有预见到的 (not seen or perceived)
volatile [ˈvɔlətail] a. 易变的 (subject to rapid or unexpected change)，不稳定的
speed-of-light 光速 (speed at which light waves propagate through different materials)
science-fiction works 科幻作品
drafting trade 制图行业
computer chip 计算机芯片，电脑芯片 (a small integrated circuit of a kind used in computers)
T-square 丁字尺 (a T-shaped ruler used in mechanical drawing)
drawing compass 绘图圆规 (a drawing tool used to draw circles or arcs)，亦为 compass
drafting machine 绘图机 (一种综合性的、具有多功能的手工绘图机器或设备)
monitor 显视器 (an electronic device with a screen used for display information)
video display screen 视频显示屏，显示器
video display terminal 视频显示终端，显示屏终端
computer-aided drafting 计算机辅助绘图，亦为 computer-aided drawing
acronym [ˈækrənim] n. 首字母缩写词 (an abbreviation formed from initial letters)
exponential [ˌekspəuˈnenʃəl] n. 指数；a. 指数的
engineering firm 工程公司
competitive edge 竞争优势 (a factor which gives a company an advantage over its competitors)
encyclopedia [enˌsaikləuˈpiːdjə] n. 百科全书
a fraction of 一小部分 (a very small part or amount of something)，零点几，几分之一
account for 解释 (to give a reason or explanation for)，……的原因 (to be the cause of)
reason [ˈriːzn] v. 推理 (由一个或几个已知的判断，推导出一个未知的结论的思维过程)

Lesson 26 Dimensional Tolerance

Because of the highly competitive nature of most manufacturing businesses, the question of finding ways to reduce cost is ever present. A good starting point for cost reduction is in the design of the product. The design engineer should always keep in mind the possible alternatives available to him in making his design. It is often impossible to determine the best alternatives without a careful analysis of the probable production cost. Designing for function, interchangeability, quality, and economy requires a careful study of tolerances, surface roughness, processes, materials, and equipment.

The engineer in industry is constantly faced with the fact that no two machine parts can ever be made exactly the same. He learns that the small variations that occur in repetitive production must be considered in the design so that the tolerances placed on the dimensions will restrict the variations to acceptable limits. The tolerances provide zones in which the outline of finished part must lie. Proper tolerancing practice ensures that the finished product functions in its intended manner and operates for its expected life.

A designer is well aware that the cost of a finished product can increase rapidly as the tolerances on the components are made smaller. Designers are constantly admonished to use the widest tolerances possible. Situations may arise, however, in which the relationship between the various tolerances required for proper functioning has not been fully explored. Under such conditions the designer is tempted to specify part tolerances that are unduly tight in the hope that no difficulty will arise at the time of assembly. This is obviously an expensive substitute for a more thorough analysis of the tolerancing situation.

The allocation of proper production tolerances is therefore a most important task if the finished product is to achieve its intended purpose and yet be economical to produce. The size of the tolerances, as specified by the designer, depends on the many conditions pertaining to the design as well as on past experience with similar products if such experience is available. A knowledge of shop processes and machine tool capabilities is of great assistance in helping to determine the tolerances in the most effective manner. A revision of the design may be needed if the tolerances are too small to be maintained by the equipment available for producing the dimension.

The manufacture of machine parts is founded on the engineering drawing. Everyone engaged in manufacturing has a direct or indirect interest in understanding the meaning of the drawings on which the entire production process is established.

Ambiguities in engineering drawing can be cause of much confusion and expense. When specifying the tolerances, the designer must keep in mind that the drawing must contain all requisite information if the designer's intent is to be fully realized. The drawing must therefore give complete information and at the same time be as simple as possible. The detail of drawing must be capable of being universally understood. The drawing must have one and only one meaning to everyone who will use it—the design, purchasing, tool design, production, inspection, assembly, and servicing departments.

Tolerances may be placed on the drawing in a number of different ways. In the unilateral system one tolerance is zero and all the variation of the dimension is given by the other tolerance. In bilateral dimensioning a basic dimension is used with plus and minus variations extending each way from the basic dimension.

A part is said to be at the maximum material condition (MMC)[1] when the dimensions are all at the limits that will give a part containing the maximum amount of material. For a shaft or external dimension, the fundamental dimension is the largest value permitted, and all the variation, as permitted by the tolerance, serves to reduce the dimension. For a hole or internal dimension, the fundamental dimension is the smallest value permitted, and the variation as given by the tolerance serves to make the dimension larger.

A part is said to be at the least material condition (LMC)[2] when the dimensions are all at the limits that give a part with the smallest amount of material (see Fig. 26.1). For LMC the fundamental value is the smallest for an external dimension and the largest for an internal dimension. The tolerances thus provide parts containing larger amounts of material.

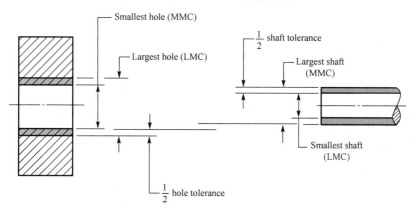

Figure 26.1 Maximum and least material conditions

Maximum material tolerances have a production advantage. For an external dimension, should the worker aim at the fundamental or largest value but form something small, the parts may be rework to bring them within acceptable limits. A worker keeping the mean dimension in mind would have smaller margins for any errors. These terms do, however, provide convenient expressions for denoting the different methods for specifying the tolerances on drawings.

Dimensional variations in manufacturing are unavoidable despite all efforts to keep production conditions as constant as possible. The reasons for the variation in a chosen dimension on parts all made by the same process are of interest. The reasons can usually be grouped into two general classes: assignable cause and chance causes.

Assignable causes A small modification in the process can cause variations in a dimension. A slight change in the properties of the raw material can cause a dimension to vary. Tools will wear and must be reset. Changes may occur in the speed, the lubricant, the temperature, the operator, and other conditions. A systematic search will generally bring such causes to light and steps can then be taken to have them eliminated.

Chance causes Chance causes, on the other hand, occur at random and are due to vague and unknown forces which can neither be traced nor rectified. They are inherent in the process and occur even though all conditions have been held as constant as possible.

When the variations due to assignable causes have been located and removed one by one, the desired state of stability or control is attained. If the variations due to chance causes are too great, it is usually necessary to move the operation to more accurate equipment rather than spend more effort in trying to improve the process.

Words and Expressions

dimensional tolerance　尺寸公差（允许的尺寸变动量），简称"公差"
interchangeability [ˌintətʃeindʒə'biliti]　*n.* 互换性
function ['fʌŋkʃən]　*v.* 起作用，行使职责；*n.* 功能，作用，职责，函数
finished product　成品（在一个企业内完成全部生产过程，可供销售的制品）
admonish [əd'mɔniʃ]　*v.* 劝告（to give friendly advice），警告（to warn gently but seriously）
unduly ['ʌn'djuːli]　*adv.* 不适当地，过度地（to an extreme or unnecessary degree）
allocation [ˌæləu'keiʃən]　*n.* 分配，配置
be founded on　建立在……之上，以……为依据
ambiguity [ˌæmbi'gjuːiti]　*n.* 含糊不清，模棱两可，不确定
requisite ['rekwizit]　*a.* 需要的，必不可少的，必备的
unilateral ['juːni'lætərəl]　*a.* 单向的，单方面的，单边的
bilateral [bai'lætərəl]　*a.* 双向的，有两面的，双边的
dimensioning [di'menʃəniŋ]　*n.* 标注尺寸，dimension 的现在分词
shaft [ʃɑːft]　*n.* 轴（主要指圆柱形外表面，也包括其他外表面中由单一尺寸确定的部分）
hole [həul]　*n.* 孔（主要指圆柱形内表面，也包括其他内表面中由单一尺寸确定的部分）
assignable [ə'sainəbl]　*a.* 可分配的，可指定的
reset [ˌriː'set]　*v.* 重新调整（to change something so that it can be used again）
systematic [ˌsisti'mætik]　*a.* 系统的，体系的，有计划的
bring…to light　发现，揭示
rectify ['rektifai]　*v.* 改正，校正，纠正

Notes

［1］ maximum material condition（MMC）意为"最大实体状态（孔或轴具有允许的材料量为最多时的状态），最大实体条件"。

［2］ least material condition（LMC）意为"最小实体状态（孔或轴具有允许的材料量为最少时的状态）"。

Lesson 27 Fundamentals of Manufacturing Accuracy

Manufacturing can be defined as the transformation of raw materials into useful products through the use of the easiest and least-expensive methods. It is not enough, therefore, to process some raw materials and obtain the desired product.

It is, in fact, of major importance to achieve that goal through employing the easiest, fastest, and most efficient methods. If less efficient techniques are used, the production cost of the manufactured part will be high, and the part will not be as competitive as similar parts produced by other manufacturers. Also, the production time should be as short as possible to enable capturing a larger market share. [1]

Modern industries can be classified in different ways. These include classification by process, classification by product, and classification based on the production volume and the diversity of products. The classification by process is exemplified by casting industries, stamping industries, and the like. When classifying by product, industries may belong to the automotive, aerospace, and electronics groups. The third method, i. e., classification based on production volume, identifies three main distinct types of production, mass, job shop, and moderate. [2] Let us briefly discuss the features and characteristics of each type.

Mass production is characterized by the high production volume of the same (or very similar) parts for a prolonged period of time. An annual production volume of less than 50,000 pieces cannot certainly be considered as mass production. The typical example of mass-produced goods is automobiles.

Job shop production is based on sales orders for a variety of small lots. Each lot may consist of 20 up to 200 or more similar parts, depending upon the customers' needs. It is obvious that this type of production is most suitable for subcontractors who produce varying components to supply various industries. The machine tools employed must be flexible to handle variations in the configuration of the ordered components, which are usually frequent. [3] Also, the employed personnel must be highly skilled in order to handle a variety of tasks, which differ for the different parts that are manufactured.

Moderate production is an intermediate phase between the job shop and the mass production types. The production volume ranges between 10,000 to 20,000 parts, and the machine tools employed are flexible and multipurpose. This type of production is gaining popularity in industry because of an increasing market demand for customized products. [4]

A very important fact of the manufacturing science is that it is almost impossible to obtain the desired nominal dimension when processing a workpiece. This is actually caused by the inevitable, though very slight, inaccuracies inherent in the machine tool as well as by various complicated factors like the elastic deformation and recovery of the workpiece and/or the fixture, temperature effects during processing, and sometimes the skill of the operator. Since it is very difficult to analyze and completely eliminate the effects of these factors, no part can be manufactured to exact dimensions.

Fortunately, exact sizes are not needed. The need is for varying degrees of accuracy according to functional requirements. A manufacturer of children's tricycles would soon go out of business if the parts were made with jet-engine accuracy—no one would be willing to pay the price. So what is wanted is a means of specifying the permissible variation in dimension with whatever degree of accuracy is required. The answer to the problem is the specification of tolerance on each dimension. [5]

The tolerance is the difference between the permitted maximum and minimum sizes of a part. It is a permissible degree of inaccuracy or a permissible deviation from the nominal dimension that would not affect the proper functioning of the manufactured part in a detrimental way. According to the ISO system, the nominal dimension is referred to as the basic size of the part. The toleranced size of the part is designated by the basic size followed by a letter and a number, such as 20H6 or 30k8 (see Fig. 27.1). [6]

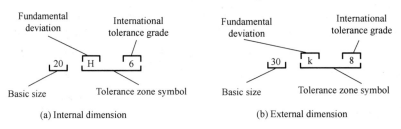

Figure 27.1 Metric tolerance symbols

Before two components are assembled together, the relationship between the dimensions of the mating surfaces must be specified. This determines the degree of tightness or freedom for relative motion between the mating surfaces. [7] There are basically three types of fits, namely, clearance fit, transition fit, and interference fit. In all cases of clearance fit, the upper limit of the shaft is always smaller than the lower limit of the mating hole. This is not the case in interference fit, where the lower limit of the shaft is always larger than the upper limit of the hole. The transition fit, as the name suggests, is an intermediate fit. According to ISO, the internal enveloped part is referred to as the shaft, whereas the surrounding surface is referred to as the hole. Accordingly, from the fits point of view, a key is referred to as the shaft and the keyway as the hole. [8]

There are two ways for specifying and expressing the various types of fits, the shaft basis and the hole basis systems. The location of the tolerance zone with respect to the zero line is indicated by a letter, which is always capital for holes and lowercase for shafts, whereas the tolerance grade is indicated by a number, as previously explained. [9] Therefore, a fit symbol can be H7/h6, F6/g5, or any other similar form.

When the service life of an electric bulb is over, all you do is buy a new one and replace the bulb. This easy operation would not be possible without two main concepts, interchangeability and standardization. Interchangeability means that identical parts must be interchangeable, i.e., able to replace each other, whether during assembly or subsequent maintenance work. As you can easily see, interchangeability is achieved by establishing a permissible tolerance, beyond which any further deviation from the nominal dimension of the part is not allowed.[10]

Words and Expressions

stamping ['stæmpiŋ] *n.* 冲压（使板料经分离或成型而得到制件的工艺），冲压件
aerospace group 航空航天集团，航空航天工业集团
mass production 大量生产（the production of large amounts of standardized products）
job shop production 单件生产，小批生产
moderate production 中批生产，成批生产
prolonged [prə'lɔŋd] *a.* 持续很久的，长期的，长时间的
lot [lɔt] *n.* 批（在一致条件下生产的一定数量的个体），也可写为 batch
lot size 批量（批中包含的个体数量），也可写为 batch size
order 订单（an instruction to buy something），下订单（to request that something be supplied）
subcontractor ['sʌbkən'træktə] *n.* 第二次转包的工厂，转包人，小承包商
multipurpose ['mʌlti'pəpəs] *a.* 多用途的（having more than one use or purpose）
nominal ['nɔminl] *a.* 标定的，额定的，规定的，名义上的，铭牌上的
(be) inherent in 为……所固有，是……的固有性质
deviation [ˌdi:vi'eiʃən] *n.* 尺寸偏差（某一尺寸减其基本尺寸所得的代数差），简称"偏差"
basic size 基本尺寸（设计给定的尺寸）
toleranced size 注有公差的尺寸，尺寸及其公差（size and its tolerance）
fundamental deviation 基本偏差（用以确定公差带相对于零线位置的上偏差或下偏差。一般指靠近零线的那个偏差）
international tolerance grade 国际公差等级，标准公差等级（standard tolerance grade）
fit 配合（基本尺寸相同的，相互结合的孔和轴公差带之间的关系，用以表达结合松紧程度的功能要求）
fit symbol 配合代号
mating surface 配合表面，啮合表面
tightness ['taitnis] *n.* 紧密性，松紧度，密封性
clearance fit 间隙配合（具有间隙的配合，其最小间隙可以等于零）
transition fit 过渡配合（可能具有间隙或过盈的配合）
interference [ˌintə'fiərəns] *n.* 过盈（孔的尺寸减去相配合的轴的尺寸所得的代数差为负）
interference fit 过盈配合（具有过盈的配合，其最小过盈可以等于零）
envelop [in'veləp] *v.* 包装，包围，包络
shaft basis system 基轴制（采用公制单位时），采用英制单位时为 basic shaft system
hole basis system 基孔制（采用公制单位时），采用英制单位时为 basic hole system
zero line 零线（公差与配合中确定偏差的一条基准直线。通常零线表示基本尺寸）
tolerance zone 公差带（限制实际要素变动量的区域）
tolerance grade 公差等级（确定尺寸精确程度的等级）
standardization [ˌstændədai'zeiʃən] *n.* 标准化，规格化，标定，校准

Notes

[1] market share 意为"市场份额,市场占有率"。全段可译为:事实上,通过采用最容易、最快和最有效的方法来实现这个目标是非常重要的。如果没用采用最有效的方法,零件的生产成本就会增加,所生产零件的竞争力就不如其他厂家生产的类似零件。此外,为了获取更大的市场份额,生产时间应该尽可能缩短。

[2] types of production 意为"生产类型"。全句可译为:第三种方法,也就是根据产量来分类的方法,确定了三种主要的生产类型:大批量生产、单件小批生产、中批生产。

[3] flexible 意为"柔性(机床适应加工不同种类工件的能力)的",ordered component 意为"订购的零件"。全段可译为:单件小批生产建立在各种各样小批量销售订单的基础上。根据客户需求,每批可由 20 至 200 或更多的相似零件组成。很明显,这种生产类型最适合那些为不同工业行业生产各种零件的小企业。所采用的机床必须具有足够的柔性,以适应订购的零件的外形变化,而这种变化是经常发生的。此外,这种企业必须雇用那些能够处理在制造不同种类零件过程中出现的各种问题的高技能人才。

[4] customized product 意为"定制产品(为满足客户特定需要而制造的产品)"。全句可译为:由于定制产品的市场需求日益增长,这种生产类型的应用越来越多。

[5] permissible variation in dimension 意为"允许尺寸的变动量"。这两句可译为:因此,我们应该有一种能够根据需要的精度等级来确定允许尺寸的变动量的方法。这个问题的解决方案是确定每一个尺寸的公差。

[6] ISO (International Organization for Standardization),国际标准化组织。其中,"ISO"并不是其全称首字母的缩写,而是一个词,它来源于希腊语,意为"相等",从"相等"到"标准",内涵上的联系使"ISO"成为组织的名称。nominal dimension 意为"公称尺寸"。这两句可译为:在 ISO(国际标准化组织)体系中,公称尺寸被称为零件的基本尺寸。注有公差的零件尺寸用基本尺寸和其后面的字母与数字来表示,如 20H6、30k8 等(见图 27.1)。

[7] degree of tightness or freedom 意为"松紧程度"。这两句可译为:在对两个零件进行装配之前,必须确定配合表面间的尺寸关系。这实际上是确定配合件相对运动的松紧程度。

[8] internal enveloped part 意为"被包容件",point of view 意为"观点"。这两句可译为:根据 ISO 规定,被包容件被称为轴,而周围表面则被称为孔。因此,从配合的观点来说,键被称为轴,键槽被称为孔。

[9] tolerance zone 意为"公差带"。全句可译为:公差带相对于零线的位置是由一个字母来表示的,总是用大写字母表示孔,用小写字母表示轴;而公差等级则如前所述用数字表示。

[10] 全段可译为:当一个电灯泡坏了之后,你所要做的就是买一个新的并替换坏灯泡。如果没有互换性和标准化这两个主要概念,这种简单的操作是不可能完成的。互换性意味着在相同的零件之间必须是可互换的,即无论是在装配时还是在其后的维修时,这些零件都能够互相替换。正如你可以很容易地看到的那样,互换性是通过建立一个允许的公差来实现的。相对于公称尺寸,超出这个范围的任何偏差都是不允许的。

Lesson 28　Tolerances and Fits

Today's technology requires that parts be specified with increasingly precise dimensions. Each dimension is allowed a certain degree of variation within a specified zone, or tolerance. For example, a part's dimension might be expressed as 20±0.06, which allows a tolerance (variation in size) of 0.12 mm. A tolerance should be as large as possible without interfering with the function of the part to minimize production costs. Manufacturing costs increase as tolerances become smaller.

The definitions of some terms related to tolerances and fits are given below.

Nominal size　size of a feature of perfect form[1] as defined by the drawing specification. In former times, this was referred to as "basic size".

Actual size　measured size of the finished part.

Limits of size　two extreme permissible sizes of a feature of size. To fulfill the requirement, the actual size shall lie between the upper and lower limits of size; the limits of size are also included.

Upper limit of size　largest permissible size of a feature of size.

Lower limit of size　smallest permissible size of a feature of size.

Deviation　algebraic difference between an actual size and its nominal size. The deviation is a signed value and may be positive, negative or zero.

Upper limit deviation　upper limit of size minus its nominal size. This was referred to as "upper deviation".

Lower limit deviation　lower limit of size minus its nominal size.

Fundamental deviation　one of the limit deviations closest to the nominal size. The fundamental deviation is identified by a letter (e.g. H, k).

Tolerance　difference between the upper limit of size and the lower limit of size. The tolerance is an absolute value without a sign. The tolerance is also the difference between the upper limit deviation and the lower limit deviation.

Standard tolerance grade　group of tolerances for linear sizes characterized by a common identifier. The standard tolerance grade identifier consists of IT followed by a number (e.g. IT6), the letters in the abbreviated term "IT" stand for "International Tolerance". There are 20 IT grades: IT01, IT0, IT1, ..., IT18.

Tolerance interval　variable values of the size between and including the tolerance limits.[2] In former times, this was referred to as "tolerance zone". The tolerance interval is contained between the upper and lower limits of size. It is defined by the magnitude of the tolerance and its placement relative to the nominal size.

Tolerance class　combination of a fundamental deviation and a standard tolerance grade. The tolerance class consists of the fundamental deviation identifier followed by the tolerance grade number (e.g. H6, k8, etc.).

Fit   relationship between an internal feature of size and an external feature of size (the hole and shaft of the same nominal size) which are to be assembled.

Clearance fit   fit that always provides a clearance between the hole and the shaft when assembled, i. e. the lower limit of size of the hole is either larger than or, in the extreme case, equal to the upper limit of size of the shaft.

Interference fit   fit that always provides an interference between the hole and the shaft when assembled, i. e. the upper limit of size of the hole is either smaller than or, in the extreme case, equal to the lower limit of size of the shaft.

Transition fit   fit which may provide either a clearance or an interference between the hole and the shaft when assembled.

Hole-basis fits system   fits where the fundamental deviation of the hole is zero, i. e. the lower limit deviation is zero. Possible examples of hole-basis fits are: H7/m6 and H6/k5. Standard broaches (Fig 28.1), reamers (Fig 28.2), and other standard tools are often used to produce precise holes, and standard plug gages (Fig 28.3) are used to check the hole sizes. On the other hand, shafts can easily be machined to any size desired. Therefore, the hole-basis fits system is recommended for general use. In this fit system, the lower limit of size of the hole is identical to the nominal size because a hole can be enlarged by machining but not reduced in size.

Figure 28.1   A round broach

Figure 28.2   Reamers   Figure 28.3   A plug gage

Shaft-basis fits system   fits where the fundamental deviation of the shaft is zero, i. e. the upper limit deviation is zero. Possible examples of shaft-basis fits are: G7/h6 and M6/h6. Fits are sometimes required on the shaft-basis fits system. For example, it is advantageous when several parts having different fits, but one nominal size, are required on a single shaft. In this fit system, the upper limit of size of the shaft is identical to the nominal size because shafts can be machined to smaller size but not enlarged.

Selective assembly[3]   a method of selecting and assembling parts by trial and error and by hand, allowing parts to be made with greater tolerances at less cost as a compromise between a high manufacturing accuracy and ease of assembly.

In selective assembly, all parts are measured and classified into several grades according to actual sizes, so that "small" shafts can be matched with "small" holes, "medium" shafts with "medium" holes, and so on.

Because the surface texture of a part affects its function, it must be precisely specified. Surface texture is the variation in a surface, including roughness, waviness, lay and imperfections.

Surface roughness　the finest of the irregularities in the surface caused by the manufacturing process used to smooth the surface. Roughness height is measured in micrometers (μm) or microinches (μin).

Surface waviness　a widely spaced variation that exceeds the roughness width cutoff measured in millimeters or inches; roughness may be regarded as a surface variation superimposed on a wavy surface.

Surface lay　the direction of the surface pattern caused by the production method used.

Surface imperfections　surface defects occurring infrequently or at widely varying intervals on a surface, including cracks, blowholes, checks, scratches, and the like.

Words and Expressions

feature [ˈfiːtʃə]　n. 要素（构成零件几何特征的点、线和面的统称）
feature of size　尺寸要素（由一定大小的线性尺寸或角度尺寸确定的几何形状）
limit of size　极限尺寸
tolerance class　公差带代号，曾称"symbol for tolerance zone"或"tolerance zone symbol"
abbreviated term　缩略术语，缩写词
broach [brəutʃ]　n. 拉刀（在拉力作用下进行切削的刀具），亦为 pull broach
round broach　圆拉刀（加工圆柱形孔的拉刀）
reamer [ˈriːmə]　n. 铰刀（一种孔的精加工刀具）
plug gage　塞规（检验孔用的专用量规。可分为通规 go gage 和止规 no go gage）
trial and error　试错，反复试验（the trying of one thing or another until something succeeds）
surface texture　表面结构（表面粗糙度、表面波纹度、表面纹理和表面缺陷等的总称）
irregularity [iˌregjuˈlæriti]　n. 不规则，不平整（something that is unusual, uneven）
superimpose [ˈsjuːpərimˈpəuz]　v. 叠加（to lay or place something on or over something else）
surface waviness　表面波纹度（固体表面主要因机械加工系统的振动而形成的有一定周期性的形状和起伏的特征量度）
cutoff [ˈkʌtɔːf]　n. 界限（a designated limit）
surface lay　表面纹理（工件表面上的微观机械加工痕迹）
imperfection [ˌimpəˈfekʃən]　n. 缺陷，瑕疵（something imperfect; a defect or flaw）
blowhole　气孔（a hole in metal caused by a bubble of gas captured during solidification）
check [tʃek]　n. 微细裂纹，微小裂纹（a small crack）

Notes

[1] feature of perfect form 意为"理想形状要素"。
[2] tolerance limits 意为"公差极限（确定允许值上界限和/或下界限的特定值）"。
[3] selective assembly 意为"选择装配法（将零件的制造公差放大到经济可行的程度，然后选择合适的零件进行装配，以保证装配精度）"。

Lesson 29　Introduction to Mechanical Design

Mechanical design is the application of science and technology to devise new or improved products for the purpose of satisfying human needs. It is a vast field of engineering technology which not only concerns itself with the original conception of the product in terms of its size, shape and construction details, but also considers the various factors involved in the manufacture, marketing and use of the product.

People who perform the various functions of mechanical design are typically called designers, or design engineers. Mechanical design is basically a creative activity. However, in addition to being innovative, a design engineer must also have a solid background in the areas of mechanical drawing, kinematics, dynamics, materials engineering, strength of materials and manufacturing processes.

As stated previously, the purpose of mechanical design is to produce a product which will serve a need for humankind. Inventions, discoveries and scientific knowledge by themselves do not necessarily benefit people; only if they are incorporated into a designed product will a benefit be derived. It should be recognized, therefore, that a human need must be identified before a particular product is designed.

Mechanical design should be considered to be an opportunity to use innovative talents to envision a design of a product, to analyze the system and then make sound judgments on how the product is to be manufactured. It is important to understand the fundamentals of engineering rather than memorize mere facts and equations. There are no facts or equations which alone can be used to provide all the correct decisions required to produce a good design. On the other hand, any calculations made must be done with the utmost care and precision. For example, if a decimal point is misplaced, an otherwise acceptable design may not function.

Good designs require trying new ideas and being willing to take a certain amount of risk, knowing that if the new idea does not work the existing method can be reinstated. Thus a designer must have patience, since there is no assurance of success for the time and effort expended. Creating a completely new design generally requires that many old and well-established methods be thrust aside[1]. This is not easy since many people cling to familiar ideas, techniques and attitudes. A design engineer should constantly search for ways to improve an existing product and must decide what old, proven concepts should be used and what new, untried ideas should be incorporated.

New designs generally have "bugs" or unforeseen problems which must be worked out before the superior characteristics of the new designs can be enjoyed. Thus there is a chance for a superior product, but only at higher risk. It should be emphasized that, if a design does not warrant radical new methods, such methods should not be applied merely for the sake of[2] change.

During the beginning stages of design, creativity should be allowed to flourish without a great number of constraints. Even though many impractical ideas may arise, it is usually easy to eliminate them in the early stages of design before firm details are required by manufacturing. In this way, innovative ideas are not inhibited. Quite often, more than one design is developed, up to the point where they can be compared against each other. It is entirely possible that the design which is ultimately accepted will use ideas existing in one of the rejected designs.

Another important point which should be recognized is that a design engineer must be able to communicate ideas to other people if they are to be incorporated. Communicating the design to others is the final, vital step in the design process. Undoubtedly many great designs, inventions, and creative works have been lost to mankind simply because the originators were unable or unwilling to explain their accomplishments to others. Presentation is a selling job. The engineer, when presenting a new solution to administrative, management, or supervisory persons, is attempting to sell[3] or to prove to them that this solution is a better one. Unless this can be done successfully, the time and effort spent on obtaining the solution have been largely wasted.

Basically, there are only three means of communication available to us. These are the written, the oral, and the graphical forms. Therefore the successful engineer will be technically competent and versatile in all three forms of communication. A technically competent person who lacks ability in any one of these forms is severely handicapped. If ability in all three forms is lacking, no one will ever know how competent that person is!

The competent engineer should not be afraid of the possibility of not succeeding in a presentation. In fact, occasional failure should be expected because failure or criticism seems to accompany every really creative idea. There is a great deal to be learned from a failure, and the greatest gains are obtained by those willing to risk defeat. In the final analysis, the real failure would lie in deciding not to make the presentation at all. To communicate effectively, the following questions must be answered:

(1) Does the design really serve a human need?

(2) Will it be competitive with existing products of rival companies?

(3) Is it economical to produce?

(4) Can it be readily maintained?

(5) Will it sell and make a profit?

Only time will provide the true answers to the preceding questions, but the product should be designed, manufactured and marketed only with initial affirmative answers. The design engineer also must communicate the finalized design to manufacturing through the use of detail and assembly drawings.

Quite often, a problem will occur during the manufacturing cycle. It may be that a change is required in the dimensioning or tolerancing of a part so that it can be more readily produced. This falls in the category of engineering changes which must be approved by the design engineer so that the product function will not be adversely affected. In other cases, a deficiency in the design may appear during assembly or testing just prior to shipping. These realities simply bear out the fact that design is a dynamic process. There is always a better way to do it and the designer should constantly strive towards finding that better way.

Words and Expressions

mechanical drawing　机械制图（用图样确切表示机械的结构形状、尺寸大小、工作原理和技术要求的学科）
manufacturing process　工艺过程（改变生产对象的形状、尺寸、相对位置和性质等，使其成为成品或半成品的过程）
envision [en'viʒən]　v. 想象，预见，展望
innovative ['inəuveitiv]　a. 革新的，创新的（introducing or using new ideas or methods）
make sound judgment　做出正确判断（to weigh the evidence and come up with the right answer）
utmost ['ʌtməust]　a.; n. 极度（的），极端（的），最大限度（的），最大可能
decimal point　小数点（a dot between a whole number and a decimal fraction）
misplace ['mis'pleis]　v. 放错位置（to put in a wrong or inappropriate place）
reinstate [ˌriːin'steit]　v. 复原，修复，恢复，使正常，使恢复原状
cling to　不愿失去某物或停止做某事（to be unwilling to lose sth or stop doing sth）
bug [bʌg]　n. 故障，缺陷（an unexpected defect, fault, or imperfection）
unforeseen ['ʌnfɔː'siːn]　a. 想不到的，未预见到的，偶然的
warrant ['wɔrənt]　v. 需要（to need），有必要（to make a particular activity necessary）
constraint [kən'streint]　n. 限制，制约，约束，束缚
inhibit [in'hibit]　v. 防止，阻止，禁止（to prohibit; forbid）
originator [ə'ridʒəneitə]　n. 创新者（one who creates or introduces something new），发明人
presentation [ˌprezen'teiʃən]　n. 介绍，讲述（a talk giving information about something）
graphical ['græfikəl]　a. 用图表示的（of or relating to pictorial representation），图解的
in the final analysis　总之，归根结底（after considering everything），亦为 in the last analysis
affirmative [ə'fəːmətiv]　a. 肯定的，正面的，赞成的；n. 肯定，确认
deficiency [di'fiʃənsi]　n. 缺乏，不足之处，缺陷，不足额，亏空
dimensioning [di'menʃəniŋ]　n. 标注尺寸（writing dimensions on a drawing）
tolerancing ['tɔlərənsiŋ]　n. 给（机器零部件等）规定公差，标注公差
reality [ri'æliti]　n. 事实，真实（the true situation that exists; the real situation）
bear out　证明，证实

Notes

[1] thrust aside 意为"推开，把……搁置一边"。
[2] for the sake of 意为"为了（for the purpose of; for the motive of）"。
[3] sell 这里指"说服（to persuade someone to accept or approve of something）"。

Lesson 30 Engineering Design

What differentiates engineering from many other fields is that it attempts to go from theory into practice for the purpose of developing products and processes instead of merely observing the phenomena of that science or art. For example, a physicist studies and records findings in order to understand some phenomenon or physical process better. On the other hand, an engineer utilizes scientific information to make a particular process or product for use by consumers.

Engineering design is a systematic process by which solutions to the needs of humankind are obtained. The process is applied to problems (needs) of varying complexity. For example, mechanical engineers will use the design process to find an effective, efficient method to convert reciprocating motion to rotary motion for the drivetrain in an internal combustion engine; electrical engineers will use the process to design electrical generating systems using falling water as the power source; and materials engineers use the process to design ablative materials which enable astronauts to safely reenter the earth's atmosphere.

The vast majority of complex problems in today's high technology society depend for solution not on a single engineering discipline, but on teams of engineers, scientists, environmentalists, economists, sociologists, and legal personnel. [1] Solutions are not only dependent upon the appropriate applications of technology but also upon public sentiment, government regulations and political influence. [2] As engineers we are empowered with the technical expertise to develop new and improved products and systems, but at the same time we must be increasingly aware of the impact of our actions on society and the environment in general and work conscientiously toward the best solution in view of all relevant factors.

A formal definition of engineering design is found in the curriculum guidelines of the Accreditation Board for Engineering and Technology (ABET). ABET accredits curricula in engineering schools and derives its membership from the various engineering professional societies. Each accredited curriculum has a well-defined design component which falls within the ABET guidelines. The ABET statement on design reads as follows:

Engineering design is the process of devising a system, component, or process to meet desired needs. It is a decision making process (often iterative), in which the basic sciences, mathematics, and engineering sciences are applied to convert resources optimally to meet a stated objective. Among the fundamental elements of the design process are the establishment of objectives and criteria, synthesis, analysis, construction, testing, and evaluation.

The engineering design component of a curriculum must include most of the following features: development of student creativity, use of open-ended problems, development and use of modern design theory and methodology, formulation of design problem statements and specifications, consideration of alternative solutions, feasibility considerations, production processes, and detailed system descriptions. Further, it is essential to include a variety of realistic constraints such as economic factors, safety, reliability, aesthetics, ethics, and social impact. [3]

If anything can be said about the last half of the twentieth century, it is that we have had an explosion of information. The amount of data that can be uncovered on most subjects is overwhelming. People in the upper levels of most organizations have assistants who condense most of the things that they must read, hear, or watch. When you begin a search for information, be prepared to scan many of your sources and document their location so that you can find them easily if the data subsequently appear to be important.

Some of the sources that are available include the following:

1. Existing solutions Much can be learned from the current status of solutions to a specific need if actual products can be located, studied and, in some cases, purchased for detailed analysis. [4] An improved solution or an innovative new solution cannot be found unless the existing solutions are thoroughly understood.

2. Your library Many universities have courses that teach you how to use your library. Such courses are easy when you compare them with those in chemistry and calculus, but their importance should not be underestimated. There are many sources in the library that can lead you to the information that you are seeking. You may find what you need in an index such as the *Engineering Index*. There are many other indexes that provide specialized information. The nature of your problem will direct which ones may be helpful to you. Don't hesitate to ask for assistance from the librarian.

3. Professional organizations The American Society of Mechanical Engineers is a technical society that will be of interest to students majoring in mechanical engineering. [5] Each major in your college is associated with not one but often several such societies. The National Society of Professional Engineers is an organization that most engineering students will eventually join, as well as at least one technical society such as the Society of Manufacturing Engineers, the American Society of Civil Engineers (ASCE), the Institute of Electrical and Electronics Engineers (IEEE), or any one of dozens that serve the technical interests of the host of specialties with which professional practices seem most closely associated. Many engineers are members of several associations and societies. [6]

Money and economics are part of engineering design and decision making. We live in a society that is based on economics and competition. It is no doubt true that many good ideas never get tried because they are deemed to be economically infeasible. Most of us have been aware of this condition in our daily lives. We started with our parents explaining why we could not have some item that we wanted because it cost too much. Likewise, we will not put some very desirable component into our designs because the value gained will not return enough profit in relation to its cost. [7]

Industry is continually looking for new products of all types. Some are desired because the current product is not competing well in the marketplace. Others are tried simply because it appears that people will buy them. How do manufacturers know that a new product will be popular? They seldom know with certainty. Statistics is an important consideration in market analysis. Most of you will find that probability and statistics are an integral part of your chosen engineering curriculum. The techniques of this area of mathematics allow us to make inferences about how large groups of people will react based on the reactions of a few.

Words and Expressions

systematic [ˌsistiˈmætik]　*a.* 有系统的，成体系的，有计划的
reciprocating motion　往复运动（motion alternately backward and forward, or up and down）
drivetrain [ˈdraivtrein]　*n.* 动力传动系统，亦为 driveline
internal combustion engine　内燃机
falling water　水位差，水流下落
astronaut [ˈæstrənɔːt]　*n.* 宇航员（a person who travels beyond the earth's atmosphere）
ablative [ˈæblətiv]　*a.* 烧蚀的，脱落的；*n.* 烧蚀材料
reenter [ˌriːˈentə]　*v.* 重新进入，重返大气层，重新加入
sentiment [ˈsentimənt]　*n.* 感情，情绪，意见，感想
empower [imˈpauə]　*v.* 授权给，准许，授予……的权利（资格）
conscientiously [ˌkɔnʃiˈenʃəsli]　*ad.* 认真地，凭良心办事地，有责任心地，尽责地
culmination [ˌkʌlmiˈneiʃən]　*n.* 顶点，最高潮，极点
salient [ˈseiljənt]　*a.* 突出的，显著的
curriculum [kəˈrikjuləm]　（*pl.* curricula）*n.* 全部课程，学习计划，课程设置
guideline [ˈgaidlain]　*n.* 方针，准则，指导方针
accreditation [əˌkrediˈteiʃən]　*n.* 鉴定合格，任命，认证
accreditation board　认证委员会
Accreditation Board for Engineering and Technology（ABET）　美国工程技术认证委员会（是美国大学在应用科学、计算机、工程和技术专业认证领域最具权威的机构）
derive [diˈraiv]　*v.* 得到，取得，衍生出，引出
iterative [ˈitərətiv]　*a.* 反复的，迭代的
optimally [ˈɔptiməli]　*ad.* 最佳地，最优地，最恰当地，最适宜地
criterion [kraiˈtiəriən]　（*pl.* criteria）*n.* 标准，准则，依据，判据
synthesis [ˈsinθisis]　*n.* 合成，综合，机构综合
open-ended [ˈəupənˈendid]　*a.* 可扩展的，未确定的
open-ended problem　开放性问题（a problem that has several or many correct answers, and several ways to the correct answers）
methodology [ˌmeθəˈdɔlədʒi]　*n.* 方法（学，论），分类法
formulation [ˌfɔːmjuˈleiʃən]　*n.* 列方程式，列出公式，（有系统的）阐述
specification [ˌspesifiˈkeiʃən]　*n.* 规格，规范，技术要求，说明书，详细说明
feasibility [ˌfiːzəˈbiləti]　*n.* 可行性（the possibility that can be made, done, or achieved）
aesthetics [iːsˈθetiks]　*n.* 美学
ethics [ˈeθiks]　*n.* 道德规范（moral principles that control or influence a person's behavior）
impact [ˈimpækt]　*n.* 冲击，影响（the effect of one thing on another）；*v.* 对……产生影响
uncover [ʌnˈkʌvə]　*v.* 发现，展现，显示（to make known, bring to light）
overwhelming [ˌəuvəˈwelmiŋ]　*a.* 压倒性的，不可抵抗的，优势的

condense [kən'dens] v. 浓缩，精简，压缩，简要叙述
scan [skæn] v. 浏览，匆忙地阅读或翻阅（to look over or leaf through hastily）
existing solution 已有的解决方案，现有的解决方案
current status 当前状态，目前状况
innovative ['inəuveitiv] a. 创新的（introducing or using new ideas or methods）
calculus ['kælkjuləs] n. 微积分
underestimate [ˌʌndər'estimeit] v. 低估，看轻
index ['indeks] n. 索引，检索，目录
database 数据库（a large collection of data organized for rapid search and retrieval）
ASCE 美国土木工程师学会
IEEE 电气电子工程师学会
professional organization 专业组织（an association that is formed to further the interests of people engaged in a specific profession and serve the public good）
trade journal 行业刊物（a publication that is targeted to people in a very specific industry）
infeasible [in'fi:zəbl] a. 不能实行的，办不到的（not feasible；impracticable）
marketplace ['mɑ:kit'pleis] n. 市场（a location where public sales are held），集会场所
statistics [stə'tistiks] n. 统计，统计学，统计数字
inference ['infərəns] n. 推论（a conclusion reached on the basis of evidence and reasoning）

Notes

[1] engineering discipline 意为"工程学科"。全句可译为：在当今这个高技术社会中，绝大多数复杂问题的解决不能仅依靠单一的工程学科，而是需要依靠由工程师、科学家、环境学家、经济学家、社会学家和法律工作者组成的团队。

[2] government regulation 意为"政府法规"。全句可译为：解决方案不只是一个技术问题，还要考虑公众情感、政府法规和政治影响。

[3] realistic constraint 意为"实际的约束条件"，social impact 意为"社会影响"。全句可译为：此外，有必要包括各种实际的约束条件，如经济因素、安全性、可靠性、美学、道德规范和社会影响。

[4] much can be learned 意为"了解关于……更多的信息，具有重要的启示和借鉴意义"，current status of solution 意为"现有的解决方案"，亦可写为 existing solution。全句可译为：如果一些能够满足特定需要的实际产品可以被发现、被研究，在某些情况下还可以被购买用来进行详细的分析，那么这些现有的解决方案对于当前的设计工作具有重要的启示和借鉴意义。

[5] major in 意为"主修（to specialize in a certain subject in college），专业为"。全句可译为：美国机械工程师学会是机械工程专业的学生们感兴趣的技术性学会。

[6] association and society 意为"协会和学会"。全句可译为：有很多工程师同时是几个协会和学会的会员。

[7] 全句可译为：同样，在我们的设计中不能采用某些性能优越的零件，这是因为相对其成本而言，增加的价值不能带来足够的利润。

Lesson 31 Some Rules for Mechanical Design

The old saying that "necessity is the mother of invention" is still true. Designing starts with a need, real or imagined. Existing apparatus may need improvements in durability, efficiency, weight, speed, or cost. New apparatus may be needed to perform a function previously done by men, such as material handling or product assembly. With the objective wholly or partly defined, the next step in design is the conception of mechanisms and their arrangements that will perform the needed functions. For this, freehand sketching is of great value, not only as a record of one's thoughts and as an aid in discussion with others, but particularly for communication with one's own mind, as a stimulant for creative ideas.

When the general shape and a few dimensions of the several components become apparent, analysis can begin in earnest.[1] The analysis will have as its objective satisfactory or superior performance, plus safety and durability with minimum weight, and a competitive cost. Optimum proportions and dimensions will be sought for each critically loaded section. Materials and their treatment will be chosen. These important objectives can be attained only by analysis based upon the principles of mechanics, such as those of statics for reaction forces; of dynamics for inertia and acceleration; of elasticity and strength of materials for stress and deflection; and of fluid mechanics for lubrication and hydraulic transmission.

Finally, a design based upon function and reliability will be completed, and a prototype may be built. If its tests are satisfactory, and if the device is to be produced in quantity, the initial design will undergo certain modifications that enable it to be manufactured in quantity at a lower cost. During subsequent years of manufacture and service, the design is likely to undergo changes as new ideas are conceived or as further analysis based upon tests and experience indicate alterations. Product quality, customer satisfaction, and manufacture cost are all related to design.

To stimulate creative thought, the following rules are suggested for the designer.

1. Apply ingenuity to utilize desired physical properties and to control undesired ones The performance requirements of a machine are met by utilizing laws of nature or properties of matter (e. g. , strength, stiffness, inertia, buoyancy, centrifugal force, principles of the lever and inclined plane,[2] friction, viscosity, fluid pressure, and thermal expansion), also the many electrical, optical, thermal, and chemical phenomena. However, what may be useful in one application may be detrimental in the next. Friction is desired at the clutch facing (see Figs. 6.4 and 31.1) but not in the clutch bearing. Ingenuity in design should be applied to utilize and control the physical properties that are desired and to

Figure 31.1 Clutch facings

minimize those that are not desired.

2. Provide for favorable stress distribution and stiffness with minimum weight On components subjected to alternating stress, particular attention is given to a reduction in stress concentration, and to an increase of strength at fillets (see Fig. 8.2), threads, holes, and fits. Stress reductions are made by modification in shape, and strengthening may be done by prestressing treatments such as surface rolling. Hollow shafts and box sections give a favorable stress distribution, together with high stiffness and minimum weight. The stiffness of shafts and other components must be suitable to avoid resonant vibrations.

3. Use basic equations to calculate and optimize dimensions The fundamental equations of mechanics and the other sciences are the accepted bases for calculations. They are sometimes rearranged in special forms to facilitate the determination or optimization of dimensions, such as the beam and surface stress equations for determining gear-tooth size. Factors may be added to a fundamental equation for conditions not analytically determinable, e.g., on thin steel tubes, an allowance for corrosion added to the thickness based on pressure. When it is necessary to apply a fundamental equation to shapes, materials, or conditions which only approximate the assumptions for its derivation, it is done in a manner which gives results "on the safe side".

4. Choose materials for a combination of properties Materials should be chosen for a combination of pertinent properties, not only for strength, hardness, and weight, but sometimes for resistance to impact, corrosion, and low or high temperatures. Cost and fabrication properties are factors, such as weldability, machinability, sensitivity to variation in heat-treating temperatures, and required coating.

5. Select carefully between stock and integral components A previously developed component is frequently selected by a designer and his company from the stocks of parts manufacturers, if the component meets the performance and reliability requirements and is adaptable without additional development costs to the particular machine being designed. However, its selection should be carefully made with a full knowledge of its properties, since the reputation and liability of the company suffer if there is a failure in any one of the machine's parts. In other cases, the strength, reliability, and cost requirements are better met if the designer of the machine also designs the component, with the advantage of compactness if it is designs integral with other components, e.g., gears to be integral with a shaft (see Fig. 31.2).

6. Provide for accurate location and noninterference of parts in assembly A good design provides for the correct locating of parts and for easy assembly and repair. Shoulders give accurate location without measurement during assembly. Shapes can be designed so that parts cannot be assembled backwards or in the wrong place. Interferences, as between screws in tapped holes (see Fig. 31.3), and between linkages must be foreseen and prevented.

Figure 31.2   Gear shafts

Figure 31.3   Tapped holes

Words and Expressions

saying ['seiiŋ] n. 话，谚语，格言（something said, especially a proverb or maxim）
apparatus [,æpə'reitəs] n. 设备（the equipment needed for a particular activity or purpose）
conception [kən'sepʃən] n. 构思，构想，设想，见解，想法
freehand sketching 徒手绘草图
an aid to 对……有所帮助，有助于
communication with one's own mind 和自己的大脑进行交流
stimulant ['stimjulənt] n. 刺激物，促进因素
have as 把……作为
hydraulic transmission 液压传动（以液体作为工作介质来传递能量和进行控制的传动方式）
alteration [,ɔːltə'reiʃən] n. 变更，改造，改变
creative thought 创造性思维
buoyancy ['bɔiənsi] n. 浮力
centrifugal [sen'trifjugəl] a. 离心的
clutch facing 离合器面片，离合器摩擦片（clutch friction plate）
alternating stress 交变应力（随时间作周期性变化的应力）
allowance [ə'lauəns] n. 余量，增加量
resonant ['rezənənt] vibration 共振
derivation [deri'veiʃən] n. 推导（an act or process by which one thing is formed from another）
on the safe side 安全可靠（avoiding potential dangers, problems, or challenges），稳妥
combination of properties 综合性能，性能组合
weldability ['weldəbiliti] n. 焊接性（在一定工艺条件下，获得优质焊接接头的难易程度）
machinability [məʃiːnə'biliti] n. 可加工性（在一定生产条件下，材料加工的难易程度）
adaptable [ə'dæptəbl] a. 能适应的，可改变的（able to adjust to new conditions）
noninterference ['nɔnintə'fiərəns] n. 不干涉；不（相互）干扰
locating of parts 零件的定位
tapped hole 螺纹孔

Notes

[1] in earnest 意为"认真地、真正地"。
[2] principle of lever and inclined plane 意为"杠杆原理和斜面原理"。

Lesson 32 Computer Applications in Design and Graphics

Computers are widely used in engineering and related fields and their use is expected to grow even more rapidly than in the past. Engineering and technology students must become computer literate, to understand the applications of computer and their advantages. Not to do so will place students at a serious disadvantage in pursuing their careers.

Computer-aided design (CAD) involves solving design problems, generating engineering drawings and other technical documents with the help of computer. In CAD, the traditional tools of graphics, such as the T-square, drawing compass (see Figs. 25.1 and 25.2), and drawing board are replaced by electronic input and output devices. When using a CAD system, the designer can conceptualize the object to be designed more easily on the computer screen and can consider alternative designs or modify a particular design quickly to meet the necessary design requirements. The designer can then subject the design to a variety of engineering analyses and can identify potential problems (such as an excessive load or deflection). The speed and accuracy of such analyses far surpass what is available from traditional methods. The CAD user usually inputs data by keyboard and/or mouse to produce graphic images[1] on the computer screen that can be reproduced as paper copies[2] with a plotter or printer (see Figs. 32.1 and 32.2).

Figure 32.1 A plotter

Figure 32.2 Printers

Draft efficiency is significantly improved. When something is drawn once, it never has to be drawn again. It can be retrieved from a library, and can be duplicated, stretched, sized, and changed in many ways without having to be redrawn. Cut and paste techniques are used as labor-saving aids.

Engineers generally agree that the computer does not change the nature of the design process but is a significant tool that improves efficiency and productivity. The designer and the CAD system may be described as a design team: the designer provides knowledge, creativity, and control; the computer generates accurate, easily modifiable graphics, performs complex design analysis at great speed, and stores and recalls design information. Occasionally, the computer may augment or replace many of the engineer's other tools, but it cannot replace the design process, which is controlled by the designer.

Depending on the nature of the problem and the sophistication of the computer system, computers offer the designer or drafter some or all of the following advantages.

1. Easier creation and correction of drawings Engineering drawings may be created more quickly than by hand and making changes and modifications is more efficient than correcting drawings made by hand.

2. Better visualization of drawings Many systems allow different views of the same object to be displayed and 3D pictorials[3] (see Fig. 32.3) to be rotated on the computer screen.

Figure 32.3　3D pictorials

3. Database of drawing aids Creation and maintenance of design databases (libraries of designs) permits storing designs and symbols for easy recall and application to the solution of new problems.

4. Quick and convenient design analysis Because the computer offers ease of analysis, the designer can evaluate alternative designs, thereby considering more possibilities while speeding up the process at the same time.

5. Simulation and testing of designs Some computer systems make possible the simulation of a product's operation, testing the design under a variety of conditions and stresses. Computer testing may improve on or replace models and prototypes.

6. Increased accuracy The computer is capable of producing drawings with more accuracy than is possible by hand. Many CAD systems are even capable of detecting errors and informing the user of them.

7. Improved filing Drawings can be more conveniently filed, retrieved, and transmitted on disks and tapes.

Computer graphics has an almost limitless number of applications in engineering and other technical fields. Most graphical solutions that are possible with a pencil can be done on a computer and usually more productively. Applications vary from 3D modeling and finite element analysis[4] to 2D drawings and mathematical calculations.

Once the domain of large computer systems advanced applications can now be done on microcomputers. An important extension of CAD is its application to manufacturing. Computer-aided design/computer-aided manufacturing (CAD/CAM) systems may be used to design a part or product, devise the essential production steps, and electronically communicate this information to and control the operation of manufacturing equipment, including robots. These systems offer many advantages over traditional design and manufacturing systems, including less design effort[5], more efficient material use, reduced lead time[6], greater accuracy, and improved inventory control[7].

Words and Expressions

computer literate ['litərit] 能熟练使用计算机的人 (individuals who have the knowledge and skills to use a computer and other related technology)
plotter ['plɔtə] n. 绘图仪，绘图机
printer ['printə] n. 打印机
offshoot ['ɔːʃuːt] n. 分支，分流
augment [ɔːg'ment] n.; v. 增加，增大
pictorial [pik'tɔːriəl] a. 图示的 (shown in the form of a picture or photograph)，图画的
alternative [ɔːl'tɜːnətiv] n. 可供选择的方案 (something that is a possible selection)
alternative design 替代设计（方案），可供选择的设计（方案）
simulation [ˌsimjuˈleiʃən] n. 仿真，模拟
filing ['failiŋ] n. 存档 (the activity of putting documents, electronic information, etc. into files)，存档档案 (an official record of something)
3D modeling 三维建模（用三维软件建立零件或装配体的三维模型的过程）
limitless ['limitlis] a. 无限的，无界限的

Notes

[1] graphic image 意为"图形、图像"。
[2] paper copy 意为"纸质文本（paper version of document），打印文本"。
[3] 3D pictorial 意为"三维图像"。
[4] finite element analysis 意为"有限元分析"。
[5] design effort 意为"设计工作"。
[6] lead time 意为"前置时间，开发周期（the time between the original design or idea for a particular product and its actual production）或交付周期（the period of time that it takes for goods to be delivered after someone has ordered them）"。
[7] inventory control 意为"库存控制（对制造业或服务业生产、经营全过程的各种物品、产成品及其他资源进行管理和控制，使其储备保持在经济合理的水平上）"。

Lesson 33 Lathes and Cutting Parameters in Turning Process

Lathes are widely used in industry to produce all kinds of machined parts.[1] Some are general purpose machines, and others are used to perform highly specialized operations.

Engine Lathes Engine lathes (see Figs. 33.1 and 34.1), of course, are general-purpose machine tools used in production and maintenance shops all over the world. Sizes range from small bench models[2] to huge heavy-duty lathes.

Copy Lathes The copy lathe is designed to produce parts automatically. The basic operation of this lathe is as follows. A template of either a flat or three-dimensional shape is placed in a holder. A guide or pointer then moves along this shape and its movement controls that of the cutting tool. Workpieces such as motor shafts, spindles, pistons, and a variety of other objects can be turned using this type of lathe.

Turret Lathes When machining a complex workpiece on a general-purpose lathe such as an engine lathe, a great deal of time is spent changing and adjusting the several cutting tools that are needed to complete the work. One of the first adaptations of the engine lathe which made it more suitable to mass production was the addition of multi-tool turret in place of the tailstock. This kind of machine tool is called turret lathe (see Fig. 33.2).

Figure 33.1 An engine lathe   Figure 33.2 A turret lathe

Turret lathes can perform multiple cutting operations, such as turning, facing, drilling, thread cutting, parting, and boring. Several cutting tools (usually as many as six) are mounted on the hexagonal turret, which is rotated after each specific operation is completed. The tools for thread cutting, parting, boring, and other operations are specially shaped for their particular purpose or are available as inserts with various shapes (see Fig. 33.3).

The principal feature of all turret lathes is that the tools can perform a consecutive serial of operations in proper sequence. Once the tools have been set and adjusted, these machines do not require highly skilled operator. Vertical turret lathes also are available; they are more suitable for short, heavy workpieces with diameters as large as 1.2 m.

(a) Thread cutting tool　　　　(b) Parting tool　　　　(c) Internal turning tool

Figure 33.3　Three types of indexable turning tools

Turning Centers　Many of today's more sophisticated lathes are called turning centers since they can perform, in addition to the normal turning operations, certain milling and drilling operations. Basically, a turning center can be thought of as being a combination of the turret lathe and milling machine.

Cutting Parameters in Turning Process　In turning, the unwanted material of a workpiece is removed by a cutting tool as chips, so that the desired geometry, tolerances, and surface roughness are obtained. This requires that the cutting tool material is harder than the workpiece material.

Turning is based on a two-dimensional surface creation, which means that two relative motions are necessary between the cutting tool and the work material. These motions are defined as the primary motion, which mainly determines the cutting speed, and the feed motion, which provides the cutting zone with new material. The primary motion is provided by the rotation of the workpiece; the feed motion is a translation of the tool. Figure 33.4 shows basic cutting parameters in turning process.

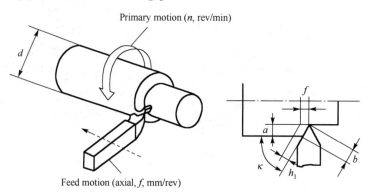

Figure 33.4　Basic cutting parameters in turning process

1. Cutting Speed　The cutting speed v is the instantaneous velocity of the primary motion of the workpiece relative to the tool (at a selected point on the cutting edge), and can be expressed as

$$v = \pi d n \tag{33.1}$$

where v is the cutting speed in m/min, d the diameter of the workpiece to be cut in meters, and n the workpiece or spindle rotation in rev/min.

2. Feed　The axial feed of a lathe may be defined as the distance the cutting tool advances along the length of the workpiece for every revolution of the spindle. The feed motion f is provided by the tool and, when added to the primary motion, leads to a repeated or continuous chip removal and the creation of the desired machined surface.

3. Depth of Cut (Back Engagement) The depth of cut a is the distance that the cutting edge engages below the original surface of the workpiece. The depth of cut determines the final dimensions of the workpiece. In turning, with an axial feed, the depth of cut is a direct measure of the decrease in radius of the workpiece; and with radial feed, the depth of cut is equal to the decrease in the length of workpiece.

4. Chip Thickness The chip thickness h_1 in the undeformed state is the thickness of the chip measured perpendicular to the cutting edge and in a plane perpendicular to the direction of cutting. The chip thickness after cutting (i.e., the actual chip thickness h_2) is larger than the undeformed chip thickness, which means that the cutting ratio or chip thickness ratio $r = h_1/h_2$ is always less than unity.[3]

5. Chip Width The chip width b in the undeformed state is the width of the chip measured along the cutting edge in a plane perpendicular to the direction of cutting.

6. Area of Cut For single-point tool operations (see Fig. 33.4), the area of cut A is the product of the undeformed chip thickness h_1 and the chip width b (i.e., $A = h_1 b$). The area of cut can also be expressed by the feed f and the depth of cut a as follows:

$$h_1 = f \sin\kappa \quad \text{and} \quad b = a/\sin\kappa \tag{33.2}$$

where κ is the tool cutting edge angle.

Consequently, the area of cut is given by

$$A = fa \tag{33.3}$$

Words and Expressions

lathe [leið] *n.* 车床,也可写为 turning machine ; *v.* 用车床加工
turning ['tə:niŋ] *n.* 车削(工件旋转做主运动,车刀做进给运动的切削加工方法)
engine lathe   普通车床,卧式车床,亦为 center lathe
general-purpose machine tool   通用机床(使用范围较广的可加工多种工件,完成多种工序的机床)
heavy-duty   重型的,耐用的(providing an unusual amount of power, durability, etc.)
workpiece ['wə:kpi:s] *n.* 工件(加工过程中的生产对象)
copy lathe   仿形车床(对工件进行仿形车削的车床),也可写为 tracer lathe
template ['templit] *n.* 样板(具有与工件轮廓相反或相同的测量边的工具)
cutting tool   刀具(能从工件上切除多余材料或切断材料的带刃工具)
spindle ['spindl] *n.* 主轴(带动工件或加工工具旋转的轴)
adaptation [ˌædæp'teiʃən] *n.* 改装(如对设备、机械装置等改装以适应新的用途),改建
holder ['həuldə] *n.* 夹持装置(a device that holds something)
turret ['tʌrit] lathe   转塔车床(具有回转轴线与主轴轴线垂直的转塔刀架,并可顺序转位切削工件的车床)

tailstock ['teilstɔk] n. 尾座（用于配合主轴箱支承工件或工具的部件）
parting ['pɑːtiŋ] n. 切断（把坯料或工件切成两段的加工方法），亦为 cutting off
drilling ['driliŋ] n. 钻孔，钻削（用钻头、扩孔钻等在工件上加工孔的方法）
thread cutting 车螺纹（用螺纹车刀切出工件的螺纹），亦为 thread turning
facing ['feisiŋ] n. 车端面，车平面，也可写为 surface turning 或 surfacing
boring ['bɔːriŋ] n. 车孔（用车削方法扩大工件的孔或加工空心工件的内表面），镗孔（镗刀旋转做主运动，工件或镗刀做进给运动的切削加工方法），镗削
hexagonal [hek'sægənəl] a. 六角形的，六边形的
insert [in'səːt] n. 刀片（装夹在刀体上的片状物体，并由它形成刀具的切削部分）
thread cutting tool 螺纹车刀（切削部分与被切螺纹牙形相同的车刀），thread turning tool
parting tool 切断刀（用于切断或切窄槽的车刀）
internal turning tool 内孔车刀（用于加工内旋转表面的车刀）
indexable turning tool 可转位车刀（一种采用机械夹固的方法，将多边形可转位刀片夹紧在刀杆上的车刀）
consecutive [kən'sekjutiv] a. 连续的（following one after another without interruption）
vertical turret lathe 立式转塔车床（主轴垂直布置的转塔车床）
chip [tʃip] n. 切屑（a small piece of something removed in the course of cutting）
work material 工件材料，亦为 workpiece material
primary motion 主运动（形成机床切削速度或消耗主要动力的工作运动）
feed [fiːd] n. 进给量（工件或刀具每转、每齿或每行程进给运动的位移量）
feed motion 进给运动（使工件的多余材料连续在相同或不同深度上被去除的工作运动）
rev/min 转/分钟，即 revolutions per minute (rpm)，r/min
cutting edge 切削刃（前后刀面的交线，它担负主要的切削工作，也叫主切削刃）
machined surface 已加工表面（刀具切除工件的加工余量后所形成的表面）
depth of cut 切削深度（指加工中工件已加工表面和待加工表面间的垂直距离）
back engagement [in'geidʒmənt] 背吃刀量
engage [in'geidʒ] v. 吃刀，切入，啮合
undeformed chip thickness 未变形切屑厚度（即将被刀具切除成为切屑的工件材料层的厚度）
perpendicular [ˌpəːpən'dikjulə] a. 垂直的（intersecting at right angles）；n. 垂直
single-point tool 单刃刀具（切削时只用一个主切削刃的刀具）
tool cutting edge angle 主偏角（主切削平面与假定工作平面间的夹角，在基面中测量）

Notes

[1] machined part 意为"经过机械加工的零件"。
[2] bench model 这里指"台式车床，也可写为 bench lathe，可安装在工作台上的小型车床"。
[3] less than unity 意为"小于 1（less than one）"。

Lesson 34 Engine Lathes

Lathes are generally considered to be the oldest machine tools. The principal form of surface produced in a lathe is the cylindrical surface. This is achieved by rotating the workpiece, while the single-point cutting tool removes the unwanted material by traversing in a direction parallel to the axis of rotation, as shown in Fig. 2.2.

The engine lathe is the most common lathe, which derives its name from the fact that early lathes were driven by steam engines. It is also called a center lathe. [1]

The essential components of an engine lathe are the bed, headstock, tailstock, carriage, leadscrew and feed rod, as shown in Fig. 34.1.

The bed is the backbone of an engine lathe. It is usually made of well-normalized or aged[2] grey or nodular cast iron and provides a heavy, rigid frame on which all the other basic components are mounted. Two sets of parallel, longitudinal ways, inner and outer, are contained on the bed, usually on the upper side (see Fig. 34.2). Some makers use an inverted V-shape for all four ways, whereas others utilize one inverted V and one flat way[3] in one or both sets. On most modern lathes the ways are surface-hardened to resist wear.

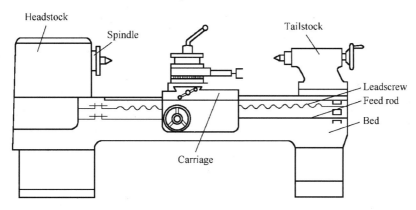

Figure 34.1 Schematic diagram of an engine lathe

The headstock is mounted in a fixed position on the inner ways, usually at the left end of the bed. It provides a powered means of rotating the work at various speeds. Essentially, it consists of a hollow spindle, mounted in accurate bearings, and a set of transmission gears—similar to a truck transmission—through which the spindle can be rotated at a number of speeds. Most lathes provide from 8 to 18 speeds, usually in a geometric ratio. An increasing trend is to provide a continuously variable speed range through electrical or mechanical drives.

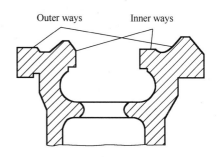

Figure 34.2 Cross section of lathe ways

Because the accuracy of a lathe is greatly dependent on the spindle, it is of heavy construction and mounted in heavy bearings, usually preloaded tapered roller or ball types (see Figs. 10.1b and 10.1c). The spindle has a hole extending through its length, through which long bar stock can be fed. The size of this hole is an important dimension of a lathe because it determines the maximum size of bar stock that can be machined when the work[4] must be fed through spindle.

The most common form of work holding device used in a lathe is the chuck. The three-jaw self-centering chuck (Fig. 34.3a) moves all of its jaws simultaneously to clamp or unclamp the work, and it is used for work with a round or hexagonal cross-section. The four-jaw independent chuck (Fig. 34.3b) can clamp on the work by moving each jaw independent of the others. This chuck exerts a stronger hold on the work and it has the ability to center non-round shapes (squares, rectangles) exactly.

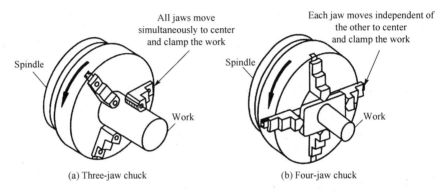

Figure 34.3 Chucks

The tailstock supports the other end of the workpiece. It is particularly useful when the workpiece is relatively long and slender. The tailstock is equipped with a center that may be fixed (dead center), or it may be free to rotate with the workpiece (live center). The tailstock is movable on the inner ways of the bed to accommodate the different lengths of workpieces. Tools such as center drill (Fig. 34.4), twist drill (Fig. 36.1), reamer (Fig. 28.2), etc. can also be mounted on the tailstock quill for making and finishing holes in the workpieces which are located in line with the axis of rotation.

Figure 34.4 Center drills

The three-jaw and four-jaw chucks are normally suitable for short workpieces. The long workpieces are machined between centers (see Fig. 34.5). Before a workpiece can be mounted between lathe centers, a center hole must be drilled in each end. Through these center holes the centers mounted in the spindle and the tailstock would locate the axis of the workpiece. However, these centers would not be able to transmit the motion to the workpiece from the spindle. For this purpose, generally a driver plate and a lathe dog would be used. In Fig. 34.5, the center located in the spindle is a dead center (Fig. 34. 6a), while that in the tailstock is a live center (Fig. 34. 6b). The shank of the center is generally finished with a Morse taper which fits into the tapered hole of the spindle or tailstock.

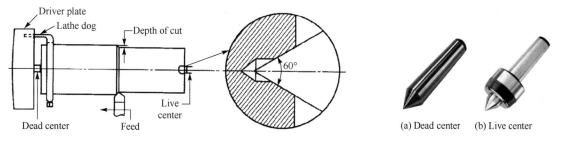

Figure 34.5 Workpiece being turned between centers in a lathe

Figure 34.6 Centers

The maximum workpiece dimensions for an engine lathe are specified by two parameters. The first is known as the swing. It is approximately twice the distance between the line connecting the lathe centers and the nearest point on the ways. The second parameter is the distance between centers. The swing thus indicates the maximum workpiece diameter that can be turned in the lathe, while the distance between centers indicates the maximum length of workpiece that can be mounted between centers.

On the lathe we can perform the following operations:

Turning Turning is the operation of removing the excess material from the workpiece to produce a cylindrical surface of required size.

Facing Facing is an operation for generating flat surfaces in lathes. The feed, in this case, is given in a direction perpendicular to the axis of rotation.

Boring Boring on a lathe is similar to turning. It is an internal turning operation used for enlarging the existing hole or cylindrical cavity in the workpiece.

Parting This operation is carried out for cutting the workpiece into two parts by using a parting tool.

Knurling Knurling is the process of embossing a straight or diamond pattern on the workpiece surface by making use of a knurling tool.

Several other operations are also performed on a lathe, such as taper turning, thread turning, grooving, drilling, centering, and chamfering.

Words and Expressions

traverse ['trævəs] *v.* 移动（to move to the side or back and forth）
axis of rotation 旋转轴（物体绕其旋转的瞬时线），回转轴线
headstock ['hedstɔk] *n.* 主轴箱（装有主轴的箱形部件），床头箱
bed 床身（机床上用于支承和连接若干部件，并带有导轨的基础零件）
carriage 溜板（在床身上使刀具作纵向移动的部件，一般由刀架、床鞍、溜板箱等组成）
leadscrew ['li:d skru:] 丝杠（a threaded rod that converts rotational motion into linear motion）
feed rod 光杠（用于传动运动部件移动的杆状零件）

schematic diagram　示意图，原理图，简图
grey cast iron　灰口铸铁（碳主要以片状石墨形式存在，断口呈灰色的铸铁），简称"灰铸铁"
nodular ['nɔdjulə] cast iron　球墨铸铁（石墨主要以球状存在的高强度铸铁）
way　导轨（引导部件沿一定方向运动的一组平面或曲面），亦为 guideway 或 slideway
surface-hardened　经过表面淬火（仅对工件表层进行淬火的工艺）的
transmission [trænz'miʃən]　n. 变速器（用于改变转速和扭矩的机构）
geometric ratio　等比（等比级数中连续项的比值）
bar stock　棒料（一种截面均匀的轧材，其截面通常为圆形、矩形或六边形）
chuck [tʃʌk]　n. 卡盘（装在机床主轴上的具有卡爪的通用夹具）
jaw [dʒɔː]　n. 卡爪（卡盘中用于夹持工件的零件）
self-centering chuck　自定心卡盘（卡爪可径向同心移动使工件自动定心的卡盘）
clamp [klæmp]　n. 夹紧（工件定位后将其固定，使其在加工过程中保持定位位置不变）
independent chuck　单动卡盘（卡爪可单独调整的卡盘）
center drill　中心钻（用于加工中心孔的一种刀具）
center ['sentə]　n. 中心，顶尖；v. 定心，对中（to adjust things so that the axes coincide）
quill [kwil]　n. 套筒（用于安装钻头、顶尖等并可轴向移动的圆柱套）
center hole　中心孔（打在工件两端中心处，承受顶尖的锥孔），顶尖孔
axis ['æksis]　n. 轴（机床部件直线运动或旋转运动的方向），轴线
driver plate　拨盘，亦为 drive plate
lathe dog　卡箍，鸡心夹头，有时也写为 dog
dead center　死顶尖（尾柄与头部锥体为一体的顶尖），固定顶尖（fixed center）
live center　活顶尖，回转顶尖（头部锥体与尾柄可相对旋转的顶尖）
Morse taper　莫氏锥度
swing [swiŋ]　n. 最大回转直径（床身上允许工件回转的最大尺寸）
distance between centers　顶尖距（前顶尖与后顶尖间的最大距离）
knurling ['nəːliŋ]　n. 滚花（用滚花工具在工件表面上滚压出花纹的加工方法）
emboss [im'bɔs]　v. 使凸出，压印花纹，滚压，在……上作浮雕图案
diamond pattern　菱形图案，由很多菱形图案构成的网纹
taper turning　车锥面（将工件车削成圆锥表面的方法）
grooving ['gruːviŋ]　n. 车槽（在工件表面上车削沟槽的方法）
centering ['sentəriŋ]　n. 钻中心孔（用中心钻在工件端面钻削带锥面定心孔的方法）
chamfering ['tʃæmfəriŋ]　n. 倒角（把工件的棱角切削成一定斜面的加工）

Notes

[1] center lathe 意为"卧式车床，普通车床，可以用两顶尖（center）装夹工件的车床"。
[2] well-normalized or aged 意为"经过充分正火或时效处理的"。
[3] one inverted V and one flat way 意为"一个凸三角形导轨和一个矩形导轨"。
[4] work 意为"工件（workpiece，加工过程中的生产对象）"。

Lesson 35　Milling Machines and Grinding Machines

Milling machines are among the most versatile and useful machine tools because they are capable of performing a variety of cutting operations. The first milling machine was built in 1818 by Eli Whitney. There is a large selection of milling machines available with numerous features.

Knee-and-column type milling machines are the most common milling machines and are used for general-purpose milling operations. The spindle, to which the milling cutter is attached, may be horizontal (Fig. 35.1a) for slab milling,[1] or vertical (Fig. 35.1b) for face milling.[2]

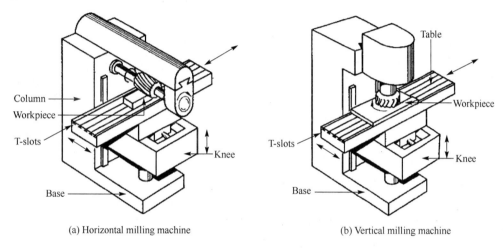

(a) Horizontal milling machine　　　　　(b) Vertical milling machine

Figure 35.1　Schematic diagram of knee-and-column type milling machines

In bed type milling machines, the table is mounted directly on the bed, which can only have horizontal movement. These machines have high stiffness and typically are used for high-volume production. Single spindle bed type machines are called simplex mills, as shown in Fig. 35.2a. The spindles of bed type milling machines may be horizontal or vertical and can be of duplex or triplex types (that is, with two or three spindles) for the simultaneous machining of two or three workpiece surfaces. Other types of milling machines are also available, such as the planer type[3] for heavy work (see Fig. 35.2b) and machines for special purposes.

Among accessories for milling machines, one of the most widely used is the universal dividing head (Fig. 35.3). This is a fixture that allows a workpiece to be easily and precisely rotated to preset angles. Typical uses are in milling parts with polygonal surfaces and in machining gear teeth. Dividing heads may also be used on many other machine tools including drilling machines and grinding machines.

Grinding is one of the most widely used methods of finishing parts to extremely close tolerances and low surface roughness. There are many types of grinding machines. The two most widely used are the cylindrical and surface grinders. Other classes of grinding machines include centerless grinders, internal grinders, and thread grinders.

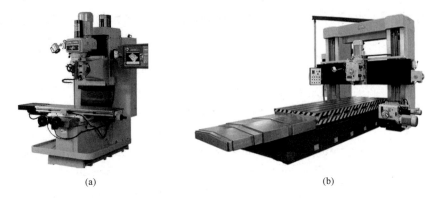

Figure 35.2 Examples of bed type (a) and planer type (b) milling machines

The cylindrical grinder is for straight cylindrical or taper work. A work is usually held between centers and rotated by a dog[4] and drive plate, as depicted in Fig. 35.4. The centers are mounted in the headstock[5] and tailstock of the grinder. A rotating grinding wheel in contact with the work removes metal from its circumference. The operation is somewhat similar to that of a milling machine. The grinding wheel replaces the milling cutter, and the thousands of abrasive particles in it may be thought of as little teeth of the cutter. They in fact produce little chips during the cutting operation.

Figure 35.3 A universal dividing head

The surface grinder is used for grinding flat surfaces. Typically, the work is held on a magnetic chuck attached to the table of the grinder; nonmagnetic materials are held by vices or some other fixtures. The table reciprocates longitudinally under the grinding wheel, as shown in Fig. 35.5. With each pass the table feeds transversely. This feed may be accomplished automatically or by hand.

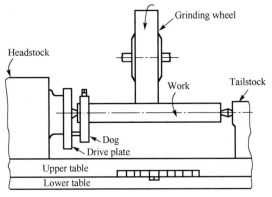
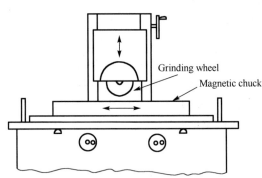

Figure 35.4 Diagram of a cylindrical grinder   Figure 35.5 Diagram of a surface grinder

The internal grinder is used for grinding of precision holes, cylinder bores, and similar operations where bores of all kinds are to be finished. Thread grinders are used for grinding precision threads for thread gages, and threads on precision parts where the concentricity between the diameter of the shaft and the pitch diameter of the thread must be held to close tolerances.

Particular design features of a part dictate to a large degree the type of grinding machine required. Where processing costs are excessive, parts redesigned to utilize a less expensive, higher output grinding method may be well worthwhile. For example, wherever possible the production economy of centerless grinding should be taken advantage of by proper design consideration.

In the grinding process, the sharp grains of the grinding wheel become rounded and hence lose their cutting ability. The cutting ability can be restored by truing and dressing (see Fig. 35.6). Truing is an operation to restore the correct geometrical shape of the wheel that has been lost due to nonuniform wear. Dressing is a sharpening operation, which removes the worn and dull grits.

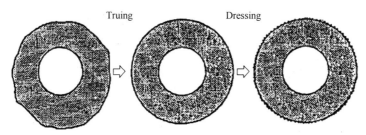

Figure 35.6 Truing makes the wheel round; dressing sharpens the wheel

With conventional grinding wheels, both truing and dressing are usually done by the same process. With superabrasive grinding wheels, separate truing and dressing processes are normally used. A properly trued and dressed wheel grinds with a minimum of grinding power, has no burn or chatter, and produces quality parts.[6]

Words and Expressions

milling ['miliŋ] *n.* 铣削（铣刀旋转做主运动，刀具或工件做进给运动的切削加工方法）
milling machine   铣床（用铣刀在工件上加工各种表面的机床）
grinding ['graindiŋ] *n.* 磨削（磨具以较高的线速度旋转，对工件表面进行加工的方法）
grinding machine   磨床（用磨具或磨料加工工件各种表面的机床），简称 grinder
knee-and-column type milling machine   升降台铣床（工作台可做纵向、横向和垂直进给运动的铣床）
table   工作台（具有工作平面，用于直接或间接装夹工件或工夹具的零部件）
column   （铣床）床柱（具有导轨，内可装主传动机构或顶部可装主传动部件的直立零件）
knee   升降台（可沿床柱导轨垂向移动，顶部可安装工作台的部件）
base   底座（用于支承和连接若干部件的基础零件）
T-slot   T形槽

horizontal milling machine 卧式铣床（装在床身上的主轴为水平布置的铣床）
vertical milling machine 立式铣床（装在床身上的主轴为垂直布置的铣床）
bed type milling machine 床身式铣床（工作台不升降，但可沿床身导轨纵向移动，主轴部件可作竖直方向移动的铣床）
simplex ['simpleks] mill 单轴床身式铣床，亦为 simplex bed type milling machine
duplex ['dju:pleks] *a.* 双的，二倍的，二重的；*n.* 由两部分组成的东西
triplex ['tripleks] *a.* 三部分的，三的，三倍的；*n.* 由三部分组成的东西
universal dividing head 万能分度头
preset ['pri:'set] *a.* 预先调整的（set in advance）；*v.* 预先设置（to set beforehand）
cylindrical grinder 外圆磨床，也可写为 cylindrical grinding machine
centerless grinder 无心磨床，也可写为 centerless grinding machine
internal grinder 内圆磨床（主要用于磨削圆柱形和圆锥形内表面的磨床）
nonmagnetic [ˌnɔnmæg'netik] material 非磁性材料，非导磁材料
magnetic chuck 磁力吸盘（安装在磨床上，用磁力吸紧工件的机床附件）
reciprocate [ri'siprəkeit] *v.* 做往复运动（to move back and forth alternately）
upper table 上工作台（安装头、尾架等部件，并可相对下工作台回转的零件）
lower table 下工作台（用于支承上工作台，与纵向运动驱动部件相连接并与床身纵向导轨相配的零件）
gage [geidʒ] *n.* 量规（一种没有刻度的量具，用量规检验工件，不能得出具体数值，只能检验工件尺寸合格与否）
thread gage 螺纹量规（包括检测外螺纹尺寸的环规，检测内螺纹尺寸的塞规，是用通端和止端综合检验螺纹的量规）
concentricity [ˌkɔnsen'trisiti] *n.* 同心，同心度
pitch diameter （螺纹）中径
redesign [ˌri:di'zain] *v.* 重新设计（to revise in appearance, function, or content）；*n.* 重新设计
truing ['tru:iŋ] *n.* 砂轮的整形（产生确定的几何形状）
dressing ['dresiŋ] *n.* 砂轮的修锐（产生锋锐的磨削面）
grit [grit] *n.* 磨粒
superabrasive grinding wheel 超硬磨料砂轮（人造金刚石或立方氮化硼砂轮）
chatter ['tʃætə] *n.* 颤振（切削过程中由于切屑动态再生引起的工艺系统自激振动现象）

Notes

[1] slab milling 意为"铣平面（用圆柱铣刀加工与机床主轴平行的平面），周铣平面"。
[2] face milling 意为"铣平面（用面铣刀加工与机床主轴垂直的平面），端铣平面"。
[3] planer type 这里指"龙门式铣床，即 planer type milling machine，工作台水平布置，两个立柱和连接梁构成龙门架式的铣床"，也可写为 gantry type milling machine。
[4] dog 这里指"磨床鸡心夹头或称为卡箍"。
[5] headstock 在磨床中通常称为"头架"，在车床中通常称为"主轴箱"。
[6] burn 这里指"磨削烧伤"，quality 意为"优质的（being of high quality）"。

Lesson 36 Drilling Operations and Machines

Drilling involves producing through or blind holes in a workpiece by forcing a tool, which rotates around its axis, against the workpiece.[1] Consequently, the range of cutting from that axis of rotation is equal to the radius of the required hole.

Cutting Tools for Drilling Operations

In drilling operations, a cylindrical cutting tool, called a drill, is employed. The drill can have either one or more cutting edges and corresponding flutes, which can be straight or helical. The function of the flutes is to provide outlet passages for the chips generated during the drilling operation and also to allow lubricants and coolants to reach the cutting edges and the surface being machined. The commonly used drills are described below.

Twist Drills The twist drill is the most common type of drill. It has two cutting edges and two helical flutes that continue over the length of the drill body. The shank of a drill can be either straight (Fig. 36.1a) or tapered (Fig. 36.1b). In the latter case, the shank is fitted into the tapered socket of the spindle and has a tang, which goes into a slot in the spindle socket, thus acting as a solid means for transmitting rotation.[2]

(a) Straight shank twist drill　　　　　　(b) Tapered shank twist drill

Figure 36.1　Twist drills

Core Drills A core drill consists of the chamfer, body, neck, and shank, as shown in Fig. 36.2. This type of drill may have either three or four flutes and an equal number of margins, which ensure superior guidance, thus resulting in high machining accuracy.[3] A core drill has flat end. The chamfer can have three or four cutting edges. Core drills are employed for enlarging previously made holes and not for originating holes.[4] This type of drill is characterized by greater productivity, high machining accuracy, and superior quality of the drilled surfaces.

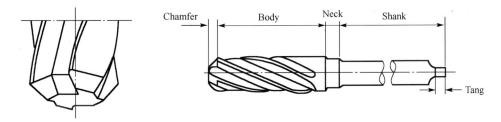

Figure 36.2　A core drill

Indexable Insert Drills Most of the drills are made of high speed steel. However, for machining hard materials as well as for large volume production, indexable insert drills are available. As shown in Fig. 36.3, the carbide inserts of suitable geometry are clamped to the end of the tool to act as the cutting edges.

Figure 36.3 Indexable insert drills

Other Types of Drilling Operations

In addition to conventional drilling, there are other operations that are involved in the production of holes in the industrial practice. Following is a brief description of each of these operations.

Counterboring Already existing holes in the parts can be further machined by counterboring as shown in Fig. 36.4a. In the counterboring operation, the hole is enlarged with a flat bottom to provide a space for the bolt head or a nut, so it would be entirely below the surface of the part.

Countersinking As shown in Fig. 36.4b, countersinking is done to enable accommodating the conical seat of a flathead screw so that the screw does not appear above the surface of the part.

Spot Facing Spot facing operation removes only a very small portion of material around the existing hole to provide a flat surface square to the hole axis, as shown in Fig. 36.4c. This has to be done only in case where existing surface is not smooth and is usually performed on castings or forgings.

Reaming Reaming is an operation used to make an existing hole dimensionally more accurate than can be obtained by drilling alone. As a result of a reaming operation, a hole has a very smooth surface. The cutting tool used in this operation is known as the reamer (see also Fig. 28.2).

Tapping Tapping is the process of cutting internal threads. The tool is called a tap (see Fig. 36.5).

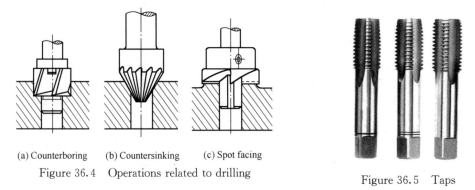

(a) Counterboring   (b) Countersinking   (c) Spot facing
Figure 36.4 Operations related to drilling

Figure 36.5 Taps

Classification of Drilling Machines

Drilling operations can be carried out by using either portable electric drills or drilling machines. The latter differ in shape and size. Nevertheless, the tool always rotates around its axis while the workpiece is kept firmly fixed. This is contrary to the drilling operation on a lathe, where the workpiece is held in and rotates with the chuck.[5] The commonly used types of drilling machines are described below.

Bench-type Drilling Machines Bench-type drilling machines are general-purpose, small machine tools that are usually placed on benches. This type of drilling machine includes an electric motor as the source of motion, which is transmitted via pulleys and belts to the spindle, where the tool is mounted (see Fig. 36.6). The feed is manually generated by lowering a feed lever, which is designed to lower (or raise) the spindle.[6] The workpiece is mounted on the machine table, although a special jig or machine vice is sometimes used to hold the workpiece.

Upright Drilling Machines Depending upon the size, upright drilling machines can be used for light, medium, and even relatively heavy workpieces. It is basically similar to bench-type machines, the main difference being a longer cylindrical column fixed to the base (see Fig. 55.3).[7] The power required for this type of drilling machine is more than that for the bench-type drilling machines.

Radial Drilling Machines A radial drilling machine is particularly suitable for drilling holes in large and heavy workpieces that are inconvenient to mount on the table of an upright drilling machine. As you can see in Fig. 36.7, a radial drilling machine has a column, which is fixed to the base. The arm, which carries the drilling head, can be raised or lowered along the column and clamped at any desired position to accommodate various heights of workpieces. The drilling head may be moved along the length of the arm and the arm may also be swung about the column, thus, enabling the drilling head to be moved rapidly to any desired location while the workpiece remained clamped in one position.[8]

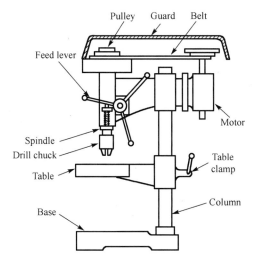

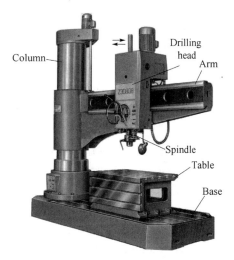

Figure 36.6 Sketch of a bench-type drilling machine

Figure 36.7 A radial drilling machine

Words and Expressions

drilling ['drilɪŋ] n. 钻削（用钻头、扩孔钻等在工件上加工孔的方法）
drilling operation 钻削加工
drilling machine 钻床（主要用钻头在工件上加工孔的机床）
flute [fluːt] n. 容屑槽（在刀体上开出的沟槽，作容屑、排屑和切削液通道用）
helical flute 螺旋槽（与刀具轴线呈螺旋状的容屑槽）
outlet ['autlet] n. 排出口，出口（a place or opening through which something is let out）
coolant ['kuːlənt] n. 冷却液，切削液（cutting fluid）

twist drill 麻花钻（容屑槽由螺旋面构成的钻头）

shank [ʃæŋk] n. 刀柄（刀具上的夹持部分）

straight shank or tapered shank 直柄（圆柱形刀柄）或锥柄（圆锥形刀柄）

tang [tæŋ] n. 扁尾（锥柄尾部用于传动的削平部分）

socket ['sɔkit] n. 孔，插口（an opening into which an inserted part is designed to fit）

core drill 扩孔钻（在钻尖中心处无切削刃，用于对已有孔扩大加工的孔加工刀具）

chamfer ['tʃæmfə] n. 倒角，斜面，（扩孔钻的）切削部分

margin ['mɑːdʒin] n. 刃带（一般指靠近切削刃且后角等于零的窄平面），也可写为 land

guidance ['gaidəns] n. 引导，制导，导向，引导装置

originate [ə'ridʒineit] v. 产生，创造（to produce or create），开始（to begin）

machining accuracy 加工精度，即工件加工后的实际几何参数（尺寸、形状和位置）与设计几何参数的符合程度

indexable insert drill 可转位钻（刀片可转位使用的镶齿钻头），可转位刀片钻

carbide ['kɑːbaid] n. 碳化物，硬质合金（cemented carbide）

large volume production 大批量生产

high speed steel 高速钢，高速工具钢，其缩写为 HSS

indexable insert 可转位刀片（采用机械方式夹固在刀体上，切削刃用钝后不重磨而转位使用的多边形刀片）

casting ['kɑːstiŋ] n. 铸件（采用铸造方法获得的有一定形状、组织和性能的金属件）

forging ['fɔːdʒiŋ] n. 锻件（金属材料经过锻造加工而得到的工件或毛坯），锻造

reaming ['riːmiŋ] n. 铰孔，铰削（用铰刀从工件预制的底孔上切除微量金属层，以提高其尺寸精度和降低表面粗糙度的切削加工方法）

tapping ['tæpiŋ] n. 攻螺纹（用丝锥加工工件的内螺纹）

tap [tæp] n. 丝锥（加工圆柱形和圆锥形内螺纹的标准工具）

counterboring [kauntə'bɔːriŋ] n. 锪沉头孔（用锪削方法加工出平底沉孔）

spot facing 锪平面（用锪削方法将工件的孔口周围切削成垂直于孔轴线的平面）

countersinking ['kauntəsiŋkiŋ] n. 锪锥孔（用锪削方法加工出锥形沉孔）

conical ['kɔnikəl] a. 圆锥形的

square to 与……成直角，垂直于

portable electric drill 手电钻（手提式钻孔用电动工具）

bench-type drilling machine 台式钻床

general purpose 通用的，多种用途的

upright ['ʌp'rait] a. 立式的；n. 立柱（a vertical part of sth that supports other parts）

upright drilling machine 立式钻床，亦为 vertical drilling machine

jig [dʒig] n. 夹具（用以装夹工件和引导刀具的装置）

special jig 专用夹具（专门为某一工件的某一工序而设计的夹具）

machine vice 机用虎钳（利用螺杆或其他机构使两钳口作相对移动夹持工件的工具）

column ['kɔləm] n. 立柱（用于支承和连接若干部件，并带有竖直导轨的直立柱状零件）

base 底座（用于支承和连接若干部件的基础零件）

medium-duty 中型的，中等的

drill chuck 钻夹头（安装在钻床主轴端部，用三个可同心移动的卡爪夹紧钻头的夹头）

guard [gɑːd] n. 防护罩，防护装置（a device that prevents injury, damage, or loss）

clamp [klæmp] n. 锁紧装置 (a device used to hold two things together tightly)
radial drilling machine 摇臂钻床（摇臂可绕立柱回转和升降，通常主轴箱在摇臂上作水平移动的钻床），亦为 radial drill
arm 摇臂（一端装在单立柱上，可绕立柱轴线回转，并可上下移动且具有水平导轨的零件）
drilling head 钻床主轴箱

Notes

[1] through or blind holes 意为"通孔或盲孔"。全段可译为：钻孔是迫使一个绕自身轴线旋转的刀具进入工件中，加工出通孔或盲孔。因此，从回转轴线开始的切削范围等于所需孔的半径。

[2] tapered socket of the spindle 意为"主轴锥孔"。这段可译为：麻花钻是最常用的钻头类型。它有两个切削刃和两个延伸至整个钻体的螺旋容屑槽。钻柄可以是直柄（见图 36.1a），也可以是锥柄（见图 36.1b）。对于后一种情况，钻柄与主轴锥孔相配合，而且钻柄上的扁尾插入主轴锥孔中的槽内，作为传递转动的一个实体工具。

[3] chamfer 这里意为"切削锥，即扩孔钻前端成斜角的切削部分"，也可写为 bevel，body 这里意为"钻体"，也可写为 drill body，margin 意为"刃带，又称棱边，钻头的圆柱或圆锥的导向面"。这两句可译为：如图 36.2 所示，扩孔钻由切削锥、钻体、颈部和柄部组成。这种类型的钻头可以有三到四个容屑槽和相同数量的刃带，它们可以保证良好的导向性，因此具有较高的加工精度。

[4] 全句可译为：扩孔钻仅能用来扩大已有的孔，而不能在实体材料上加工出孔。

[5] tool 在这里指"钻头"，contrary to 意为"与……相反"。全段可译为：可以使用手电钻或钻床进行钻削加工。钻床有不同的种类和型号。尽管如此，钻头总是绕自身轴线做旋转运动，而工件则是固定不动的。这与在车床上进行的钻孔加工相反，在车床上，工件被夹持在卡盘内并随其转动。常用的钻床类型如下所述。

[6] feed lever 意为"进给手柄"。全段可译为：台式钻床是通用的小型机床，通常安放在工作台上。这种钻床上有一个作为动力源的电动机，通过带轮和带将动力传送到装有刀具的主轴上（见图 36.6）。进给是通过用手向下搬动进给手柄而实现的，这个手柄是被设计用来使主轴向下（或向上）运动。工件装夹在机床工作台上，有时也使用专用夹具或机用虎钳来夹持工件。

[7] cylindrical column fixed to the base 意为"固定在底座上的圆形立柱"。全句可译为：它与台式钻床基本类似，主要的区别之处在于一根较长的固定在底座上的圆形立柱（见图 55.3）。

[8] arm 这里指"摇臂，是一个承载着主轴箱的悬臂"，也可写为 radial arm，clamp 意为"锁紧，固定"，swung 是 swing 的过去分词，意为"摆动，回转"。全段可译为：摇臂钻床特别适合在大而重的工件上钻孔，而这类工件不方便安放到立式钻床的工作台上。正如你在图 36.7 中所看到的那样，摇臂钻床中有一个固定在底座上的立柱。承载着钻床主轴箱的摇臂可以沿着立柱上下移动，并且可以锁紧在任何需要的位置上，以便加工不同高度的工件。钻床主轴箱可以沿着摇臂的长度方向移动，摇臂还可以绕立柱回转，这就使得当工件被夹紧后固定不动时，钻床主轴箱可以快速到达任何需要的位置。

Lesson 37 Milling Operations (*Reading Material*)

After lathes, milling machines are the most widely used for manufacturing applications. In milling, the workpiece is fed into a rotating milling cutter, which is a multi-edge tool as shown in Fig. 37.1. Metal removal is achieved through combining the rotary motion of the milling cutter and linear motions of the workpiece simultaneously.

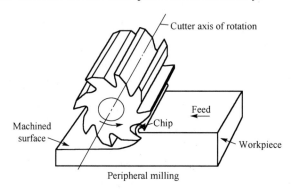

Figure 37.1 Schematic diagram of a milling operation

Each of the cutting edges of a milling cutter acts as an individual single-point cutter when it engages with the workpiece metal. Therefore, each of those cutting edges has appropriate rake and relief angles. Since only a few of the cutting edges are engaged with the workpiece at a time, heavy cuts can be taken without adversely affecting the tool life. In fact, the permissible cutting speeds and feeds for milling are three to four times higher than those for turning or drilling. Moreover, the quality of surfaces machined by milling is generally superior to the quality of surfaces machined by turning, shaping, or drilling.

As far as the directions of cutter rotation and workpiece feed are concerned, milling is performed by either of the following two methods.

1. Up milling (conventional milling) In up milling the cutting tool rotates in the opposite direction to the table movement, the chip starts as zero thickness and gradually increases to the maximum size as shown in Fig. 37.2a. This tends to lift the workpiece from the table. There is a possibility that the cutting tool will rub the workpiece before starting the removal. However, the machining process involves no impact loading, thus ensuring smoother operation of the machine tool.

The initial rubbing of the cutting edge during the start of the cut in up milling tends to dull the cutting edge and consequently have a lower tool life. Also since the cutter tends to cut and slide alternatively, the quality of machined surface obtained by this method is not very high. Nevertheless, up milling is commonly used in industry, especially for rough cuts.

2. Down milling (climb milling) In down milling the cutting tool rotates in the same direction as that of the table movement, and the chip starts as maximum thickness and goes to zero thickness gradually as shown in Fig. 37.2b.

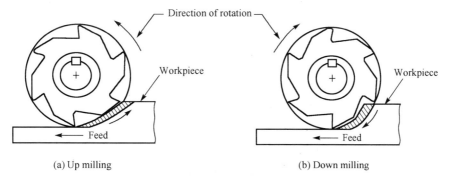

Figure 37.2 Milling methods

The advantages of this method include higher quality of the machined surface and easier clamping of workpieces, since cutting forces act downward. Down milling also allows greater feeds per tooth and longer tool life between regrinds than up milling. But, it cannot be used for machining castings or hot rolled steel, since the hard outer scale will damage the cutter.

There are a large variety of milling cutters to suit specific requirements. The cutters most generally used, shown in Fig. 37.3, are classified according to their general shape or the type of work they will do.

Plain milling cutters They are basically cylindrical with the cutting teeth on the periphery, and the teeth may be straight or helical, as shown in Fig. 37.3a. The helical teeth generally are preferred over straight teeth because the tooth is partially engaged with the workpiece as it rotates. Consequently, the cutting force variation will be smaller, resulting in a smoother operation. The cutters are generally used for machining flat surfaces.

Face milling cutters They have cutting edges on the face and periphery. The cutting teeth, such as carbide inserts, are mounted on the cutter body as shown in Fig. 37.3b.

Most larger-sized milling cutters are of inserted-tooth type. The cutter body is made of ordinary steel, with the teeth made of high speed steel, cemented carbide, or ceramics, fastened to the cutter body by various methods. Most commonly, the teeth are indexable carbide, coated carbide, or ceramic inserts.

Slitting saws They are very similar to a circular saw blade in appearance as well as function (see Fig. 37.3c). The thickness of these cutters is generally very small. The cutters are employed for cutting off operations and deep slots.

Side and face cutters They have cutting edges not only on the periphery like the plain milling cutters, but also on both the sides (see Fig. 37.3d). As was the case with the plain milling cutter, the cutting teeth can be straight or helical.

Angle milling cutters They are used in cutting dovetail grooves and the like. Figure 37.3e indicates a milling cutter of this type.

T-slot cutters T-slot cutters (Fig. 37.3f) are used for milling T-slots such as those in the milling machine table (see Fig. 35.1).

End mills There are a large variety of end mills. One of distinctions is based on the method of holding, i. e., the end mill shank can be straight or tapered. The straight shank is used on end mills of small size. The tapered shank is used for large cutter sizes. The cutter usually rotates on an axis perpendicular to the workpiece surface.

Figure 37.3g shows two kinds of end mills. The cutter can remove material on both its end and its cylindrical cutting edges.

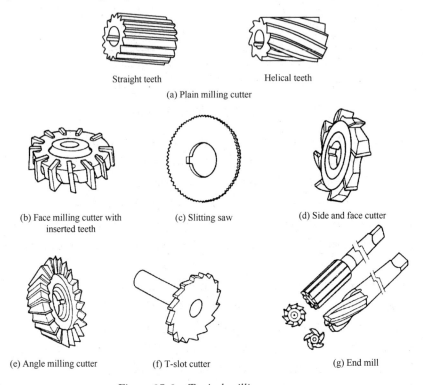

Figure 37.3　Typical milling cutters

Words and Expressions

milling cutter　铣刀（铣削加工用多齿刀具）
multi-edge tool　多刃刀具（有多个主切削刃参加切削的刀具），也可以写为 multi-point tool
peripheral milling　周边铣削（用铣刀周边齿刃进行的铣削），周铣
schematic [skiˈmætik]　a. 示意性的，图解的，图表的
schematic diagram　原理图，示意图
single-point cutter　单刃刀具（切削时只用一个主切削刃的刀具），也可以写为 single-point tool
cutting edge　切削刃（刀具前刀面上作切削用的刃）
rake angle　前角（前刀面与基面间的夹角。在法剖面中测量）
relief angle　后角（后刀面与切削平面间的夹角。在正交平面中测量），亦为 clearance angle
chip [tʃip]　n. 切屑（刀具切削产生的屑状物）
heavy cut　重切削，强力切削
shaping [ˈʃeipiŋ]　n. 刨削（刨刀与工件做水平方向相对直线往复运动的切削加工方法）
machined surface　已加工表面（工件上经刀具切削后产生的表面）
up milling　逆铣，即 conventional milling（铣刀的旋转方向与工件的进给方向相反）

down milling　顺铣，即 climb milling（铣刀的旋转方向与工件的进给方向相同）
table feed　工作台进给
tool life　刀具寿命（刀具从开始使用至达到磨钝标准时应保证的切削时间）
clamp　夹紧（工件定位后将其固定，使其在加工过程中保持定位位置不变）
feed per tooth　每齿进给量（每旋转一个刀齿时，铣刀相对工件在进给方向上的位移）
regrind [ri'graind]　v. 重新刃磨，再次刃磨
casting ['kɑ:stiŋ]　n. 铸件（将熔融金属浇入铸型，凝固后所得到的金属工件或毛坯）
hot rolled steel　热轧钢
scale　氧化层（a oxide film formed on iron or steel that has been heated to high temperatures）
outer scale　外层氧化皮（金属工件表面在高温下生成的固体腐蚀产物层）
helical teeth　螺旋齿
plain milling cutter　圆柱铣刀（在圆柱形刀体上有直齿、斜齿或螺旋齿的铣刀）
periphery [pə'rifəri]　n. 周边（the outer limits of an area or object），圆周，圆柱体表面
cutting force　切削力（在切削过程中，为使被切削材料变形、分离所需要的力）
face milling cutter　面铣刀（主要用于加工与机床主轴垂直的平面的铣刀），端铣刀
face milling cutter with inserted teeth　镶齿套式面铣刀（镶嵌刀条的面铣刀）
coated carbide insert　硬质合金涂层刀片（通过物理或化学气相沉积技术或其他方法，在硬质合金基体上涂覆一薄层耐磨性高的难熔金属化合物的刀片）
slitting saw　锯片铣刀（切削刃在外圆周上，用来下料或加工窄槽的铣刀）
circular saw blade　圆锯片
cut off　切断
side and face cutter　三面刃铣刀（外圆及两端面都带刀齿的盘形铣刀），side milling cutter
angle milling cutter　角度铣刀（加工各种角度槽用的铣刀）
T-slot cutter　T形槽铣刀（加工T形槽用的铣刀）
end mill　立铣刀（加工台阶、凹槽和各种互相垂直的平面，特别是深槽用的铣刀）
distinction [dis'tiŋkʃən]　n. 区别，差别（a noticeable difference between things or people）
cutter body　刀体（刀具上夹持刀条或刀片的部分，或由它形成切削刃的部分）
cemented carbide　硬质合金（由难熔金属碳化物和黏结金属通过粉末冶金工艺制成的烧结材料）
fasten ['fɑ:sn]　v. 固定，使坚固或稳固
carbide insert　硬质合金刀片
indexable ceramic insert　陶瓷可转位刀片
as was the case with　正如，就像
dovetail groove　燕尾槽
and the like　等等，诸如此类的东西（and other similar things）
end cutting edge　端刃
cylindrical cutting edge　周刃

Lesson 38 Gear Manufacturing Methods

The shape of the space between gear teeth is complex and varies with the number of teeth on the gear as well as gear module, so most gear manufacturing methods generate the tooth flank instead of forming.

Gear Shaping Gear shaping is the most versatile of all gear cutting processes. Although shaping is most commonly used for cutting teeth in spur and helical gears, this process is also applicable to cutting internal gears, herringbone gears, elliptical gears (see Figs. 12.2 and 19.1), and racks.

Figure 38.1 shows the principle of gear shaping with a shaper cutter. In this process, the cutter is mounted on a spindle that reciprocates axially as it rotates. The workpiece spindle is synchronized with the cutter spindle and rotates slowly as the tool meshes with and cuts the workpiece. The downward movement of the cutter represents the principal cutting motion. On the return (upward) stroke the cutter must be retracted about 1 mm to give clearance otherwise tool rub occurs on the backstroke and wear failure is rapid.

The advantages of gear shaping are that production rates are relatively high and that it is possible to cut right up to a shoulder. Unfortunately, for helical gears, a helical guide is required to impose a rotational motion on the stroking motion (as shown in Fig. 38.2); such helical guides cannot be produced easily or cheaply so the method is only suitable for high-volume production of helical gears since special cutters and guides must be manufactured for each different helix angle. A great advantage of gear shaping is its ability to cut internal gears.

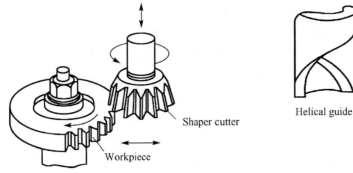

Figure 38.1 Gear shaping process   Figure 38.2 Helical gear shaping process

Gear Hobbing Gear hobbing uses the rack generating principle but avoids slow reciprocation by mounting many "racks" on a rotating cutter. The "racks" are displaced axially to form a gashed worm (see Fig. 4.9). The gear hobbing process is shown in Fig. 38.3.

Metal removal rates are high since no reciprocation of hob or workpiece is required and so cutting speeds of 40 m/min can be used for conventional hobs and up to 150 m/min for carbide hobs.

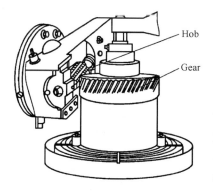

Figure 38.3 Gear hobbing process

Gear hobbing is a method widely used for cutting teeth in spur gears, helical gears, worms, and worm wheels. Gear hobbing machines are not applicable to cutting internal or bevel gears.

Gear Broaching Gear broaching is not usually used for helical gears but is useful for internal spur gears; the principal use of broaching in this context[1] is for internal splines (see Fig. 4.8) which cannot easily be made by any other method. The broaching process is rapid and produces low surface roughness with high dimensional accuracy. However, because broaches are expensive and a separate broach is required for each size of gear, this method is suitable mainly for high-volume production.

Gear Shaving This is a finishing process based on removing thin layers of chips (2 - 10 μm thick) from the profiles of the teeth by a tool called a gear shaving cutter. Shaving is currently the most widely used method of finishing spur and helical gear teeth following the gear cutting operation and prior to hardening the gear. The objective is to reduce surface roughness and improve tooth profile accuracy.

A gear shaving cutter looks like a gear which has extra clearance at the root (for swarf and coolant removal) and whose tooth flanks have been grooved to give cutting edges. It is run in mesh with the rough gear with crossed axes[2] so that there is in theory point contact with a relative velocity along the teeth giving scraping action, as indicated in Fig. 38.4. The gear shaving cutter teeth are relatively flexible in bending and so will only operate effectively when they are in double contact between two gear teeth. Cycle times can be less than half a minute and the machines are not expensive but cutters are delicate and difficult to manufacture.

Gear Grinding Gear grinding is extremely important because it is the main way hardened gears are machined. When high accuracy is required it is not sufficient to pre-correct for heat treatment distortion and grinding is then necessary.

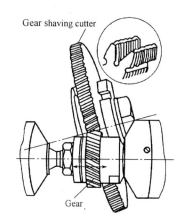

Figure 38.4 Gear shaving process

The simplest approach to gear grinding is form grinding (see Fig. 38.5). This method is fairly slow but gives high accuracy consistently. The fastest gear grinding method uses the same principle as gear hobbing but replaces the hob (Fig. 4.9) by a worm grinding wheel (see Figs. 38.6 and 38.7). Accuracy of the process is reasonably high although there is a tendency for grinding wheel and workpiece to deflect variably during grinding so the wheel form may require compensation for machine deflection effects. Generation of a worm shape on the grinding

137

wheel is a slow process since a dressing diamond must not only form the rack profile but has to move axially as the wheel rotates. Once the wheel has been trued and dressed, gears can be ground rapidly until redressing is required. This is the most popular method for high production rates with small gears.

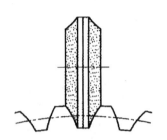

Figure 38.5 Form grinding process

Figure 38.6 Worm grinding wheels

Figure 38.7 Gear grinding with worm grinding wheels

Words and Expressions

gear teeth 轮齿（齿轮上的每一个呈辐射状排列并用于持续啮合的凸起部分）

module ['mɔdju:l] n. 模数（齿距除以圆周率 π 所得到的商）

generate ['dʒenəˌreit] v. 展成，范成

tooth flank 齿面（位于齿顶曲面和齿根曲面之间的轮齿侧表面）

forming ['fɔ:miŋ] n. 成形法（利用成形刀具对工件进行加工的方法）

gear shaping 插齿（用插齿刀按展成法或成形法加工内、外齿轮或齿条等的齿面）

helical gear 斜齿轮（齿线为螺旋线的圆柱齿轮）

helix ['hi:liks] n.; a. 螺旋，螺旋线；螺旋状的

rack [ræk] n. 齿条（具有一系列等距离分布齿的平板或直杆）

herringbone ['heriŋbəun] a. 人字形的

herringbone gear 人字齿轮（一半齿宽上为右旋齿，另一半齿宽上为左旋齿的圆柱齿轮）

elliptical [i'liptikəl] a. 椭圆的

elliptical gear 椭圆齿轮（分度曲面为椭圆柱面的非圆齿轮）

shaper cutter 插齿刀（齿轮状的插齿刀具，用于在插齿机上按展成法加工齿形）

reciprocate [ri'siprəkeit] v. 做往复运动（to move back and forth alternately）

synchronize [ˈsiŋkrənaiz]　v. 同步（to happen at the same time and speed），同时发生
stroke [strəuk]　n. 行程（刀具或工件在运动过程中相对移动的距离）
reciprocation [riˌsiprəˈkeiʃən]　n. 往复运动（an alternating back-and-forth movement）
return stroke　返回行程（工作结束后，刀具或工件返回原始位置的行程）
retract [riˈtrækt]　v. 缩回，缩进，收缩
backstroke [ˈbækstrəuk]　n. 返回行程，回程
helical guide　螺旋导轨
high-volume production　大批量生产
gear hobbing　滚齿（用齿轮滚刀或蜗轮滚刀按展成法加工齿轮或蜗轮等的齿面）
gear hob　齿轮滚刀（按齿轮啮合原理加工件齿形的一种展成刀具）
gear hobbing machine　滚齿机（主要用滚刀按展成法加工圆柱齿轮、蜗轮等齿面的齿轮加工机床）
gash [gæʃ]　n. 深长切口（a long deep cut）；v. 造成深长切口（to make a long deep cut in）
carbide hob　硬质合金滚刀（切削部分或整体采用硬质合金的滚刀）
bevel gear　锥齿轮（分度曲面为圆锥面的齿轮）
gear broaching　拉齿（用拉刀或拉刀盘加工内、外齿轮等的齿面）
broach [brəutʃ]　n. 拉刀（在拉力作用下进行切削的拉削工具）
gear shaving　剃齿（用剃齿刀对齿轮或蜗轮等的齿面进行精加工）
tooth profile　齿廓
gear shaving cutter　剃齿刀，也可写为 shaving cutter
swarf [swɔːf]　n. 细小切屑（metallic particles removed by a cutting or grinding tool）
coolant [ˈkuːlənt]　n. 冷却液（起冷却作用的调温流体），切削液（为了提高切削加工效果，如增加切削润滑，降低切削区温度而使用的液体，亦为 cutting fluid）
rough [rʌf]　a.；ad. 粗糙的（地），粗加工的（地）；v. 粗加工
scraping [ˈskreipiŋ]　n. 刮削（刮除工件表面薄层的加工方法）
gear grinding　磨齿（用砂轮按展成法或成形法磨削齿轮或齿条等的齿面）
form grinding　成形磨削（a specialized type of cylindrical grinding where the grinding wheel has the exact shape of the final product）
dress [dres]　v. 修整，砂轮的修锐（产生锋锐的磨削面）
true [truː]　v. 砂轮的整形（产生确定的几何形状）
dressing diamond　金刚石修整工具（通过钎焊或压入的方法镶嵌到钢制载体上的金刚石）
redressing [riˈdresiŋ]　n. 重新修整（砂轮）
worm grinding wheel　蜗杆砂轮（具有螺旋工作表面，以滚切法磨齿的砂轮）

Notes

［1］in this context 意为"由于这个原因，在这个意义上，在这种情况下"。
［2］crossed axes 意为"交错轴（两轴线既不平行，也不相交）"。

Lesson 39　Laser-Assisted Machining and Cryogenic Machining Technique

To meet the demand for production and precision, researchers and equipment builders are looking outside the bounds of conventional milling, drilling, turning, and grinding. New cutting and machining processes continue to emerge along with methods for modifying the conventional techniques.

Laser-assisted machining　Laser-assisted machining is a process that has long been researched and is now ready to come out of the lab. In this operation, a laser beam is projected onto the part through optical fibers or some other optical-beam delivery units, just ahead of the tool (see Fig. 39.1). [1] CO_2 lasers are usually used with power levels from 200 to 500W. The laser-induced heat softens the workpiece and makes it easier to cut.

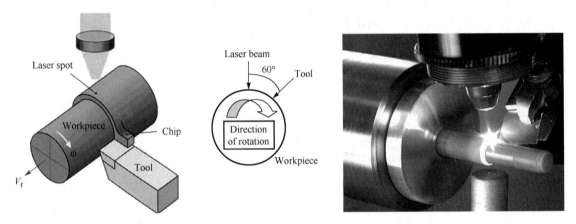

Figure 39.1　Laser-assisted machining process

This process can be used for very difficult-to-machine materials such as superalloy or ceramics. For example, with such process you can cut through ceramics like butter. It also offers a low surface roughness, usually 0.5 μm in Ra or lower.

Because the heating laser beam is tightly focused, heating is localized around the actual cutting zone. [2] Heat is carried off in the chips and there are no changes in the physical properties of the workpiece material.

There are two problems with the process. There is a physical problem initially of establishing the cutting parameters (laser power, cutting speeds and feeds, etc.) and the psychological problem of convincing people it works well and has many advantages.

Cryogenic machining technique　Water or oil based coolants have traditionally been used to reduce the heat generated during the machining process. [3] However, they have several drawbacks, including mist collection, workpiece contamination, and disposal costs. A unique use of liquid nitrogen is offering a modification of traditional turning to improve manufacturing operations. The cryogenic coolant system delivers a jet of -195°C liquid nitrogen directly to the insert during turning operations (see Fig. 39.2). [4] This removes the heat from the cutting zone more quickly and efficiently, allowing the insert to retain its hardness at high cutting temperatures. In Fig. 39.2, the white on the indexable turning tool is frost produced by the low temperature, despite the cutting heat. [5]

The typical construction of an indexable turning tool is shown in Fig. 39.3. Inserts are available in a variety shapes, such as square, triangle, diamond, and round. [6] Each insert has a number of cutting edges and, after one edge is worn, it is indexed to make another cutting edge available.

Figure 39.2 Cryogenic machining process

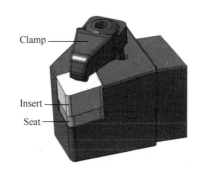

Figure 39.3 An indexable turning tool[7]

Cubic boron nitride (CBN) and polycrystalline cubic boron nitride (PCBN) inserts (see Fig. 39.4) have traditionally been the tools of choice for hard turning applications. But, they are considered too costly for many operations. The cryogenic machining technique also allows greater use of low-cost ceramic inserts (see Fig. 39.5) for hard turning operations.

Figure 39.4 CBN inserts

Figure 39.5 Ceramic inserts

CBN and ceramic inserts tend to wear unevenly and are prone to fracturing when hard turning dry or with water or oil based coolants. [8] Increased fracture toughness resulting from low-temperature liquid nitrogen cooling provides more predictable, gradual flank wear (see Fig. 39.6) for ceramic inserts, as well as increasing cutting speeds up to 200%.

By dissipating heat efficiently, cryogenic machining increases cutting tool life and material removal rates. For example, compared with conventional coolant

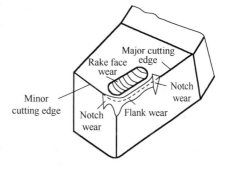

Figure 39.6 Features of tool wear

machining methods, the use of liquid nitrogen to machine titanium alloy can result in up to a tenfold increase in cutting tool life and a twofold increase in material removal rates.[9] And in addition to being cost-effective, nitrogen is a safe, breathable, nongreenhouse and nonflammable gas. It quickly evaporates after contact with the insert, and returns back into the atmosphere, leaving no residue to contaminate the workpiece, chips, machine tool, or operator, and it also eliminates disposal costs.[10] This is particularly helpful for porous powder metal parts, which often require subsequent part cleaning operations to remove coolant residue.

Now these two processes have become commercially available.

Words and Expressions

laser-assisted machining 激光加热辅助切削加工（通过激光加热软化切削区材料，再利用刀具进行切削加工），简称 LAM
cryogenic [ˌkraiəuˈdʒenik] a. 深冷（通常指-150℃以下的温度）的
cryogenic machining technique 深冷加工技术
laser beam 激光束
project onto 投射到……上
optical fibe 光导纤维（能把光闭合在纤维中而产生导光作用的纤维），又称光学纤维
soften [ˈsɔ(ː)fn] v. （使）变柔软，软化（to make soft or softer）
superalloy [ˌsjuːpəˈæloi] n. 超合金，又称"高温合金（high temperature alloy）"
physical problem 实际问题
focus [ˈfəukəs] n. 焦点，焦距；v. 聚焦，调焦
cutting parameter 切削用量
localize [ˈləukəlaiz] v. 使局限于某一区域（to keep something within a limited area）
coolant [ˈkuːlənt] n. 冷却液，切削液（cutting fluid）
mist collection （切削液）雾收集（the process of collecting coolant mist during machining）
disposal cost 废弃物处置成本，清理成本
insert 刀片（装夹在刀体上的片状物体，并由它形成刀具的切削部分）
hard turning 硬态车削（通常指对淬硬钢进行车削加工），硬车加工
cubic boron nitride (CBN) insert 立方氮化硼刀片（切削部分材料为立方氮化硼的刀片）
polycrystalline cubic boron nitride (PCBN) 聚晶立方氮化硼
hard turning dry 硬态干式车削（在不使用任何切削液的情况下进行硬态车削）
fracture toughness 断裂韧度，又称断裂韧性（构件材料应力强度因子的临界值）
unevenly [ʌnˈiːvənli] ad. 不均衡地，不规则地，参差不齐地
gradual [ˈgrædjuəl] a. 逐渐的，逐步的，渐进的
flank wear 后刀面磨损（刀楔后面上的磨损）
rake face wear 前刀面磨损，又称月牙洼磨损（crater wear）
notch [nɔtʃ] wear 边界磨损（在切削刃靠近工件外表面处产生较大的沟状磨损）
major cutting edge 主切削刃
minor cutting edge 副切削刃
features of tool wear 刀具磨损特征

titanium alloy 钛合金（以金属钛为基体元素、加入一种或多种合金元素组成的合金）
material removal rate 材料去除率，材料切除率（单位时间里所切除材料体积）
tenfold ['tenfəuld] a. 十倍的；n. 十倍
cost-effective 有成本效益的，划算的（producing good results without costing a lot of money）
breathable, nongreenhouse and nonflammable 可以吸入、不产生温室效应和不可燃烧的
residue ['rezidju:] n. 残留物（matter remaining after something has been removed）
powder metal part 粉末金属零件（通过粉末冶金方式得到的金属零件）
commercially available 作为商品提供的，市场上可买到的（buyable, purchasable）

Notes

[1] optical-beam delivery unit 意为"光束传输装置"。全句可译为：在这种加工过程中，激光束通过光导纤维或一些其他种类的光束传输装置投射到工件上靠近刀具前方的位置。

[2] tightly focused 意为"紧聚焦的"。全句可译为：由于加热时采用了紧聚焦的激光束，因此加热仅限于实际切削区周围。

[3] water or oil based coolants 意为"水基或油基冷却液，水基或油基切削液"。全句可译为：通常采用水基或油基切削液来减少切削过程中产生的热量。

[4] liquid nitrogen 意为"液氮"。全句可译为：在车削加工过程中，通过低温切削液传输系统将-195℃的液氮直接喷射到刀片上（见图39.2）。

[5] cutting heat 意为"切削热"。这两句可译为：这种方法可以将切削区的热量更快速、更有效地传导出去，使刀片在很高的切削温度下仍然可以保持其硬度。尽管有切削热，图39.2中可转位车刀上的白颜色显示了低温仍然会使其结霜。

[6] diamond 意为"菱形，菱形的"。全段可译为：可转位车刀的典型结构如图39.3所示。刀片可以有多种形状，如正方形、三角形、菱形和圆形。每种刀片上都有多个切削刃，当一个切削刃磨钝后，可将刀片转位到另一个切削刃。

[7] 在图39.3中，clamp 意为"压板"，seat 意为"刀垫"，亦可为 shim。

[8] fracturing 意为"破损，破碎"。全段可译为：无论是在进行硬态干式切削时，还是在切削加工时使用水基或油基切削液，CBN与陶瓷刀片都容易产生不均匀磨损和易于发生破损（见图39.6）。低温液氮冷却可以提高陶瓷刀片的断裂韧度，产生可以预测和逐渐增大的后刀面磨损，切削速度的提高最多可以达到200%。

[9] machining 意为"机械加工"。全句可译为：例如，与使用传统切削液的机械加工方法相比，加工钛合金时使用液氮可以使刀具寿命增加10倍，材料去除率增加两倍。

[10] contaminate 意为"污染"。全句可译为：它会在接触到刀片后快速蒸发，返回空气中，没有任何残留物，不会对工件、切屑和机床造成污染，不会对操作者造成危害，同时也节省了与切削液有关的废弃物处理费用。

Lesson 40　Machine Tool Motors

Electric motors are the prime movers for most machine tool functions. They are made in a variety of types. Most of them use 3-phase ac power supplied at 380 V.

The Basics

All electric motors use the principle that like magnetic poles repel and unlike poles attract. [1] Current through a coil or permanent magnets creates magnetic fields. Motors deliver torque by shifting the magnetic fields within the motor, so the rotor is constantly drawn around.

Initially, all motors used direct current (dc). This current creates magnetic fields in the stator and rotor, then mechanically energizing and de-energizing stator coils cause a moving field that draws the rotor around. [2] With alternating current (ac) motors, the current itself switches in polarity from positive to negative. [3] Sending ac to the stator windings creates a rotating magnetic field that draw the rotor around.

For most machine tool applications, versions of the ac asynchronous motor are used for spindle drives, while feed motors are generally synchronous.

A Little History

The design problem through the years with machine tools and motors has been how to get high torque at a variety of speeds. Initially, mechanical transmissions consisting of gears, belts, and gear/belt combinations gave speed changing capability. Up to 36 speed ranges were common at one time. But all this extra hardware is costly and needs maintenance. In the last decade, as speed requirements have risen the inaccuracy caused by vibration, which was not a problem at lower speeds, made the complex mechanical transmission unacceptable for some applications. Machine tool builders still use mechanical transmission for many applications but, because of more versatile motor speed control, three-speed transmissions are more common.

For today's operation, consider a spindle speed of 3,600 rpm as low, with high speed generally 10,000 rpm and greater. Some spindles, such as small-diameter grinding spindles, operate at 200,000 rpm. At the same time, motor design and control technology have progressed dramatically. Now, thanks to computer technology, it's possible to quickly modify motor speed and torque. [4]

Spindle Motors

A spindle is a motor-driven shaft that both positions and transmits power to a tool or holds a workpiece. [5] Spindle motors, the major motors on a machine tool, drive the spindle shafts. Spindle motors are rated by kilowatt, which generally ranges from 3 to 10 kW with the average around 5 kW.

All the early spindle motors were dc brush motors. It's a powerful motor, but speed control limitations and brush maintenance caused it to fall from favor.

The ac asynchronous induction spindle motor is also called a squirrel cage motor (see Fig. 40.1).[6] Alternating current—fluctuating from positive to negative in a sine wave (see Fig. 40.2)—is sent to three stator windings causing a rotating magnetic field. At the same time, current induced into the rotor sets up another field. Forces set up by these two fields cause rotor rotation. To stop, a control reverses current flow.

 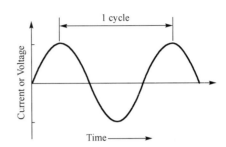

Figure 40.1　A squirrel cage motor　　　　Figure 40.2　Diagram of sinusoidal alternating current

Lower cost motors of this type control speed by varying frequency. Larger, and more precise motors, use vector control that requires a microprocessor with a math algorithm to control current accurately, both the current magnetizing the coil and the current producing torque. The motor can therefore have maximum torque at any speed.

Vector control also provides both position and current feedback in real time.[7] Current feedback gives a good indication of what the spindle and its tool are doing (spinning in air or cutting metal). This information also determines how much time a given tool has been used, which is important for automatic tool management. In some automatic tool changers, spindle position control is important for tool holder orientation. It is necessary to know where the tool is for automatic tool changers to function properly.

Another motor design now gaining acceptance as a spindle drive is the dc brushless motor with permanent magnet rotor. This motor has high torque at lower speed which is important when using larger tools. Direct current brushless motors don't develop as much heat as comparable ac motors, but are larger, more costly, and have speed limitations.

Feed Motors

As with spindle motors, positioning motors were initially dc brush type. Then ac induction motors became popular, followed by hydraulic motors (see Fig. 68.1).[8] Hydraulic motors have good acceleration and power to weight ratio because of low mass. Although still used in some high repeatability applications, they were generally displaced by ac servo motors that offer longer life, higher efficiency, and less heat generation.

Accurate Positioning

Position control is the key feature in feed motor operation. It is important to know that accurate positioning depends on feedback, or a closed-loop system. Most early systems were open loop, or nonfeedback. In this arrangement, a controller commands a motor to "turn four revolutions" or "only just turn 90°". After that, the motor stops. Accuracy depends on how well the system was made and if anything interfered with system motion.

With feedback or closed loop, the motor obeys the command, but has a sensor that sends back the signal, "Here is where I am." If it is not the right place, the control warns the operator or sends a corrective signal. This "conversation" between the motor and control takes place almost instantaneously.

Motors tell where they are with encoders and resolvers.[9] An encoder is a device attached to the motor shaft. It generates a digital signal that notes how many turns or partial turns the motor shaft made. A resolver is a device on a motor that generates a sine wave as the motor turns. A controller senses the sine wave, counting both the number of waves and sine angle to establish where the motor is.[10]

Words and Expressions

prime mover　原动力，原动机（转化自然能使其做功的机器或装置）
3-phase ac　三相交流电
energize ['enədʒaiz]　v. 使通电（to supply with an electric current; connect to a source of electricity）
de-energize　v. 切断，断开，释放（继电器、电磁铁等），去（解除）激励
direct current　直流，直流电（方向和量值不随时间变化的电流）
alternating current　交流，交流电（量值和方向作周期性变化且平均值为零的时变电流）
polarity [pəu'læriti]　n. 极性（尤指磁极或电极，especially magnetic or electrical poles）
stator ['steitə]　n. 定子（由定子铁芯、定子绕组和机座组成，主要作用是产生旋转磁场）
winding ['waindiŋ]　n. 绕组，线圈
rotor ['rəutə]　n. 转子（电机中的旋转部件）
asynchronous [ei'siŋkrənəs]　a. 不同时的，异步的
synchronous ['siŋkrənəs]　a. 同步的，同时发生的，同时出现的
feed motor　进给电动机
inaccuracy [in'ækjurəsi]　n. 误差，不准确
transmission [trænz'miʃən]　n. 传动装置（传递运动和动力的装置）
spindle ['spindl]　n. 主轴（带动工件或加工工具旋转的轴）
spindle speed　主轴转速（机床主轴在单位时间内的转数）
grinding spindle　砂轮主轴（带动砂轮旋转的轴）
fall from favor　成为不受欢迎的（to become unpopular），受到冷落
rpm (revolutions per minute)　转数/分
squirrel cage motor　鼠笼式电动机
automatic tool changer　自动换刀装置（能自动更换加工中所用刀具的装置），简称 ATC
tool holder　刀夹（夹持刀片、刀具或刀杆并与机床刀架连接的工具）
fluctuate ['flʌktjueit]　v. 脉动，振荡，变化
sine wave　正弦波形（a wave whose waveform resembles a sine curve）
sinusoidal [ˌsainə'sɔidəl]　a. 正弦的（having a magnitude that varies as a sine curve）
induction [in'dʌkʃən]　n. 感应，感应现象
algorithm ['ælgəriðəm]　n. 算法（求解数学计算问题的具体方法）
feedback ['fi:dbæk]　n. 反馈（将输出量通过恰当的检测装置返回到输入端并与输入量进行比较的过程）

induce [in'dju:s] v. 感应
magnetize ['mægnitaiz] v. 使磁化（to induce magnetic properties in）
real time 实时（在数据发生的同时处理该项数据，并在所需的响应时间内获得必要的结果）
positioning [pə'ziʃəniŋ] n. 定位，位置控制
dc brushless motor 无刷直流电动机
worktable ['wə:kteibl] n. 工作台
induction motor 感应电动机
dc brush motor 有刷直流电动机
hydraulic [hai'drɔ:lik] a. 水力的，液压的
activate ['æktiveit] v.；n. 开动，启动，驱动，激发
interfere with 干涉，干扰，妨碍

Notes

[1] like magnetic pole 意为"相同的磁极"。这两句可译为：所有的电动机都应用相同的磁极相斥、不同的磁极相吸的原理。通过线圈的电流或永久磁铁会产生磁场。

[2] energizing and de-energizing 这里指"通电和断电"。全句可译为：这个电流在定子和转子中产生磁场，然后采用机械的方式使定子线圈通电和断电，产生一个运动的磁场带动转子旋转。

[3] alternating current 意为"交流电"。全句可译为：在交流电动机中，电流本身从正到负地改变极性。

[4] 全句可译为：由于采用计算机技术，使得快速改变电动机的转速和扭矩成为可能。

[5] motor-driven 意为"由电动机驱动的"。全句可译为：主轴是一个由电动机驱动的轴，它既要确定刀具的位置又要将动力传给刀具，或者夹持工件。

[6] ac asynchronous induction spindle motor 意为"交流异步感应主轴电动机"。全句可译为：交流异步感应主轴电动机也被称为鼠笼式电动机（见图40.1）。

[7] vector control 意为"矢量控制"。全句可译为：矢量控制还能够实时地提供位置和电流的反馈。

[8] as with 意为"正如，与……一样"，hydraulic motor 意为"液压马达（做连续回转运动并输出转矩的液压执行元件）"。这两句可译为：如同主轴电动机一样，位置控制电动机最初采用有刷直流电动机。随后，交流感应电动机得到了普遍的应用，其后出现了液压马达（见图68.1）。

[9] encoder and resolver 意为"编码器（一种能提供位置反馈和速度反馈的测量装置）和分解器（一种能将旋转的和线性的机械位移转换为模拟电信号的变换器）"。全句可译为：电动机采用编码器和分解器来告诉人们它们所处的位置。

[10] sine angle 意为"正弦角"。全句可译为：控制器对正弦波进行检测，根据波的个数和正弦角来确定电动机的位置。

Lesson 41 Development of Metal Cutting
(*Reading Material*)

Before the middle of the 18th century the main material used in engineering structures was wood. The lathe and a few other machine tools existed, mostly constructed in wood and most commonly used for shaping wooden parts. The boring of cannon and the production of metal screws and small instrument parts were the exceptions. It was the steam engine, with its requirement of large metal cylinders and other parts of unprecedented dimensional accuracy, which led to the first major developments in metal cutting.

The materials of which the first steam engines were constructed were not very difficult to machine. Grey cast iron, wrought iron, brass and bronze were readily cut using hardened carbon steel tool. The methods of heat treatment of tool steel had been evolved by centuries of craftsmen, and reasonably reliable tools were available, although rapid failure of the tools could be avoided only by cutting very slowly. It required 27 working days to bore and face one of Watt's large cylinders.

At the inception of the steam engine, no machine tool industry existed—the whole of this industry is the product of the last two hundred years. The century from 1760 to 1860 saw the establishment of enterprises devoted to the production of machine tools. Henry Maudslay, Joseph Whitworth, and Eli Whitney, among many other great engineers, devoted their lives to perfecting the basic types of machine tools required for generating, in metallic components, the cylindrical and flat surfaces, threads, grooves, slots and holes of many shapes required by the developing industries. The lathe, the planer, the shaper, the milling machine, the drilling machine, and the sawing machine were all developed into rigid machines capable, in the hands of good craftsmen, of turning out large numbers of very accurate parts and structures of sizes and shapes that, had never before been contemplated. By 1860 the basic problems of how to produce the necessary shapes in the existing materials had largely been solved. There had been little change in the materials, which had to be machined—cast iron, wrought iron and a few copper based alloys. The quality and consistency of tool steels had been greatly improved, but the limitations of carbon steel cutting tools were becoming an obvious constraint on speeds of production.

From 1860 to the present day, the emphasis has shifted from development of the basic machine tools and the know-how of production of the required shapes and accuracy, to the problems of machining new metals and alloys and to the reduction of machining costs. With the Bessemer and Open Hearth steel making processes, steel rapidly replaced wrought iron as a major construction material. The tonnage of steel soon vastly exceeded the earlier output of wrought iron and much of this had to be machined. Alloy steels in particular proved more difficult to machine than wrought iron, and cutting speeds had to be lowered even further to achieve a reasonable tool life. Towards the end of the 19th century the costs of machining were becoming very great in terms of manpower and capital investment. The incentive to reduce costs by cutting faster and automating the cutting process become more intense, and, up to the present time, continues to be the mainspring of the major developments in the metal cutting field.

The technology of metal cutting has been improved by contributions from all the branches of industry with an interest in machining. Development of cutting tool materials has held a key position. Productivity could not have been increased without the replacement of carbon tool steel by high-speed steel and cemented carbide which allowed cutting speeds to be increased by many times. Tool designers and machinists have optimized the shapes of tools to give long tool life at high cutting speed. Lubricant manufacturers have developed many new coolants and lubricants to reduce surface roughness and permit increased rates of metal removal.

The producers of those metallic materials which have to be machined played a double role. Many new alloys were developed to meet the increasingly severe conditions of stress, temperature and corrosive atmosphere imposed by the requirements of our industrial civilization. Some of these, like aluminum and magnesium, are easy to machine, but others, like high-alloy steels and nickel-based alloys, become more difficult to cut as their useful properties improve. On the other hand, metal producers have responded to the demands of production engineers for metals which can be cut faster. New heat treatments have been devised, and the introduction of alloys like the free-machining steels and leaded brass has made great savings in production costs.

Today metal cutting is a very large segment indeed of our industry. The motorcar industry, electrical engineering, railways, shipbuilding, aircraft manufacture, and the machine tool industry itself —all these have large machine shops with many thousands of employees engaged in machining. An estimate shows that something like 10% of all the metal produced is turned into chips. Thus metal cutting is a very major industrial activity, employing tens of millions of people throughout the world. The wastefulness of turning so much metal into low-grade scrap has directed attention to methods of reducing this loss. Much effort has been devoted to the development of ways of shaping components in which metal losses are reduced to a minimum, employing processes such as cold-forging, precision casting and powder metallurgy, and some success has been achieved. So far there are few signs that the numbers of components machined or the money expended on machining are being significantly reduced. In spite of its evident wastefulness, it is still the cheapest way to make very many shapes and is likely to continue to be so for many years.

Progress in the technology of machining is achieved by the ingenuity and experiment, the intuition, and logical thought of many thousands of practitioners engaged in metal cutting. The worker operating the machine, the tool designer, the lubrication engineer, the metallurgist, are all constantly probing to find answers to new problems created by the necessity to machine novel materials, and by the incentives to reduce costs, by increasing rates of metal removal, and to achieve greater precision or improved surface roughness.

Words and Expressions

bore [bɔː] *v.* 镗孔，加工深孔（to make a long deep hole with a tool）；*n.* 孔
cylinder [ˈsilində] *n.* 汽缸（与活塞构成工作容积的部件）
unprecedented [ʌnˈpresidəntid] *a.* 空前的，史无前例的（not done or experienced before）
wrought [rɔːt] iron 熟铁（an iron alloy with very less content of carbon）
hardened [ˈhɑːdənd] *a.* 硬化的，经过淬火的
craftsman [ˈkrɑːftsmən] *n.* 工匠，手艺精巧的人
Henry Maudslay 亨利·莫兹利（1771—1831），英国机械发明家，英国机床工业之父
Joseph Whitworth 约瑟夫·惠特沃思（1803—1887），英国机械发明家，工程教育家
Eli Whitney 伊莱·惠特尼（1765—1825），美国发明家和企业家
sawing machine 锯床（用圆锯片或锯条等将材料锯断或加工成所需形状的机床）
turn out 生产，制造（to produce rapidly or regularly by machine）
bore and face 加工孔和端面
contemplate [ˈkɔntempleit] *v.* 思考（to consider carefully），想到（to have in mind）
the present day 目前，当代（the period of time that exists now; the present time）
know-how [ˈnəuhau] *n.* 专门技能（the knowledge and skill required to do something correctly）
Henry Bessemer [ˈbesimə] 亨利·贝塞麦（1813—1898），英国发明家和冶金学家，有多项专利发明，以其1856年发明的贝塞麦炼钢法最为著名
Open Hearth [hɑːθ] 平炉（又称"马丁炉"，法国冶金学家马丁1865年发明炼钢平炉）
tonnage [ˈtʌnidʒ] *n.* 吨数（total weight in tons shipped, carried, or produced）
manpower [ˈmænpauə] *n.* 劳动力（the number of people working or available for work）
mainspring [ˈmeinspriŋ] *n.* 主要因素，主要动机
carbon tool steel 碳素工具钢（适宜于制作各种小型工模具的高碳非合金钢）
tool steel 工具钢（适宜于制造刃具、模具和量具等各式工具用的钢）
high-speed steel 高速钢（一种具有高硬度、高耐磨性和高耐热性的工具钢），又称锋钢
cemented carbide 硬质合金（以一种或几种难熔金属碳化物或氮化物、硼化物等为硬质相和金属黏结剂相组成的烧结材料）
coolant [ˈkuːlənt] *n.* 冷却液，切削液（为了提高切削加工效果，如增加切削润滑、降低切削区温度而使用的液体，亦为 cutting fluid）
lubricant [ˈluːbrikənt] *n.* 润滑剂（为减少表面磨损和摩擦力而施加在摩擦表面上的物质）
nickel-based alloy 镍基合金（an alloy generally contains 50% nickel or more）
free-machining steel 易切削钢（加入某些元素使切削性能改善的钢），亦为 free-cutting steel
leaded [ˈledid] brass 铅黄铜，易切削黄铜（free-cutting brass 或 free-machining brass）
motorcar [ˈməutəkɑː] *n.* 汽车（automobile）
low-grade 低劣的（of inferior grade or quality），低等的，低级的
cold forging 冷锻（在室温下进行的锻造工艺）
powder metallurgy 粉末冶金（制取金属、合金、金属化合物等粉末，以及将这些粉末或粉末混合料经过成型和烧结，制成材料或制品的冶金技术）
novel [ˈnɔvəl] *a.* 新型的（new and original, not like anything seen before），新颖的

Lesson 42 History of Machine Tools
(*Reading Material*)

The history of tools began during the Stone Age (over 50,000 years ago), when the only tools were hand tools made of wood, animal bones, or stone.

Around the end of the Stone Age, people began to experiment with new materials to make tools. This is how we got the field of metallurgy, the study of extracting metals from their ores and modifying the metals so as to give them certain desired shapes or properties.

For a brief time, they tried working with copper. While that metal was great for cooking pots, it was much too soft to use for farming equipment and weapons.

Through experimentation, someone discovered that copper could be made into a stronger material by adding other metals or nonmetals. A metal made by melting and mixing two or more metals or a metal and another material together is called an alloy. At first, they tried arsenic. This made the first form of bronze, but arsenic is poisonous. Next, they added tin to the copper to produce the perfect bronze. Mixing 90% copper and 10% tin, the new metal came out much stronger.

With the introduction of bronze, people could make much better tools. Bronze axes could hold a sharper edge than stone axes and last much longer. It made cutting trees much faster and allowed farmers to clear more land for planting.

Humans may have smelted iron sporadically throughout the Bronze Age. Iron tools and weapons weren't as hard or durable as their bronze counterparts during that period.

The Iron Age was a period in human history that started between 1200 B.C. and 600 B.C., depending on the region, and followed the Stone Age and Bronze Age. During the Iron Age, people across much of Asia, Europe and parts of Africa began making tools and weapons from iron and steel.

The use of iron became more widespread after people learned how to make steel, a hard and tough metal made by treating iron with great heat and mixing carbon with it. Tools and weapons were greatly improved, and animals were domesticated to provide power for some of these tools, such as the plow.

Before the Industrial Revolution of the 18th century, hand tools were used to cut and shape materials for the production of goods such as wagons, ships, furniture, and other products. After the advent of the steam engine, goods were produced by power-driven machines that could only be manufactured by machine tools. Machine tools (capable of producing dimensionally accurate parts in large quantities) and jigs and fixtures (for holding the work and guiding the tool) were the indispensable innovations that made mass production and interchangeable parts realities in the 19th century.

The earliest steam engines suffered from the imprecision of early machine tools, and the large cast cylinders of the engines often were bored inaccurately by machines powered by waterwheels and originally designed to bore cannon. Within 50 years of the first steam engines, the basic machine tools, with all the fundamental features required for machining heavy metal parts, were designed and developed. Some of them were adaptations of earlier woodworking machines; the metal lathe derived from woodcutting lathes used in France as early as the 16th century. In 1775 John Wilkinson of England built a precision machine for boring engine cylinders. In 1797 Henry Maudslay, also of England and one of the great inventive geniuses of his day, designed and built a screw-cutting engine lathe. The outstanding feature of Maudslay's lathe was a lead screw for driving the carriage (see Fig. 34. 1). Geared to the spindle of the lathe, the lead screw advanced the tool at a constant rate of speed and guaranteed accurate screw threads. By 1800 Maudslay had equipped his lathe with 28 change gears that cut threads of various pitches by controlling the ratio of the lead-screw speed to the spindle speed.

The shaper (see also Fig. 2. 8) was invented by James Nasmyth, who had worked in Henry Maudslay's shop in London. In Nasmyth's machine, a workpiece could be clamped horizontally to a table and worked by a cutter using a reciprocating motion to machine straight-line surfaces or cut keyways. A few years later, in 1839, Nasmyth invented the steam hammer for forging heavy workpieces. Another disciple of Maudslay, Joseph Whitworth, invented or improved a great number of machine tools and came to dominate the field; at the International Exhibition of 1862, his firm's exhibits took up a quarter of all the space devoted to machine tools.

Britain tried to keep its lead in machine-tool development by prohibiting exports, but the attempt was foredoomed by industrial development elsewhere. British machine tools were exported to the European continent and to the United States despite the prohibition, and new machine tools were developed outside Britain. Notable among these was the milling machine invented by Eli Whitney, produced in the United States in 1818, and used to manufacture firearms. The first fully universal milling machine was built in 1862 by Joseph Brown of the United States and was used to cut helical flutes in twist drills(Fig. 36. 1). The turret lathe, also developed in the United States in the middle of the 19th century, was fully automatic in some operations, such as making screws. Various gear-cutting machines reached their full development in 1896 when Edwin Fellows, an American, designed a gear shaper that could rapidly turn out almost any type of gear.

The production of artificial abrasives in the late 19th century opened up a new field of machine tools, that of grinding machines. Charles Norton of Massachusetts dramatically illustrated the potential of the grinding machine by making one that could grind an automobile crankshaft(Fig. 20. 5) in 15 minutes, a process that previously had required five hours.

By the end of the 19th century a complete revolution had taken place in the working and shaping of metals that created the basis for mass production and an industrialized society. The 20th century has witnessed the introduction of numerous refinements of machine tools, such as numerical control machine tools. Yet even today the basic machine tools remain largely the legacy of the 19th century.

Words and Expressions

hand tool   手工工具（手动操作的工具，a tool used with workers' hands）
metallurgy [me'tælədʒi]   n. 冶金学（研究从矿石等原料中提取金属或金属化合物并加工成具有一定性能和应用价值的金属材料的学科）
for a brief time   在很短的时间内（a very brief period of time）
extract [iks'trækt]   v. 提炼（to separate a metal from an ore），选取，提取
nonmetal ['nɔn'metl]   n. 非金属（a chemical element that lacks the characteristics of a metal）
arsenic ['ɑːsənik]   n. 砷（a solid poisonous chemical element），砒霜（As_2O_3）
planting ['plɑːntiŋ]   n. 栽培，种植（the activity of putting plants into the ground）
smelt [smelt]   v. 熔炼，冶炼（to melt ore in order to separate the metal contained）
sporadically [spə'rædikəli]   ad. 偶发地，零星地（not regularly or constantly）
counterpart ['kauntəpɑːt]   n. 对应物（something that has the same purpose as another）
domesticate [də'mestikeit]   v. 驯养（驯化动物从而使其在人类环境中生存并造福于人）
jigs and fixtures   夹具（用以装夹工件和引导刀具的装置）
indispensable [,indis'pensəbl]   n. 不可缺少之物；a. 不可缺少的，绝对必要的
imprecision [,impri'siʒən]   n. 不精确（the state or quality of being not accurate or exact）
inaccurately [in'ækjuritli]   ad. 不精密地，不精确地
waterwheel ['wɔːtəwiːl]   n. 水轮（以水流为动力的旧式装置，用以带动机器中的轴转动，a large wheel driven by flowing water, used to work machinery）
adaptation [,ædæp'teiʃən]   n. 改装（如对设备、机械装置等改装以适应新的用途），改建
John Wilkinson   约翰·威尔金森（1728—1808），英国发明家，第一台镗床的发明者
gear [giə]   v. 通过齿轮进行连接（to connect by gears）
screw thread   螺纹，通常简称为 thread（often shortened to thread）
change gears   交换齿轮（用以改变机床切削速度或进给量的可更换齿轮组）
pitch [pitʃ]   n. 螺距（相邻两牙在中径线上对应两点间的轴向距离）
spindle speed   主轴转速（机床主轴在单位时间内的转数）
James Nasmyth   詹姆斯·内史密斯（1808—1890），英国发明家，其最著名的发明为蒸汽锤
disciple [di'saipl]   n. 弟子，门生，追随者（a person who is a student of another; follower）
foredoomed [fɔː'duːmd]   a. 注定失败的（going to fail or extremely unlucky from the beginning）
notable ['nəutəbl]   a. 值得注意的（worthy of note or notice），显著的（remarkable），著名的
firearm ['faiərɑːm]   n. 枪械（尤指手枪或步枪，a weapon, especially a pistol or rifle）
Joseph Brown   约瑟夫·布朗（1810—1876），美国发明家，研制并生产出万能铣床等机床
helical flute   螺旋槽（与刀具轴线呈螺旋状的容屑槽）
Edwin Fellows   埃德温·费洛斯（1865—1945），美国发明家，企业家
gear shaper   插齿机（用插齿刀按展成法插削内、外齿轮齿面的机床），gear shaping machine
Charles Norton   查尔斯·诺顿（1851—1942），美国发明家，磨床的发明者
grind [graind]   v. 磨削（to perform the operation of grinding）

Lesson 43 Nontraditional Manufacturing Processes

The human race has distinguished itself from all other forms of life by using tools and intelligence to create items that serve to make life easier and more enjoyable. Through the centuries, both the tools and the energy sources to power these tools have evolved to meet the increasing sophistication and complexity of mankind's ideas.

In their earliest forms, tools primarily consisted of stone instruments. Considering the relative simplicity of the items being made and the materials being shaped, stone was adequate. When iron tools were invented, durable metals and more sophisticated articles could be produced. The twentieth century has seen the creation of products made from the most durable and, consequently, the most difficult-to-machine materials in history. In an effort to meet the manufacturing challenges created by these materials, cutting tools have now evolved to include materials such as high speed steel, cemented carbide, coated cemented carbide, cubic boron nitride, diamond, and ceramics.

A similar evolution has taken place with the methods used to power our tools. Initially, tools were powered by muscles; either human or animal. However as the powers of water, wind, steam, and electricity were harnessed, mankind was able to further extended manufacturing capabilities with new machines, greater accuracy, and faster machining rates.

Every time new tools, tool materials, and power sources are utilized, the efficiency and capabilities of manufacturers are greatly enhanced. However as old problems are solved, new problems and challenges arise so that the manufacturers of today are faced with tough questions such as the following: How do you drill a 2 mm diameter hole 670 mm deep without experiencing taper or runout? Is there a way to efficiently deburr passageways inside complex castings and guarantee 100% that no burrs were missed? Is there a welding process that can eliminate the thermal damage now occurring to my product?

Since the 1940s, a revolution in manufacturing has been taking place that once again allows manufacturers to meet the demands imposed by increasingly sophisticated designs and durable, but in many cases nearly unmachinable, materials. This manufacturing revolution is now, as it has been in the past, centered on the use of new tools and new forms of energy. The result has been the introduction of new manufacturing processes used for material removal, forming, and joining, known today as nontraditional manufacturing processes.

The conventional manufacturing processes in use today for material removal primarily rely on electric motors and hard tool materials to perform tasks such as sawing, drilling, and broaching. Conventional forming operations are performed with the energy from electric motors, hydraulics, and gravity. Likewise, material joining is conventionally accomplished with thermal energy sources such as burning gases and electric arcs[1].

In contrast, nontraditional manufacturing processes harness energy sources considered unconventional by yesterday's standards. Material removal can now be accomplished with electrochemical reactions, electrical sparks (see Fig. 43.1), high-temperature plasmas, and high-velocity jets of liquids and abrasives. Materials that in the past have been extremely difficult to form, are now formed with magnetic fields, explosives, and the shock waves from powerful electric sparks. Material-joining capabilities have been expanded with the use of high-frequency sound waves[2] and beams of electrons.

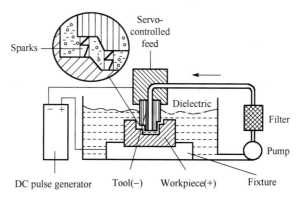

Figure 43.1 Schematic diagram of EDM process

In the past 70 years, over 20 different nontraditional manufacturing processes have been invented and successfully implemented into production. The reason there are such a large number of nontraditional processes is the same reason there are such a large number of conventional processes; each process has its own characteristic attributes and limitations, hence no one process is best for all manufacturing situations.

For example, nontraditional processes are sometimes applied to increase productivity either by reducing the number of overall manufacturing operations required to produce a product or by performing operations faster than the previously used method.

In other cases, nontraditional processes are used to reduce the number of rejects experienced by the old manufacturing method by increasing repeatability, reducing in-process breakage of fragile workpieces, or by minimizing detrimental effects on workpiece properties.

Because of the aforementioned attributes, nontraditional manufacturing processes have experienced steady growth since their introduction. An increasing growth rate for these processes in the future is assured for the following reasons:

1. Currently, nontraditional processes possess virtually unlimited capabilities when compared with conventional processes, except for volumetric material removal rates. Great advances have been made in the past few years in increasing the removal rates of some of these processes, and there is no reason to believe that this trend will not continue into the future.

2. Approximately one half of the nontraditional manufacturing processes are available with computer control of the process parameters. The use of computers lends simplicity to processes that people may be unfamiliar with, and thereby accelerates acceptance. Additionally, computer control assures reliability and repeatability, which also accelerates acceptance and implementation.

3. Most nontraditional processes are capable of being adaptively-controlled through the

use of vision systems, laser gages, and other in-process measurement techniques. If, for example, the in-process measurement system determines that the size of holes being produced in a product are becoming smaller, the size can be modified without changing hard tools, such as drills.

Words and Expressions

sophistication [sə‚fisti'keiʃən] n. 复杂化，完善，采用先进技术
durable ['djuərəbl] a. 耐用的，耐久的 (able to withstand wear or damage for a long time)
difficult-to-machine material 难加工材料，亦为 difficult-to-cut material
coated cemented carbide 涂层硬质合金（表面涂有难熔金属化合物，如 TiC、TiN、TiC-TiN 等薄膜的硬质合金），亦为 coated carbide
cubic boron nitride 立方氮化硼（一种人造的超硬材料），其缩写为 CBN
runout ['rʌn'aut] n. 偏摆，偏斜，径向振摆，径向跳动
deburr [di'bəː] v. 去毛刺（清除工件已加工部位周围所形成的刺状物或飞边）
passageway ['pæsidʒ'wei] n. 通道，通路
burr [bəː] n. 毛刺
sawing ['sɔːiŋ] n. 锯削（锯切工具旋转或往复运动，把工件加工成所需形状的切削方法）
unconventional [‚ʌnkən'venʃənl] a. 非传统的，非常规的，不一般的
electrochemical [i‚lektrəu'kemikl] a. 电化学的
plasma ['plæzmə] n. 等离子体，等离子
dielectric [‚daii'lektrik] n. 电介质，（特种加工中指 dielectric fluid）工作液
DC (direct current) 直流，直流电（方向和量值不随时间变化的电流）
pulse generator 脉冲发生器，（在特种加工中通常称为）脉冲电源
pump [pʌmp] n. 泵（改变容积内流体的压力或输送流体的机器）
EDM (electrical discharge machining) 电火花加工
reject [ri'dʒekt] n. 废品（不符合验收条件，且不能修复使用的零件），不合格品
repeatability [ri‚piːtə'biliti] n. 重复精度（在同一条件下，操作方法不变，进行规定次数操作所得到的连续结果的一致程度）
in process 过程中的，在进行中 (begun, and not completed)
breakage ['breikidʒ] n. 破损，断裂，损坏
detrimental [‚detri'mentl] a. 有害的，有损的，不利的；n. 有害的东西
aforementioned [ə‚fɔː'menʃnd] a. 上述的，前面提到的
thereby ['ðɛə'bai] ad. 因此，所以，在那方面，大约
adaptive [ə'dæptiv] control 自适应控制（按照预先给定的评价指标自动改变加工系统的参数，使之达到最佳工作状况的控制）
in-process measurement 加工过程中测量（在加工的过程中，为控制加工量所进行的测量）

Notes

[1] burning gases and electric arcs 意为"燃烧的气体和电弧"。
[2] high-frequency sound wave 这里指"超声波，即 ultrasonic wave"。

Lesson 44   Overview of Nontraditional Manufacturing Processes

Many advanced materials cannot be machined by traditional means, or at best they are machined with excessive tool wear and at high costs. [1] In addition, the complexity and surface quality of machined parts, tools, and dies have dramatically increased. Traditional methods also cause undesired changes in the properties of a workpiece, such as deformation and residual stresses, which require further processing to remove the effects. [2]

Many new workpiece materials are either harder than conventional cutting tools or cannot withstand the high cutting forces involved in traditional machining. There is an acute need to develop new manufacturing methods and improve existing techniques that are capable of economically machining advanced materials such as superalloys, ceramics, plastics, and fiber-reinforced composites. Nontraditional manufacturing processes can machine precision components from these advanced materials.

Economic considerations require that technologies such as flexible manufacturing system (FMS), computer integrated manufacturing systems (CIMS), computer-aided engineering (CAE), computer numerical control (CNC), robotics, and artificial intelligence be employed not only on the shop floor but throughout all facets of the corporate structure. Many nontraditional manufacturing processes are especially well suited to on-line monitoring and adaptive control and thus can easily be interfaced with a manufacturer's database. [3]

Nontraditional manufacturing processes may be broadly categorized in two ways: processes in which there is a nontraditional mechanism of interaction between the tool and workpiece and processes in which nontraditional media are used to transfer of energy from the tool to the workpiece. [4] Nontraditional interaction mechanisms include chemical, electrochemical, thermal, and mechanical with high impact velocity. Nontraditional media include solids, abrasives, aqueous and nonaqueous liquids, gas, plasma, ions, electrons, and photons. [5] Also, in nontraditional manufacturing processes, the energy can be transferred in discrete pulses or continuously, and the energy at a given time can be applied to localized portion of the workpiece or over a broad machining area. [6] Electrical discharge machining (see Fig. 43.1), ultrasonic machining (see Fig. 44.1), water jet machining (see Fig. 44.2), and electrochemical machining are some of commonly used nontraditional manufacturing processes.

Nontraditional manufacturing processes are currently employed by many diverse industries, yet some manufacturers are not familiar with these processes. In addition, those people who believe that they are already familiar with nontraditional processes are often unaware of all of the process options available to them or are lacking information on the most recent advances which can drastically increase the cost-effectiveness and capabilities of these processes.

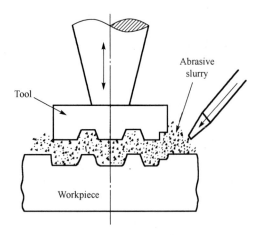

Figure 44.1 Schematic diagram of an ultrasonic machining operation

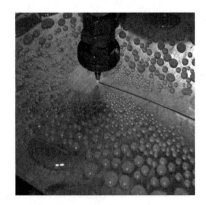

Figure 44.2 Water jet machining

Numerous articles, technical papers, and reports have addressed various nontraditional manufacturing processes in recent years. [7] It is nearly impossible, however, for those who are interested to be aware of all the available information sources as well as to find the time to stay abreast of all the developments concerning this very broad and rapidly changing subject.

If you feel that a nontraditional process should be investigated further for possible implementation at your facility, minimize the chances of encountering problems, and assure yourself success by following these common sense guidelines. [8]

Before deciding that one particular process is correct for your application, investigate and experiment with as many alternative processes as possible. If the option is available, select and investigate several applicable processes, preferably by processing your own hardware on the equipment under consideration.

To determine a potential supplier's level of experience with your prospective equipment, request a list of customers who have purchased similar machines. This list will reflect repeat orders, which is a good indication of customer satisfaction. Be extremely cautious if the equipment being proposed to you is one of the first to be built by a particular manufacturer. You could end up debugging the equipment at the expense of your production quotas. [9]

Contact several of the manufacturer's customers and confirm directly the level of operator and maintenance skills that they require to properly support the equipment, and inquire about the most frequent causes of downtime. Question also the long-term repeatability of the equipment. Are constant tuning or replacement of components required to maintain repeatability? Will your application place greater demands on the equipment than it was designed for?

If your facility is part of a multidivisional corporation, check with the manufacturing personnel at the other divisions to determine if they have had experience with the process or manufacturer in question. If they have had experience and are willing to share their information with you, your company and your parent corporation will save time and money by avoiding a duplication of efforts. [10]

When you are ready to make your equipment purchase, have the manufacturer supply you with a list of the recommended spare parts that you should have on hand to minimize downtime when problems arise. Also, take advantage of any operator, maintenance, or programmer training courses offered by the manufacturer. Inadequate training can severely affect your chances for what should be a successful and profitable implementation.

Words and Expressions

at best　充其量，至多，最好的情况下（under the most favorable circumstances）
die　模具（用以限定生产对象的形状和尺寸的装置），亦为 mould
residual stress　残余应力（金属加工过程中由于不均匀的应力场、应变场、温度场，以及不均匀的组织，在加工后，工件或产品内保留下来的应力）
cutting force　切削力（在切削过程中，为使被切削材料变形、分离所需要的力）
acute [ə'kju:t]　*a*. 敏锐的，尖锐的，急切的，剧烈的
fiber-reinforced composite　纤维增强复合材料
electrochemical [i,lektrəu'kemikəl]　*a*. 电化学的
aqueous ['eikwiəs]　*a*. 水的（made by water; relating to water），含水的（containing water）
nonaqueous ['nɔn'eikwiəs]　*a*. 非水的（made from or with a liquid other than water）
photon ['fəutɔn]　*n*. 光子
electrical discharge machining　电火花加工（在一定的介质中，通过工具电极和工件电极之间的脉冲放电的电蚀作用，对工件进行加工的方法），简称 EDM
ultrasonic machining　超声加工（利用超声振动的工具带动工件和工具间的磨料悬浮液，冲击和抛磨工件的被加工部位，使其局部材料被蚀除的加工方法），简称 USM
water jet machining　高压水射流加工（利用通过高压喷嘴的具有很高动能的高速水射流，对工件材料进行切割加工的方法），简称 WJM
electrochemical machining　电化学加工（利用金属在电解液中产生的阳极溶解的原理去除工件材料的特种加工），又称电解加工，简称 ECM
slurry ['slə:ri]　*n*. 悬浮液（固体颗粒悬浮在液体中所形成的混合物）
abrasive slurry　磨料悬浮液（a mixture of abrasive particles suspended in water or oil）
address [ə'dres]　*v*. 提出，致力于解决（to deal with something usually skillfully or efficiently）
stay abreast of　了解最新进展（to know most recent information），亦为 keep abreast of
prospective [prəs'pektiv]　*a*. 预期的，预见的，未来的，盼望的
debug [di:'bʌg]　*v*. 调试，发现并排除故障，审查
quota ['kwəutə]　*n*. 份额，分担部分，定额，限额，定量
downtime　*n*. 停机时间，发生故障时间
duplication [,dju:pli'keiʃən]　*n*. 复制，重复
profitable ['prɔfitəbl]　*a*. 有利可图的，有用的，有益的

Notes

[1] advanced material 意为"先进材料"。全句可译为：许多先进材料不能采用传统的方式加

工，或者即使能够加工，刀具也会产生很严重的磨损，而且加工成本也会很高。

[2] deformation and residual stresses 意为"变形和残余应力"。这两句可译为：此外，被加工零件、工具和模具的复杂程度和表面质量都已经显著提高。传统方法还会使工件上出现人们所不希望的性能变化，如变形和残余应力，这些都需要进一步加工以去除其影响。

[3] shop floor 意为"车间，（工厂的）生产区"，on-line monitoring 意为"在线监控"。这段可译为：从经济角度考虑，诸如柔性制造系统（FMS）、计算机集成制造系统（CIMS）、计算机辅助工程（CAE）、计算机数字控制（CNC）、机器人技术和人工智能等项技术不仅要在工厂的车间中得到应用，还要应用于公司结构的所有方面。许多特种加工工艺非常适合于在线监控和自适应控制，因此可以很容易地与制造厂的数据库相连接。

[4] mechanism of interaction 意为"相互作用机理"。全句可译为：特种加工工艺大致可分为两类：一类加工工艺在工具和工件之间采用非传统的相互作用机理，另一类加工工艺采用非传统的介质将能量从工具传到工件上。

[5] aqueous and nonaqueous liquids 意为"水和非水液体"。全句可译为：非传统介质包括固体、磨料、水和非水液体、气体、等离子、离子、电子和光子。

[6] discrete pulse 意为"离散脉冲"。全句可译为：此外，在特种加工工艺中，能量可以以离散脉冲或连续的形式传递。在给定时间内，能量可以作用于零件上的很小的区域中或大的加工面积上。

[7] technical paper 意为"技术论文"。全句可译为：近年来，很多文章、技术论文和报告致力于解决各种各样的特种加工工艺问题。

[8] facility 这里指"部门"，可用 department 代替，或者为 manufacturing facility 制造部门（A place where an industrial or manufacturing process takes place）。全句可译为：如果你认为某种特种加工工艺可能会被用于你所在的部门中而对其做进一步调研时，那么遵循下列常识性准则会将可能遇到的问题减至最少，以确保获得成功。

[9] supplier 意为"供应商（提供商品或服务的公司或个人）"，repeat order 意为"后续订单（another order from a customer who has ordered the same thing before）"，propose to 意为"提出（to suggest that some action be taken）"，production quota 意为"生产定额"。这几句可译为：为了确定潜在的供应商在你想要购买的设备方面的经验水平，索要一张购买过类似产品的客户名单，这张名单上的那些后续订单的数量将会表明客户们对产品的满意程度。如果将要向你提供的设备是某个企业所生产的第一批此类产品中的一个，那么你就应该非常谨慎。你最终得到的结果可能是需要不断地对设备进行调试，其代价是减少生产定额。

[10] parent corporation 意为"母公司，控股公司"，大大小小的公司都可以称为 company，用英语说就是 a group of people who work together，而 corporation 就更为正式，主要是指大公司（a large business organization）。此段可译为：如果你所在的部门是一个由多个部门组成的大公司的一部分，则你可以与其他部门的制造人员联系，看看他们对于在你考虑中要采用的加工工艺或选择的厂家是否有所了解。如果他们有这方面的经验，并且愿意与你分享他们的信息，你所在的公司及其母公司将会因为避免了重复性的工作而节省时间和金钱。

Lesson 45　Machining of Engineering Ceramics

Engineering ceramic materials have attractive properties: high hardness, high thermal resistance, chemical inertness, and low thermal or electrical conductivity, to name a few. However, these properties make ceramics extremely difficult to machine, whether by abrasive or nonabrasive processes. For the latter, material removal rates must be increased in current processes to improve productivity rates for economic justification. Other requirements include: improved surface roughness, control of geometric features, and reduced capital equipment[1] costs.

Grinding involves a complex interaction between a number of variables: workpiece material properties, wheel specifications, and machine tools selection. This interaction, called "grindability", can be quantified in terms of material-removal rate, power or grinding force required, surface roughness, tolerances, and surface integrity[2].

Coarse grained grinding wheels are usually used for fast removal of materials. During grinding flaws are introduced on the ground surface (see Fig. 45.1) that vary from piece to piece, affecting the surface integrity. This translates into a variation in fracture strength. Residual stresses are also generated on the surface and subsurface, due to the mechanical process of chip removal. Defective regions within the surface layer can have linear dimensions between 10 and 100 μm. Such damage often must be removed by finish machining[3] with a finer-grit wheel.

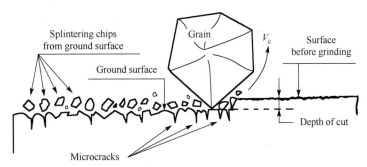

Figure 45.1　Model of chip formation in grinding of engineering ceramics

For conventional grinding of ceramics, the selection of the grinding machine and the grinding wheel also influence grindability. In general, only diamond grinding wheels can be used for ceramics. Resin-bonded diamond wheels provide better results than either metal or vitrified-bonded wheels for fired ceramics. Resin-bonded wheels produce lower grinding forces and wear away faster, exposing new cutting edges which help to minimize surface flaw size of the ceramic workpiece.

Since the ceramic's properties play a major role in the final grinding results, their interaction with machining variables must be understood completely to determine costs. Coolant application becomes critical for low thermal-conductivity materials to prevent thermally induced cracks. Porosity, grain size, and microstructure affect surface roughness and surface quality; high porosity contributes to poor surface roughness. Generally, ceramics require higher grinding forces and power, which lead to low wheel life.

Several abrasive techniques do not require expensive diamond wheels and the problems that go with them. Water jet machining (see Fig. 44.2) uses a high-velocity fluid jet[4], either alone or with abrasive particles, to erode workpiece material. The jet can be used in the pulsed, or continuous modes—the latter for most ceramic applications. This type of machining is best for cutting slots and grooves or trepanning large holes.

There are no heat effects nor large mechanical forces in water-jet machining, so that even the most thermal-shock sensitive, weakest ceramics can be cut without damage. Chatter, vibration, surface distortion, and subsurface damage of the workpiece also are eliminated.

Ultrasonic machining (see Fig. 44.1) is another abrasive process that has certain advantages over conventional grinding. Sometimes called impact grinding, it is a mechanical process that uses an ultrasonic transducer. As the electrical energy is converted into mechanical motion, a tool of desired shape vibrates longitudinally at an ultrasonic frequency (20–25 kHz) with an amplitude of around 15–30 μm over the workpiece.

A constant stream of abrasive slurry passes between the tool and the workpiece. Commonly used abrasives include diamond, boron carbide, silicon carbide and alumina, and the abrasive grains are suspended in water or a suitable chemical solution. In addition to providing abrasive grain to the cutting zone, the slurry is used to flush away debris. The size of abrasive grain determines the workpiece surface roughness, the size of the cavity in relation to the tool, and the cutting rate. The vibrating tool, combined with the abrasive slurry, abrades the workpiece material uniformly, leaving a precise reverse image of the tool shape.

Ultrasonic machining can be used to machine almost any hard, brittle material, though it is more effective in materials of 40 HRC or more. These include silicon, glass, quartz, fiber-optic materials, structural ceramics such as SiC, and electronic substrates such as alumina.

Abrasionless machining, such as laser-beam and electron-beam machining (LBM, EBM), is not limited by the hardness of ceramics and therefore offers an alternative to conventional grinding. In LBM, a laser focuses an intense beam of light onto the workpiece to vaporize material. Because the laser beam is mechanically positioned, it is not as fast as EBM. However, unlike EBM, no vacuum chamber is required so workpiece loading is faster, and there is no limit on size. Another advantage is less-expensive equipment. Usually a pulsed mode is used for drilling, while a continuous mode is used for cutting. LBM can machine just about any hard material, including diamond.

Machining rate is controlled by the rate at which material, which is melted and vaporized, is removed by the beam. Removal occurs by thermal convection[5] and beam pressure. The rate may be greatly increased if a gas jet is used to blow away the material. Thermal gradients can cause cracking so that thick ceramics may have to be machined in the unfired state. Conversely, laser machined ceramics may be stronger than diamond-machined ceramics[6] since heating can produce beneficial residual stresses. Laser power requirements are determined by the amount of stock to be removed; one 150 W laser is sufficient for thin ceramics, while a 15,000 W laser usually is required for thick ceramics.

Words and Expressions

inertness [i'nə:tnis]　*n.*　惰性，不活泼，无自动力
material removal rate　材料去除率，材料切除率（单位时间里所切除的材料体积）
justification [ˌdʒʌstifi'keiʃən]　*n.*　可接受的理由（an acceptable reason for doing something）
grindability [ˌgraində'biliti]　*n.*　可磨削性，磨削加工性（材料进行磨削加工的难易程度）
fracture strength　断裂强度（材料发生断裂时所承受的应力）
grain [grein]　*n.*　磨粒（单颗粒磨料），晶粒（多晶体材料内以晶界分开的晶体）
ground surface　磨削表面，经过磨削加工后的表面，这里 ground 是 grind 的过去分词
splinter ['splintə]　*n.*　碎片；*v.*　裂成碎片（to break something into small pieces）
splintering chip　崩碎切屑（形状不规则的切屑，使加工后的表面凹凸不平）
microcrack ['maikrəukræk]　*n.*　微裂纹，显微裂纹（指微观尺度的裂纹）
chatter ['tʃætə]　*n.*；*v.*　颤振，自激振动（self-excited vibration）
vitrified bond　陶瓷结合剂（由低熔点的玻化料制成的一种物质，用于磨具的制造）
resin　树脂（a liquid organic compound, that under certain circumstances, will harden）
porosity [pɔː'rɔsiti]　*n.*　孔隙率（一材料中孔隙的总体积与整个材料的体积之间的比率）
grain size　晶粒尺寸（多晶体内的晶粒大小），晶粒度，粒度（磨粒平均直径）
trepanning [tri'pæniŋ]　*n.*　套料加工（能在钻削的内孔中套出一段棒料的孔加工方法）
erode [i'rəud]　*v.*　冲蚀（材料表面与流体之间由于高速相对运动引起的损伤）
ultrasonic transducer [trʌnz'djuːsə]　超声换能器（一种能把高频电能转化为机械能的装置）
amplitude ['æmplitjuːd]　*n.*　振幅（振动物体离开平衡位置的最大距离）
boron ['bɔːrɔn] carbide　碳化硼（化学式为 B_4C，chemical formula B_4C）
alumina [ə'ljuːminə]　*n.*　氧化铝，刚玉（以三氧化二铝 Al_2O_3 为主体的磨料）
debris ['deibriː]　*n.*　碎片，碎屑（scattered pieces of unwanted material），微粒
size of abrasive grain　粒度（磨粒平均直径），也可写为 grain size
convection [kən'vekʃən]　*n.*　对流（依靠流体的宏观运动进行的热量传递），传递
thermal gradient ['greidiənt]　*n.*　热梯度，温度梯度（temperature gradient）

Notes

[1] capital equipment 意为"生产设备（equipment used to manufacture a product）"。
[2] surface integrity 意为"表面完整性"。
[3] finish machining 意为"精加工（使工件达到预定的精度和表面质量的加工）"。
[4] high-velocity fluid jet 意为"高速液体射流"。
[5] thermal convection 意为"热对流"。
[6] diamond-machined ceramics 意为"采用金刚石加工过的陶瓷"。

Lesson 46 Definitions and Terminology of Vibration

All matter—solid, liquid and gaseous—is capable of vibration, e.g. vibration of gases occurs in tail ducts of jet engines causing troublesome noise and sometimes fatigue cracks in the metal. A typical machine has many moving parts, each of which is a potential source of vibration or shock excitation.[1] Designers face the problem of compromising between an acceptable amount of vibration and noise, and costs involved in reducing excitation.

Vibration of a body is periodic change in position or displacement from a static equilibrium position. Associated with vibration are the interrelated physical quantities of acceleration, velocity and displacement—e.g. an unbalanced force causes acceleration ($a = F/m$) in a system which, by resisting, induces vibration as a response. The vibratory motion may be classified broadly as (a) transient; (b) continuing or steady-state; and (c) random.

Transient Vibration Transient vibrations die away and are usually associated with irregular disturbances, e.g. shock or impact forces, initial engagement of cutting tools—i.e. forces which do not repeat at regular intervals. Although transients are temporary components of vibrational motion, they can cause large amplitudes initially and consequent high stress but, in many cases, they are of short duration and can be ignored leaving only steady-state vibrations to be considered.

Steady-State Vibration Steady-state vibrations are often associated with the continuous operation of machinery and, although periodic, are not necessarily simple harmonic or sinusoidal. Since vibrations require energy to produce them, they reduce the efficiency of machines and mechanisms because of dissipation of energy, e.g. by friction and consequent heat transfer to surroundings, sound waves and noise, stress waves through frames and foundations, etc. Thus, steady-state vibrations always require a continuous energy-input to maintain them.

Random Vibration Random vibration is the term used for vibration which is not periodic, i.e. has no cyclic basis and is not regularly repetitive.

In the following paragraphs certain terms and definitions used in vibration analysis are made clear—several of which are probably known to engineering students already.

Period, Frequency and Amplitude A steady-state mechanical vibration is the motion of a system repeated after an interval of time known as the period (see Fig. 2.7). The number of vibrations per unit of time is called the frequency. The maximum displacement of any part of the system from its static-equilibrium position is the amplitude of the vibration of that part—the total travel being twice the amplitude. Thus, "amplitude" is not synonymous with "displacement" but is the maximum value of the displacement from the static-equilibrium position.

Free Vibration A free vibration occurs without any external force except gravity (see Fig. 46.1), and normally arises when an elastic system is displaced from a position of stable equilibrium and released, i.e. free vibration occurs under the action of restoring forces inherent in an elastic system, and natural frequency[2] is a property of the system.

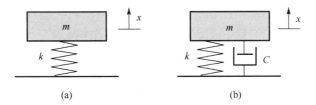

Figure 46.1　Free vibration (a) without damping and (b) with viscous damping

Forced Vibration　A forced vibration takes place under the excitation of an external force which is usually a function of time, e. g. in unbalanced rotating parts, imperfections in manufacture of gears. The frequency of forced vibration is that of the exciting or applied force, i. e. the frequency is a quantity independent of the natural frequency of the system.

Self-excited Vibration　Generally called chatter, self-excited vibration is caused by the interaction of the chip-removal process with the structure of the machine tool. Self-excited vibration typically begins with a disturbance in the cutting zone.

The most important type of self-excited vibration is regenerative chatter, which is caused when a tool is cutting a surface that has a roughness or geometric disturbances developed from the previous cut. Consequently, the depth of cut varies, and the resulting variations in the cutting force subject the tool to vibrations. The process continues repeatedly (hence the term regenerative). This type of vibration can be observed easily while driving a car over a rough road. Self-excited vibrations usually have a very high amplitude.

Resonance　Resonance occurs when the frequency of an applied force coincides with, or is near to a natural frequency of the system. In this critical condition, dangerously large amplitudes and stresses may occur in mechanical systems but, electrically, radio and television receivers are designed to respond to resonant frequencies. The calculation or estimation of natural frequencies is, therefore, of great importance in all types of vibrating systems. When resonance occurs in rotating shafts and spindles, the speed of rotation is known as the critical speed. Hence, the prediction and avoidance of a resonant condition in mechanisms is of vital importance since, in the absence of damping or other amplitude limiting devices, resonance is the condition at which a system gives an infinite response to a finite excitation.

Damping　Damping is the dissipation of energy from a vibrating system, and thus prevents excessive response. It is observed that a free vibration diminishes in amplitude with time and, hence, eventually ceases owing to some restraining or damping influence. Thus, if a vibration is to be sustained, the energy dissipated by damping must be replaced from an external source.

The dissipation is related in some way to the relative motion between the components or elements of the system, and is caused by frictional resistance of some sort, e. g. in structures, internal friction in material, and external friction caused by air or fluid resistance called viscous damping (see Fig. 46.1b) if the drag force is assumed proportional to the relative velocity between moving parts. One device assumed to give viscous damping is the dashpot. A dashpot cannot store energy but can only dissipate it.

Words and Expressions

terminology [ˌtəːmiˈnɔlədʒi] *n.* 术语（the technical terms used in a particular field）
duct [dʌkt] *n.* 导管，输送管
jet engine 喷气发动机
excitation [ˌeksiˈteiʃən] *n.* 激励（能使系统产生响应的某种外作用力或其他输入）
static equilibrium position 静平衡位置
interrelated [ˌintəriˈleitid] *a.* 相互关联或影响的（having a mutual or reciprocal relation）
transient vibration 瞬态振动（非稳态、非随机、持续时间短暂的振动）
response [risˈpɔns] *n.* 响应（激励作用引起的反应，系统的输出量）
steady-state vibration 稳态振动（连续的周期振动）
simple harmonic vibration 简谐振动（自变量为时间的正弦函数的振动）
random vibration 随机振动（在未来任意一个给定时刻，其瞬时值不能精确预知的振动）
impact force 碰撞力（接触物体间因碰撞引起的力）
sinusoidal [ˌsainəˈsɔidəl] *a.* 正弦的
dissipation [ˌdisiˈpeiʃən] *n.* 耗散，消耗
free vibration 自由振动（激励或约束去除后出现的振动）
forced vibration 受迫振动（由外激励作用导致系统产生的振动）
viscous [ˈviskəs] *a.* 黏性的，黏稠的
damping [ˈdæmpiŋ] *n.* 阻尼（能量随时间或距离而耗散）
disturbance [disˈtəːbəns] *n.* 扰动，扰乱，紊乱，故障
viscous damping 黏性阻尼（振动系统的运动受大小与运动速度成正比而方向相反的阻力所引起的能量损耗）
self-excited vibration 自激振动（机械加工过程中，加工系统本身引起的交变切削力反过来加强和维持系统自身振动的现象。因其振动频率较高，又称颤振）
chatter [ˈtʃætə] *n.* 颤振，自激振动
regenerative chatter 再生颤振，再生自激振动
rough road 起伏不平的道路（an uneven and not smooth road）
resonance [ˈrezənəns] *n.* 共振（外加力的频率与系统固有频率一致时所产生的振动）
critical speed 临界转速（系统产生共振的特征转速），又称共振转速（resonant speed）
in some way 在某种程度上（to some degree），以某种方式（in some manner）
dashpot [ˈdæʃˌpɔt] *n.* 阻尼器（a device for damping vibrations），缓冲器

Notes

[1] shock excitation 意为"冲击激励（作用于系统并产生机械冲击的激励）"。
[2] natural frequency 意为"固有频率（结构系统在自由振动下所具有的振动频率，the frequency at which a system tends to vibrate in the absence of any driving or damping force）"。

Lesson 47　Mechanical Vibrations

Engineering systems possessing mass and elasticity are capable of relative motion. If the motion of such systems repeats itself after a given interval of time, the motion is known as vibration. Vibration, in general, is a form of wasted energy and undesirable in many cases. This is particularly true in machinery; for it generates noise, breaks down parts, and transmits unwanted forces and movements to close-by objects.

To eliminate the adverse effects of most vibration, one of the approaches is to make a complete study of the equation of motion of the system in question.[1] The system is first simplified in terms of mass, spring, and dashpot (see Fig. 46.1b), which represent the body, the elasticity, and the friction of the system respectively. The equation of motion, then, expresses displacement as a function of time or will give the distance between any instantaneous position of the mass during its motion and the equilibrium position. The important property of the vibrating system, the natural frequency, is then obtained from the equation of motion.

In both the rectilinear and torsional types of vibration analysis, the period is the time required for a periodic motion to repeat itself; and the frequency is the number of cycles per unit time. Due to the similarities between rectilinear and torsional types of vibration, the discussion and analysis of one type apply equally well to the other type.[2]

Free vibration is the periodic motion observed as the system is displaced from its static equilibrium position. The forces acting are the spring force, the friction force, and the weight of the object. Due to the presence of friction, the vibration will diminish with time.

When external forces are acting on the system during its vibration motion, it is termed forced vibration. At forced vibration, the system will tend to vibrate at its own natural frequency as well as to follow the frequency of the excitation force. In the presence of friction, that portion of motion not sustained by the sinusoidal excitation force will gradually die out.[3] As a result, the system will vibrate at the frequency of the excitation force regardless of the initial conditions or the natural frequency of the system.

In actuality, most engineering systems during their vibratory motion encounter friction or resistance in the form of damping. Damping, in its various forms such as air damping, fluid friction, Coulomb dry friction, magnetic damping, internal damping, etc., will always slow down the motion, and cause the eventual dying out of the oscillation.[4]

Resonance occurs when the frequency of the excitation is equal to the natural frequency of the system. If a resonance is not damped, the displacement of the mass and hence the stretching of the spring element will tend towards infinity.[5] The spring component will fracture and for this reason undamped resonances must be avoided for the protection of equipment and instruments.

Long-term exposure of a mechanical system to vibrations of frequencies away from resonance can also cause damage through the mechanism of fatigue. Thus, if a mechanical component such as a spring is subjected to repetitive or cyclical applications of stress levels much lower than the ultimate strength, it will fracture after a large number of repetitions of this stress.[6] Indeed, if the number of cycles of stress is increased, the amplitude of the stress needed eventually to cause fracture becomes lower. The underlying mechanism in fatigue appears to be the gradual unzipping of intermolecular bonds starting from a defect or weakness in the molecular structure.[7]

Another undesirable side-effect of vibrating structures is the generation of audible noise. Such noise can be psychologically annoying to human beings working in the environment and can render normal voice communication impossible.[8] If extreme, noise can irreparably damage human hearing. The most thorough way of suppressing such noise is to reduce or eliminate the vibrations causing it.

Many systems can vibrate in more than one manner and direction. If a system is constrained so that is can vibrate in only one mode or manner, or if only one independent coordinate is required to specify completely the geometric location of the masses of the system in space, it is single-degree-of-freedom system.[9]

For an object in rectilinear motion (see Fig. 46.1), if its acceleration is always proportional to the displacement of the object from an equilibrium position and is directed toward the equilibrium position, then the object is said to have simple harmonic motion or simply SHM. SHM is the simplest form of periodic motion. The periodic motion of vibration, whether simple or complex, may be considered to be composed of SHM, or a number of SHM of various amplitudes and frequencies by means of a Fourier series.[10]

Many vibration problems in engineering are nonlinear in nature, i.e. the restoring forces are not proportional to the displacements and the damping forces are not proportional to the first power of the velocity.[11] This is also true when motions of appreciable magnitude in linear vibrating systems are of concern.

An essential difference in the study of nonlinear systems is that a general solution cannot be obtained by superposition as in the case of linear systems.[12] Moreover, in many instance, new phenomena occur in nonlinear systems, which cannot occur in linear systems.

In general, advanced mathematics is required for the analysis and solution of nonlinear systems due to their complicated configurations and nonlinear differential equations of motion.

Words and Expressions

  mechanical vibration 机械振动（描述机械系统运动或位置的量值绕其平衡位置做往复运动的现象）
  close-by 邻近的，附近的
  instantaneous [ˌinstən'teinjəs] *a.* 瞬间的，即刻的，即时的
  excitation force 激振力
  actuality [ˌæktju'æliti] *n.* 现实，实际情况

side effect 副作用（a secondary and usually adverse effect）
psychologically [ˌsaikəˈlɔdʒikəli] ad. 心理上地，心理学地，从心理学的角度
fatigue 疲劳（材料在低于其断裂应力的循环加载下，经一定循环次数后发生断裂的现象）
mechanism of fatigue 疲劳机理
irreparably [iˈrepərəbli] ad. 不能挽回地，不能恢复地
suppress [səˈpres] v. 抑制，使止住
nonlinear [ˈnɔnˈliniə] a. 非线性的
superposition [ˌsjuːpəpəˈziʃən] n. 重叠，重合，叠加

Notes

[1] equation of motion 意为"运动方程"。全句可译为：对于大部分振动而言，避免由其产生的有害影响的一种方式是全面地研究这个系统的运动方程。

[2] rectilinear and torsional vibration 意为"直线振动和扭转振动"。全句可译为：由于直线振动和扭转振动存在着相似性，对于一种类型的讨论和分析同样适用于另一种类型。

[3] sinusoidal excitation force 意为"正弦激振力"。全句可译为：在有摩擦存在的情况下，不是由正弦激振力支持的那部分振动将逐渐停止。

[4] magnetic damping, internal damping 意为"磁阻尼，内阻尼"。全句可译为：以诸如空气阻尼、流体摩擦、库仑干摩擦、磁阻尼、内阻尼等形式存在的阻尼将会使运动减慢，并且最终使振动停止。

[5] displacement of the mass 意为"质量的位移"。全句可译为：如果产生共振的系统是无阻尼的，那么质量的位移和弹簧的伸长量将趋于无穷大。

[6] repetitive or cyclical applications of stress 意为"重复或周期性应力的作用"。全句可译为：因为像弹簧这样的机械零件，在比其极限强度低得多的重复应力或周期性应力下，经过很多次的重复作用之后也会发生断裂。

[7] underlying mechanism 意为"基本理论"。全句可译为：疲劳方面的基本理论认为，分子之间的键是从分子结构有缺陷或薄弱的地方逐渐拉开的。

[8] 全句可译为：这种噪声会使处于这个环境中工作的人在精神上感到烦躁，并会使人不能以正常的声音进行交流。

[9] single-degree-of-freedom 意为"单自由度"。

[10] Fourier series 意为"傅里叶级数"。全段可译为：当一个物体做直线运动时（见图46.1），如果其加速度的大小始终与物体相对平衡位置的位移成正比，而且加速度的方向总是指向平衡位置，那么这个物体就是在做简谐运动或简称为SHM。简谐运动是一种最简单的周期性运动。周期性的振动，无论是简单的还是复杂的，都可以认为是由简谐运动组成的，或者是由用傅里叶级数表示的很多振幅不同和频率不同的简谐运动组成的。

[11] restoring force 意为"回复力（振动物体所受的总是指向平衡位置的合外力）"，first power 意为"一次方"。全句可译为：工程中的许多问题在其本质上是非线性的，也就是说，回复力并不与位移成正比，而且阻尼力也不与速度的一次方成正比。

[12] general solution 意为"（微分方程的）通解"。全句可译为：非线性系统与线性系统的本质差别是它的微分方程的通解不能像线性系统那样通过叠加来获得。

Lesson 48　Automated Assembly

Over the past decade, advances in automation technology have significantly boosted production capabilities throughout the world despite conflicting demands for increased quality, efficiency, and flexibility to accommodate smaller lot sizes and greater variety. [1]

Automated assembly solves the quantity and quality problems and gradually is being adapted to smaller lot sizes. However, one inherent aspect of automated assembly is the extremely rapid production of scrap when any one of production factors goes out of control and is not detected immediately. This is a major problem in high-volume production and in small-lot automated production as well, where even one off-spec unit means heavy financial loss because of the high value added at each step. Periodic or end-of-line inspection cannot prevent this, and continuous 100% manual inspection is both prohibitively expensive and notoriously prone to human error. [2]

Automated inspection can detect problems immediately, increase throughput and yield, cut downtime and improve process control. [3]

Vision systems today measure dimensions, count quantities, inspect for correct assembly and surface damage, and even compete with sensors in applications requiring fast, foolproof, and largely automatic changeover between related products. Vision systems commonly inspect marking or printing on containers and products, comparing each acquired image with a standard, and accepting or rejecting the product.

In most vision applications, correct lighting is critical to reliable operation. It often requires more detailed engineering effort than any other part of the vision system and presents difficult challenges to the machine designer. To avoid difficulties, locations for lamps, mirrors and other lighting elements should be considered carefully in the basic machine design.

Vision applications also depend heavily on the quality of the algorithms used for control and analysis. [4] While the basic camera and vision system determines the time to acquire an image, the algorithms determine the time to analyze and manipulate the data. The time to perform the analysis can exceed the acquisition time substantially, and may even require investment in multiple parallel inspection stations, more expensive faster vision systems, or other undesirable options.

Smaller vision cameras support the trend toward integrated inspection. For some simple applications, all necessary controls can be installs in the camera housing. For more complex applications, vision systems typically can control from one to four cameras. Even as vision systems have became more powerful and versatile, prices have steadily fallen, with vision systems suitable for integration now ranging from \$2,500 on the low end to more than \$50,000 for sophisticated, powerful systems that often include several cameras performing different complex inspection.

Only a few years ago, laser sensors were considered exotic, expensive device; however, lasers are now available for prices well below those of even simple vision systems.[5] Lasers offer key advantages of high speed, extreme accuracy and ease of access. Simple laser sensors give results almost instantaneously, and can keep up with virtually any assembly system.

The laser sensor's high speed also facilitates inspection of moving targets, often saving machine dwell positions, which reduces machine size and cost. Laser sensors can inspect to an accuracy of 2 or 3 microns and, unlike vision systems, lasers are essentially independent of lighting conditions.

Lasers have proven highly effective in on-line inspection of height differences and dimensions, accurate positioning, and detecting specified product faults.[6] Lasers also have proven uniquely effective in measuring the thickness of a moving part that is too thin for reliable measurement by other means—down to the level of 3 or 4 microns. The thickness of a part presented at a variable angle can still be measured accurately, because the laser spot is so small that the angle effect is virtually meaningless.

A variety of simple sensors can confirm completion of motions or myriad other factors. In recent years, sensor capabilities have virtually exploded, with sensor size decreasing drastically, accuracy improving and prices falling. Sensors have become economically justifiable as means of achieving very high yield, detecting malfunctions immediately, preventing equipment damage and increasing net throughput. Today's automatic assembly machine typically uses from dozens to hundreds of sensors.

Sensors can measure displacement to any desired accuracy. The linear variable differential transformer (see Fig. 74.1), which has been used for decades, is extremely accurate, highly reliable, relatively inexpensive, and can collect data for analysis and corrective action.[7] Noncontact optical, inductive, capacitive, and other types of sensors provide a wide range of inspection functions. Typical applications include detecting presence and position, and measuring displacement, speed, orientation, and temperature.[8] Creative application of simple, inexpensive sensors can solve complex problems economically.

Words and Expressions

automated assembly 自动装配（以机械的动作代替人工操作，自动地完成各种装配作业的生产过程）
out of control 失去控制的，失控
high-volume production 大量生产
small-lot production 小批量生产，小批生产
off-spec 质量不合格（failure to meet the prescribed specifications or standards）
end-of-line 终点
notoriously [nəu'tɔːriəsli] ad. 出了名地，众人皆知地
prone [prəun] a. 倾向于；有……倾向，易于
throughput ['θruːput] n. 生产量（output or production over a period of time），生产能力（the quantity or amount of raw material processed within a given time）
downtime ['dauntaim] n. 故障时间，停机时间

foolproof [fu:lpru:f] *a.* 不会被误用的，安全的（very well designed and easy to use so that it cannot fail and you cannot use it wrongly）

changeover ['tʃeindʒ'əuvə] *n.* （生产方法，装备等的）完全改变，转变，大变更（a conversion to a different purpose or from one system to another）

substantially [səb'stænʃəli] *ad.* 充分地

exotic [ig'zɔtik] *a.* 外来的（from another country），新奇的（very different or unusual）

dwell [dwel] *v.* 居住，占据

myriad ['miriəd] *n.* 无数；*a.* 无数的，种种的

malfunction [mæl'fʌŋkʃən] *n.* 故障（the state of something that functions wrongly or does not function at all），不能正常工作

noncontact [ˌnɔn'kɔntækt] *n.* 无触点，非接触

inductive [in'dʌktiv] *a.* 诱导的，感应的，电感的

capacitive [kə'pæsitiv] *a.* 电容的

orientation [ɔrien'teiʃən] *n.* 方向，方位，倾向性

Notes

[1] small lot size and greater variety 意为"小批量，多品种"。全句可译为：在过去的十年中，尽管为了适应小批量、多品种生产且又要面对提高质量、效率和柔性这些相互矛盾的要求，自动化技术的进步仍然极大地提高了世界各地的生产能力。

[2] prohibitively expensive 意为"费用高得令人望而却步"。全句可译为：定期检验或在生产线末端进行检验不能避免这种现象，而连续的100%人工检验既是费用高得令人难以接受，又是众所周知地易于出现人为错误。

[3] throughput and yield 意为"生产能力和产量"。全句可译为：自动检验能够立即发现问题，增加生产能力和产量，减少故障时间和改进过程控制。

[4] vision application 意为"视觉应用"。全句可译为：视觉应用也在很大程度上依赖用于控制和分析的算法的质量。

[5] lasers 这里指"laser sensors，即激光传感器"。全句可译为：仅仅几年前，激光传感器还被认为是新奇的、昂贵的装置。然而，现在激光传感器甚至比简单视觉系统的价格还要低得多。

[6] on-line inspection 意为"在线检测"。全句可译为：激光传感器已被证明在高度差、尺寸的在线检测、精确定位和发现产品的特定缺陷方面非常有效。

[7] linear variable differential transformer 意为"线性可变差动变压器，简称LVDT"。全句可译为：已被人们使用了几十年的线性可变差动变压器（见图74.1）的精度和可靠性都很高，而且比较便宜，能够为分析和校正工作收集数据。

[8] presence and position 意为"存在和位置"。全句可译为：典型的应用包括检测零件的存在和位置，以及测量位移、速度、方向和温度。

Lesson 49 Roles of Engineers in Manufacturing

Many engineers have as[1] their function the designing of products that are to be brought into reality through the processing or fabrication of materials. They are a key factor in the selection of materials and manufacturing processes. A design engineer, better than any other person, should know what he or she wants a design to accomplish. He knows what assumptions he has made about service loads and requirements, what service environment the product must withstand, and what appearance he wants the final product to have. In order to meet these requirements he must select and specify the material(s) to be used. In most cases, in order to utilize the material and to enable the product to have the desired form, he knows that certain manufacturing processes will have to be employed. In many instances, the selection of a specific material may dictate what processing must be used. At the same time, when certain processes are to be used, the design may have to be modified in order for the process to be utilized effectively and economically. Certain dimensional tolerances can dictate the processing.

In any case, in the sequence of converting the design into reality, such decisions must be made by someone. In most instances they can be made most effectively at the design stage, by the designer if he has a reasonably adequate knowledge concerning materials and manufacturing processes. Otherwise, decisions may be made that will detract from[2] the effectiveness of the product, or the product may be needlessly costly. It is thus apparent that design engineers are a vital factor in the manufacturing process, and it is indeed a blessing to the company if they can design for production—that is, for efficient production.

Manufacturing engineers select and coordinate specific processes and equipment to be used, or supervise and manage their use. Some of them design special tooling that is used so that general purpose machine tools can be utilized in producing specific products. These engineers must have a broad knowledge of machine and process capabilities and of materials, so that desired operations can be done effectively and efficiently without overloading or damaging machines and without adversely affecting the materials being processed. These manufacturing engineers also play an important role in manufacturing.

A relatively small group of engineers design the machines and equipment used in manufacturing. They obviously are design engineers and, relative to their products, they have the same concerns of the interrelationship of design, materials, and manufacturing processes. However, they have an even greater concern regarding the properties of the materials that their machines are going to process and the interreaction of the materials and the machines.

Still another group of engineers—the materials engineers—devote their major efforts toward developing new and better materials. They, too, must be concerned with how these materials can be processed and with the effects the processing will have on the properties of the materials.

Although their roles may be quite different, it is apparent that a large proportion of engineers must concern themselves with the interrelationship between materials and manufacturing processes. Low-cost manufacture does not just happen. There is a close and interdependent relationship between the design of a product, selection of materials, selection of processes and equipment, and tooling selection and design. Each of these steps must be carefully considered, planned, and coordinated before manufacturing starts. This lead time[3], particularly for complicated products, may take months, even years, and the expenditure of large amount of money may be involved. Typically, the lead time for a completely new model of an automobile is about 2 years, for a modern aircraft it may be 4 years.

With the advent of computers and machines that can be controlled by computers (see Fig. 69.1), we are entering a new era of production planning[4]. The integration of the design function and the manufacturing function through the computer (see Figs. 49.1 and 49.2) is called CAD/CAM (computer aided design/computer aided manufacturing). The design is used to determine the manufacturing process planning and the programming information for the manufacturing processes themselves. Detailed drawings can also be made from the database used for the design and manufacture, and programs can be generated to make the parts as needed. In addition, extensive computer aided testing and inspection (CATI) of the manufactured parts is taking place. There is no doubt that this trend will continue at ever-accelerating[5] rates as computers become cheaper and smarter.

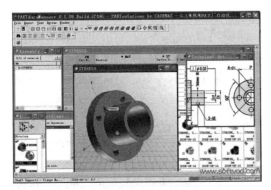 

Figure 49.1　3D CAD models

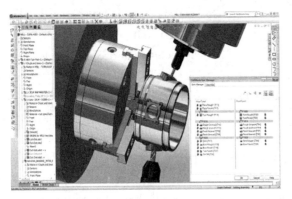 

Figure 49.2　Machining simulation using CAM softwares

Words and Expressions

fabrication [ˌfæbriˈkeiʃən] *n.* 制造，生产，加工，装配
assumption [əˈsʌmpʃən] *n.* 假设，设想，前提
dictate [dikˈteit] *v.* 指示，命令，规定，限定，确定
procedure [prəˈsiːdʒə] *n.* 程序，工序，生产过程，方法，措施
sequence [ˈsiːkwəns] *n.* 顺序，程序，工序，序列，连续
in any case 无论如何，不管怎样（no matter what happens）
detract [diˈtrækt] *v.* 贬低，减损（to undergo reduction in value, importance, or quality）
coordinate [kəuˈɔːdinit] *n.* 坐标，一致；*v.* 使协调，调整，整理
tooling [ˈtuːliŋ] *n.* 工艺装备（产品制造过程中所用的各种工具的总称，包括刀具、夹具、模具、量具、检具、辅具和钳工工具等），简称"工装"
process capability 工序能力
effectively [iˈfektivli] *ad.* 有效地（producing the intended result）
efficiently 高效地（producing a satisfactory result without wasting time or energy）
general purpose machine tool 通用机床（使用范围较广，可加工多种工件、完成多种工序的机床）
interdependent [ˌintədiˈpendənt] *a.* 相互依赖的，相互影响的，相互关联的
interrelationship [ˌintəriˈleiʃənʃip] *n.* 相互关系，相互联系，相互影响
advent [ˈædvənt] *n.* 到来，出现，来临
with the advent of... 随着……的到来（出现）
integration [ˌintiˈgreiʃən] *n.* 积分，集成，综合，整体化
database [ˈdeitəbeis] *n.* 数据库（a large amount of information stored in a computer system in such a way that it can be easily looked at or changed）
accomplish [əˈkɔmpliʃ] *v.* 完成，达到目的，实行
modify [ˈmɔdifai] *v.* 变更，改变，改进，改善，改良，调整
needlessly [ˈniːdlisli] *ad.* 不需要地，无用地，多余地
overload [ˌəuvəˈləud] *v.*；*n.* （使）超载，超重，过负荷，使负担过重
expenditure [iksˈpenditʃə] *n.* 消费，支出，经费，开支，消费额
inspection [inˈspekʃən] *n.* 检验（a checking or testing of sth against established standards）

Notes

[1] have as 意为"把……作为"。
[2] detract from 意为"有损于（to impact someone or something negatively）"。
[3] lead time 意为"前置时间，从产品设计到生产所需要的时间（the time between the design of a product and its production）"。
[4] production planning 意为"生产规划（the process of deciding how a product will be manufactured before the manufacturing process begins）"。
[5] ever-accelerating 意为"不断加速的"。

Lesson 50 Manufacturing Enterprises

A manufacturing enterprise is a factory-based, profit-making organization. Business enterprises, also called companies, firms, and corporations, can be small or large. They have in common a willingness to assume risks. [1] The payoff for these risks is profit, which is the amount of money resulting when revenues exceed costs. Profit is an excess that allows the payment of dividends to owners and shareholders, the purchase of new equipment and plants, and the payment of taxes. [2] If dividends, the rent on invested capital and money, are not paid, it would lessen the faith of investors and jeopardize a source of money for growth.

Manufacturing is unable to make everything that everyone wants. Firms are free to decide which products they will manufacture. People cannot buy all the products they desire, but each buyer is free to select what he or she will buy. Both producers and the buyers have a freedom of choice.

Consumer demand serves as the stimulant to encourage business to provide products. [3] Materials, which are the minerals of nature, such as coal, ores, hydrocarbons, and many more, are converted into these products. It takes financing and money, either by loans from banks or capital investments from stockholders or from plowback of profit into the business, to sustain this activity. Working capital is money used to buy materials and pay employees. Fixed capital is the money for tools, machines, and factory buildings. A manufacturing enterprise needs money for these and other requirements.

Energy is an important input to manufacturing. Energy exists in many different forms, such as electricity, compressed air, steam, gas, or coal.

Management provides planning, organization, direction, control, and leadership of the business enterprise to make it productive and profitable. Managers have responsibilities to the owners, employees, customers, general public, and the enterprise itself. The business enterprise must make a profit; otherwise it will fail.

The design step consists of creating plans for products so that they are attractive, perform well, and give service at low cost. Manufactured products are designed before they are made.

The processes needed to manufacture a product must be designed in great detail. General plans for the processes are recognized during the design stage. Now the techniques of manufacturing engineering are used. The best combination of machines, processes, and people is selected to satisfy the objectives of the firm, shareholders, employees, and customers.

The output of manufacturing system is a product. Look around you. Products are everywhere. Goods can be classified into consumer and capital kinds. [4] Consumer goods are those products that people buy for their personal consumption such as food or cars. Capital goods are products purchased by manufacturing firms to make the consumer products. [5] Machine tools, computer-controlled robots, and plant facilities themselves are examples.

The cost of a product depends on raw materials, production costs for machines and labor, management and sales, and overhead. Machine and labor costs are inexorably related and make up, along with raw materials expenditures, the bulk of production costs. When a material is chosen, the process, including the machine, is frequently specified. Alternatively, if a machine is available, the raw material that can be processed on that machine may be utilized. One could say that the purpose of economical production is to produce a product at a profit. This infers that the cost must be acceptable and competitive; also, a demand for the product must exist or must be created.

Since the first use of machine tools, there has been a gradual trend toward making machines more efficient by combining operations and by transferring more skill to the machine, thus reducing time and labor. To meet these needs, machine tools have become complex both in design and in control. Automatic features have been built into many machines, and some are completely automatic. This technical development has made it possible to attain the high production rate with low labor cost that is essential for any society wishing to enjoy high living standards. Computer-aided design and manufacturing are significant steps of progress.

Along with the development of production machines, the quality in manufacturing must be maintained. Quality and accuracy in manufacturing operations demand that dimensional control be maintained to provide parts that are interchangeable and give the best operating service. For mass production, any one of a quantity of parts must fit in a given assembly.[6] A product made of interchangeable parts is quickly assembled, lower in cost, and easily serviced. To maintain this dimensional control, appropriate inspection facilities must be provided.

To produce parts of greater accuracy, more expensive machine tools and operations are necessary, more highly skilled labor is required, and rejected parts may be more numerous. Products should not be designed with greater accuracy than the service requirements demand. A good design includes consideration of a finishing or coating operation, because a product is often judged for appearance as well as function and operation. Many products, such as those made from colored plastics or other special materials are more saleable because of appearance. In most cases the function of the part is the deciding factor. This is particularly true where great strength, wear, corrosion resistance, or weight limitations are encountered.

People are vital resources for a manufacturing enterprise. People perform work or labor. The work may be mental, physical, or a mixture of both.[7] It is evident that educational requirements for human resources have been increasing. The skill requirements of the operator, technician, engineer, and researcher have become more complex. Special skills and training are necessary as manufacturing moves into computer-aided design and manufacturing. Many schools, colleges, and universities are adding courses in production or manufacturing.

Words and Expressions

payoff ['peiɔːf]   *n.* 回报（reward），（一系列行动的）结果（the result of a set of actions）
revenue ['revinjuː]   *n.* 收入（income），收益（yield from property or investment）
dividend   股息（an amount of a company's profits that is paid to the owners of its stock）

shareholder [ˈʃɛəhəuldə]　n. 股票持有人（one that owns or holds a share or shares of stock）
jeopardize [ˈdʒepədaiz]　v. 危及，使受危害，使遭遇危险
stimulant [ˈstimjulənt]　n. 刺激性的；n. 兴奋剂，刺激物
hydrocarbon [ˈhaidrəuˈkɑːbən]　n. 烃类，碳氢化合物
capital [ˈkæpitl]　n. 首都，资本，资金；a. 基本的，主要的，资本的
plowback [ˈplaubæk]　n. 利润再投资（reinvestment of profits in the business that earned them）
working capital　流动资金，营运资金，周转资本
fixed capital　固定资本
drafting [ˈdrɑːftiŋ]　n. 起草，绘图
infer [inˈfəː]　v. 推论，推断，猜想，意味着，指出，暗示
gradual [ˈgrædjuəl]　a. 逐渐的，渐进的，渐变的，平缓的
dimensional [diˈmenʃənl]　a. ……维的，……度的，量纲的
overhead [ˈəuvəhed]　n. 经常费用，管理费用，杂项
general plan　总体规划
consumer goods　消费品（things purchased by customers and will be used right away）
capital goods　资本货物（machines and tools used in the production of other goods）
plant　工厂，车间（a factory or workshop for the manufacture of a particular product）
plant facilities　工厂设备，工厂设施
inexorably [inˈeksərəbli]　ad. 不可逆转地，不可阻挡地（in a way that continues without any possibility of being stopped）
imake up　组成，构成，补充
the bulk of something　大多数，大部分（the majority or largest part of something）

Notes

［1］assume risk 意为"承担风险"。全句可译为：它们共同的特点是愿意承担风险。

［2］excess 意为"超出额，超出量"，这里指"收入超出成本的数额"。全句可译为：利润是收入超出成本的数额，可以用来向业主和股票持有人支付股息红利，购买新的设备和交税。

［3］consumer demand 意为"消费要求，消费者需求，市场需求"。全句可译为：消费需求可以作为激励企业提供产品的促进因素。

［4］consumer and capital kind 这里指"consumer goods and capital goods"，意为"消费品和资本货物"。全句可译为：商品可以分为消费品和资本货物。

［5］全句可译为：资本货物是制造企业为制造消费品而购买的产品。

［6］fit in 意为"装配到……中"。全句可译为：在大量生产中，一批零件中的任何一个必须能够被装配到指定的装置中。

［7］mixture of both 意为"二者兼而有之"。全句可译为：工作可以是脑力劳动或体力劳动，也可以二者兼而有之。

Lesson 51 Careers in Manufacturing

Manufacturing gives jobs to people and makes the products consumers need. Manufacturing companies hire people with many types of skills and talents to help the companies compete well in the world of business. Manufacturing companies often divide themselves into several departments. A department is a small part of the company.

Some of the more common departments found in large manufacturing companies include management, engineering, production, marketing, finance, and human resources. People with different talents, skills, and educational training can find many different types of jobs and careers in each of these departments.

Management Department For a manufacturing company to be successful, all of the departments must work together toward one goal. Management's job is to make sure all the departments work together. Workers in management, called managers, make company policies and then make sure the other departments follow these policies.

Engineering Department The engineering department designs, produces, and tests the first model of a new product. [1] These first models are called prototypes. In the engineering department, scientists conduct tests and experiments on new materials like plastics, ceramics, and metal alloys. Technicians also work in the engineering department. A technician is usually a skilled and experienced design or production worker. Only the most experienced and skilled technicians can work in the engineering department as drafters, designers, and machinists. Engineers are the problem-solvers when it comes to designing a new product or the system for making that product.

Many engineering departments have a research and development team. [2] This team is made up of the best engineers, technicians, and production workers. Their job is to create and test new product ideas.

Production Department The production department sets up and uses the tools and machines to make the product. A management worker, the supervisor, is normally in charge of this department. The supervisor tells the production workers what to do and then makes sure they do it. Supervisors also train new workers and encourage safe work habits. Frequently, supervisors begin as production workers.

Production jobs use skilled, semiskilled, and unskilled labor. Skilled workers design and make specialized tools, jigs, fixtures, and quality control devices used to manufacture the product. [3] They also adjust and maintain equipment, fixtures, or quality control devices when they break down. Because their jobs require precision skills, these workers usually have years of manufacturing experience and training. Skilled workers also set up and operate a variety of machines, including computer-controlled devices such as lathes, milling machines, and even robots. Semiskilled workers run machines and use the special tools, jigs, and fixtures that are made and set up by skilled workers. Unskilled workers perform physically demanding, routine jobs. They include material handlers, helpers, and assemblers. Unskilled workers do not have the skills to run machinery.

Marketing Department The marketing department conducts market research and advertises, packages, and sells the product. Market researchers do surveys to see what consumers want, if they like their product, and what price they will pay. They also compare their products to those of competitors.

Advertisers decide which media will be used to publicize the new product.[4] Artists, photographers, and writers prepare advertising campaigns for print media like newspapers, magazines, and direct mailing or for broadcast media like radio and television. The advertising department decides on a name and a trademark for the product as well as a theme or motto for the campaign.

Package designers and artists match the product to the best packaging system. A good package is easy to make, adds little to the product cost, displays the best product features, and attracts consumer attention. Package designers want to make their product stand out from the crowd. They use bright colors as well as unusual package shapes, sizes, and trademarks.

Salespeople identify ways to get the manufactured product into consumers' hands. The consumer could be either an individual or another industry. When an individual buys a manufactured product in a store, it is called retail sales. When another industry buys a manufactured product, it is called industrial sales. The job of salespeople, whether in retail or industrial sales, is to point out the product's best features as well as how it can be sold or used.

Finance Department Financial record-keeping is very important to manufacturing companies. The main objective of a manufacturing company is to make a profit on the product they produce and sell. The finance department keeps track of all income and expenditures. Expenditures include payroll; materials costs; utility costs for lighting, heating, air conditioning, and machines; and the costs of buying and maintaining tools, equipment, and machines. Accountants, bookkeepers, and other record-keepers work in the finance department to make sure the company develops and follows a budget.

Human Resources Department The human resources department is responsible for the following:

1. Recruitment—Recruiters identify new workers to fill vacant positions in the manufacturing company. They must match workers to the jobs by examining job applications and by conducting interviews and performance testing.[5]

2. Training —Trainers are teachers in industry. They prepare new workers for their jobs or teach experienced workers how to use new equipment.

3. Public Relations—Public relations people keep the general public aware of the most recent developments in the company.[6]

It is important that human resources workers get along well with other people. Their job is working with people to make the manufacturing company run smoothly and efficiently.

Words and Expressions

prototype　样机，原型机（在研制过程中按设计图样制造的第一批供试验用的机械）

supervisor ['sjuːpəvaizə]　n. 主管（one who is in charge of a particular department）
skilled worker　技术工人，熟练技工（a worker who has acquired special skills）
semiskilled worker　半技术工人，半熟练工人
unskilled worker　非技术工人，非熟练工人
physically demanding job　体力工作，需要体力的工作
material handler　物料管理员，仓库管理员
helper ['helpə]　n. 辅助工人（工业企业中为基本生产服务，从事辅助性工作的工人）
assembler [ə'semblə]　n. 装配工人（one that assembles）
marketing department　市场营销部门，营销部，市场部
advertising campaign　广告宣传活动
print media　平面媒体（以纸张为载体发布新闻或资讯的媒体）
trademark ['treidmɑːk]　n. 商标，标志，品种
motto ['mɔtəu]　n. 座右铭，格言
salespeople ['seilzˌpiːpl]　n. 销售人员（persons employed to sell goods or services）
retail sales　零售（the sale of consumer goods, or final goods, by businesses to end consumers）
industrial sales　工业销售（工业企业通过一定方式，将产品提供给购货单位的经济活动）
stand out from the crowd　脱颖而出，引人注目，与众不同
record-keeping　记账（process of recording transactions and events in an accounting system）
accountant　会计（someone whose job is to keep the financial records of a business or person）
bookkeeper　记账员，簿记员（a person who records the accounts or transactions of a business）
human resources department　人力资源部，人事部
recruitment [ri'kruːtmənt]　n. 补充，招收，充实
vacant ['veikənt]　a. 空位，未占的，空职的

Notes

[1] engineering department 意为"工程部门，工程设计部门"。model 这里指"样机，即为验证设计或方案的合理性和正确性，或者生产的可行性而制作的样品"。这两句可译为：工程设计部门对新产品的第一台样机进行设计、生产和试验。这些第一批的样机称为原型机。

[2] research and development 意为"研究和开发，研制"，set up 意为"调试"。这段与下一段第一句可译为：在许多工程部门中都有一个产品的研究与开发小组。这个小组由最好的工程师、技师和生产工人组成。他们的工作是提出新的产品创意，并对其进行测试。

　　生产部门　生产部门对工具和机器进行调试，并使用它们制造产品。

[3] quality control device 意为"质量检测装置"。全句可译为：技术工人设计和制造生产产品所需要的专用刀具、夹具和质量检测装置。

[4] publicize the new product 意为"对新产品的宣传和推销"。全句可译为：广告负责人员决定采用哪一种媒体来对新产品进行宣传和推销。

[5] match to 意为"与……相配"。全句可译为：他们必须通过分析工作申请表，以及进行面试和能力测试来为工人分配合适的工作。

[6] public relations 意为"公共关系，公关"。全句可译为：公关人员应该使公众知道公司最近的发展情况。

Lesson 52 Manufacturing Research Centers at U. S. Universities (*Reading Material*)

A variety of research centers devoted to manufacturing have been established at numerous universities across the country in an effort to maintain U. S. leadership in manufacturing technology. For instance, the University of Illinois has initiated a program in manufacturing research and education to serve industry's needs. This program is being carried out by the Manufacturing Research Center (MRC). The MRC is an industry-driven center of excellence in manufacturing research established to foster collaborative research initiatives between the university and industry. The Center conducts collaborative research projects with its member companies, educates students in the problems and issues of manufacturing, and provides a broader access for its members to the laboratories and programs of the university.

The research concentrates on four engineering sectors: electronics, automotive/vehicular, chemical, and machine tool. The Center funds two types of projects. Member companies designate that one-half of the funds it contributes be applied to research of a specific interest to that company. The results of the research from these company-designated projects are available on an exclusive basis to the company. The remaining funds from each company are employed collectively to support center-designated projects, the results of which are shared by all of the participating companies.

The MRC is also concentrating on robotics technology and manufacturing systems. The latter includes projects related to manufacturing systems design and control, integrated design and manufacturing, and management/systems integration. Robotics projects are in the areas of intelligent robotics, robotic vision, and path planning/control of robotic devices.

The Center for Manufacturing Engineering Systems (CMES) established at the New Jersey Institute of Technology (NJIT) has a similar goal as MRC—that of strengthening New Jersey's industrial base. Dedicated to the development and dissemination of advanced manufacturing knowledge, CMES is designed to serve industrial and government organizations. Activities focus on research and technology transfer. A particular emphasis of CMES is to aid small and medium sized businesses in adapting advanced manufacturing technology to their needs. Funding has been received from a variety of sources, including the New Jersey Commission on Science and Technology, NJIT, and industry.

CMES-supported projects generally fall into one of three categories. The first project type, the most common, is the collaborative long-term project with industry, involving the application of computer-integrated manufacturing technology. Most of these projects involve small to medium-sized firms and serve both a technology transfer and a research and development function. The second type of project is the short-duration technology extension, collaborative project which aids industry in the selection and application of advanced manufacturing technology. For instance, a company seeking to purchase a CAM software package may come to CMES for advice about which software is most appropriate for this operation. The third project type involves typical university-based research on major manufacturing problems, with support from government and industry.

A substantial by-product of the CMES' activities is the provision of hands-on educational experience for the students who work on Center projects. Many of these students are enrolled in either the undergraduate and graduate degree programs in manufacturing engineering—the master's degree is the only such degree offered in New Jersey. Students earning this degree may select from several specializations, all of which share a common core of courses and call for completion of a thesis or project. Thesis and project topics, which usually involve applied research, are developed based on input from industrial advisors who keep NJIT faculty apprised of industrial priorities. NJIT has also teamed with community colleges to offer a new degree in manufacturing technology, a two-year CIM associate degree in applied science. Many students in this program complete part of their course work at NJIT, using the university's advanced laboratories.

The Center for Manufacturing Productivity (CMP) is a collaboration between the UMass College of Engineering and the School of Management to provide a comprehensive program of assistance to small and medium-sized manufacturing companies in Western Massachusetts. CMP was founded in 1991 with a grant from the U. S. Small Business Association.

The CMP provides manufacturing businesses with comprehensive and integrated assistance in management and engineering through the expertise of UMass faculty in the School of Management and College of Engineering. Through on-site consultation and evaluation, the CMP staff, working with company management, will develop a detailed and cost-effective program specific to each company's needs. Among other strategies, CMP works with businesses to convert from defense to commercial manufacturing, develop computer-aided technologies and systems, modernize management structures, adapt to meet overseas product standards and adequately address export requirements of foreign markets, better utilize human resources, and improve production and control.

CMP offers a wealth of University expertise in engineering and management including CAD/CAM, strategic planning, design for manufacturing, market analysis, ISO9000, machining and grinding, total quality management, production process control, human resource management, and software development.

Words and Expressions

initiate [iˈniʃieit] v. 开始，创始，起始，着手，引起；n. 首创精神，主动，积极性
industry-driven 产业驱动型，工业驱动型
project 项目 (planned work that has a specific purpose and usually requires a lot of time)
collaborative research 合作研究 (research project that is carried out by at least two people)
member company 会员企业，会员公司

concentrate ['kɔnsentreit] v. 集中（to bring or direct toward a common center or objective）
designated project 指定项目
exclusive [iks'klu:siv] a. 排他的，专用的（only to be used by one particular person or group）
on an exclusive basis 独家（报道、供应）
intelligent robotics 智能机器人技术
medium-sized firm 中等规模的公司，中型企业
technology transfer 技术转移（拥有技术的一方将现有技术有偿转让给他人的行为）
technology extension 技术推广
short-duration 短期的
collaborative project 合作项目，合作课题
by-product ['baiprɔdʌkt] n. 副产品（something produced in addition to the main product）
hands-on experience 实践经验（knowledge that someone gets from doing something）
undergraduate and graduate degree 本科生和研究生学位
specialization 专业方向（a focused area of study attached to a specific major）
call for 要求，需要（to need, require, or demand something）
common core of courses 共同的核心课程（common core courses）
thesis ['θi:sis] n. 毕业论文，学位论文（a long piece of writing on a particular subject that is done to earn a degree at a university）
community college 社区学院，社区大学，两年制的社区大学
associate degree (or associate's degree) 大专学位，副学士学位
CIM＝Computer Integrated Manufacturing 计算机集成制造
comprehensive [,kɔmpri'hensiv] a. 广泛的（so large in scope or content as to include much）
integrated ['intigreitid] a. 综合的，完整的
UMass (University of Massachusetts) 马萨诸塞大学，俗称麻省大学
adapt 适合，适应（to change sth so that it functions better or is better suited for a purpose）
on-site [ɔn'sait] a. 现场的（at the place where a business or activity happens）
strategic [strə'ti:dʒik] planning 战略规划（制定组织的长期目标）
design for manufacturing 面向制造的设计（指产品设计需要满足产品制造的要求，具有良好的可制造性，使得产品以最低的成本、最短的时间、最高的质量制造出来）
total quality management 全面质量管理，简称TQM
production process control 生产过程控制（对直接或间接影响产品质量的生产、安装和服务过程所采取的作业技术和生产过程的分析、诊断和监控）
human resource management 人力资源管理
software development 软件开发
convert from defense to commercial manufacturing 将生产军工产品转为生产民用产品，军转民

Lesson 53 Developments in Manufacturing Technology
(*Reading Material*)

A manufacturing business is a production business that is intended to make a profit. Profits result from the product development cycle, quality, reliability, pricing, public image, productivity, team work, and so on. The primary objective of a manufacturing business is to convert raw materials into quality goods that have value in the marketplace and that are sold at competitive prices. The goods are produced through good management techniques in the use of such resources as capital, human labor, materials, equipment, and energy involving many activities and operations. Manufacturing may be defined as a series of interrelated activities and operations involving the design, selection of materials, planning, production, quality assurance, management, and marketing of consumable and durable goods. The interaction among these activities and operations form a total manufacturing system. A system is an organized collection of human resources; machines, tools, and equipment resources; financial resources; and methods required to accomplish a set of specific functions. Many processes are used in meeting the primary objective of a manufacturing system in transforming or converting a set of input parameters of new materials and shapes into an output of proper size, configuration, and performance according the specification for it.

People continue to seek ways to produce devices that will improve their standard of living, meet their basic needs, and control their environment. Such improved methods create demands that translate to larger quantities of devices that are produced more quickly than previously and that are of better quality at a lower cost per item.

The construction and application of simple machines for production started in Europe around 1770. These developments moved production from the home to factories, marking the beginning of the Industrial Revolution. Mechanization signified the movement from making products by hand in the home to making products by machines in factories.

Mechanization also created a system of mass production, which placed a demand on machines to duplicate parts with a high degree of accuracy. This resulted in a need for more accurate measuring tools, improved measuring techniques, and standards to help manufacturers make interchangeable parts. Fixed automation mechanisms and transfer lines of modular machine were major results of mass production. A transfer line is an organization of manufacturing facilities for faster output and shorter production time. Mass production was the first example of automated production.

Fixed automation gave way to machine tools with simple automated controls. This type of controller operates automatically to regulate a controlled variable or system. Advancement in controller technology opened the new era of automation, called programmable automation.

Programmable automation is designed to accommodate changes in a product. A new technique in automation, numerical control (NC), developed around 1952. Numerical control is based on digital computer principles, a form of programmable automation that controls manufacturing processes by numbers, letter, and symbols. Advances in computer technology extended NC to direct numerical control (DNC), computer numerical control (CNC), graphical numerical control (GNC), and voice numerical control (VNC).

Numerical control caused a revolution in the manufacture of metal parts. The success of NC led to a number of extensions such as adaptive control (AC) and industrial robots. Adaptive control determines the proper speeds and feeds during machining as a function of variations in such factors as hardness of work material, width or depth of cut, and so on. It denotes a control system that measures certain output process variables and uses these to control the speed and feed of the machine tool. Typical process variables used in AC machining systems are spindle deflection or force, torque, cutting temperature, vibration amplitude, and horsepower. Industrial robots started playing a major role in manufacturing during the late 1970s. Initially, robots were used for material handing. Today's robotic technology, however, has been developed to such an extent that robots are used to perform many high-level tasks in manufacturing.

Computers are being given an increasingly important role in manufacturing systems. A computer's ability to receive and handle large amounts of data, coupled with the fast processing speed, makes a system approach indispensable. The use of computers in manufacturing is now coming of age. Computer application in manufacturing production is typically referred to as computer-aided manufacturing (CAM). It is built on the foundation of such systems as NC, AC, robotics, automated guided vehicle (see Fig. 53.1), and flexible manufacturing system (FMS).

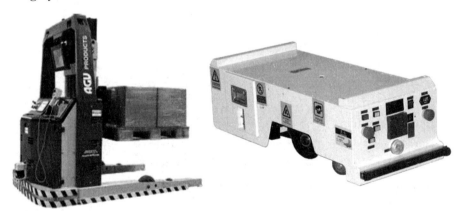

Figure 53.1 Automated guided vehicles

Computer-aided manufacturing is the effective use of computer technology in the planning, management, control, and operation of a manufacturing production facility through either direct or indirect computer interface with physical and human resources of the company. Computers play important roles in CAM systems. They integrate manufacturing data into a common database. Database management concepts are applied to CAM operations to speed up data access and to ensure that all users work from a common design. Under the CAM definition, computer applications include such systems as inventory control, scheduling, machine monitoring, and management information. These applications are primarily for transferring, interpreting, and keeping track of manufacturing data.

Words and Expressions

development cycle　开发周期

public image　公众形象（the opinion that the public has about a person or an organization）

quality　优质的，高质量的（very good or excellent; of or having superior quality）

quality goods　优质产品，高质量的商品（high-quality goods）

productivity [ˌprɔdʌkˈtiviti]　n. 生产率（the rate at which goods or services are produced）

quality assurance　质量保证（指为使人们确信某一产品、过程或服务的质量所必需的全部有计划有组织的活动）

mechanization [ˌmekənaiˈzeiʃən]　n. 机械化（the act or process of causing a task to be performed by machinery）

fixed automation　固定式自动化，刚性自动化

transfer line of modular machine　组合机床自动线（由若干台组合机床及其他辅助设备组成的自动生产线），亦为 automatic production line of modular machine

modular machine　组合机床（以系列化、标准化的通用部件为基础，配以少量的专用部件组成的专用机床），亦为 modular machine tool

programmable automation　可编程自动化

direct digital control　直接数字控制

graphical numerical control　图形数控

adaptive control　自适应控制（无论外界是否发生巨大变化或系统是否产生不确定性，控制系统都能自行调整参数或产生控制作用，使系统仍能按某一性能指标运行在最佳状态的一种控制方法）

indispensable [ˌindiˈspensəbl]　n. 不可缺少之物；a. 不可缺少的，绝对必要的

come of age　成年，成熟（something has reached its full successful development）

horsepower [ˈhɔːsˌpauə]　n. 马力（a unit of power in the U.S., equal to 746 watts）

automated guided vehicles (AGV)　自动导引车（沿标记或外部引导命令指示的，沿预设路径移动的移动平台，一般应用于工厂中）

flexile manufacturing system　柔性制造系统（在成组技术的基础上，以多台数控机床或数组柔性制造单元为核心，通过自动化物流系统将其联结，统一由主控计算机和相关软件进行控制和管理，组成多品种变批量和混流方式生产的自动化制造系统），简称 FMS

physical resources　物质资源（the material assets that a business owns, including buildings, materials, manufacturing equipment and office furniture）

inventory control　库存控制

scheduling [ˈʃedjuːliŋ]　n. 制订计划，进度安排（setting an order and time for planned events）

Lesson 54 Mechanical Engineering and Mechanical Engineers

The field of mechanical engineering is concerned with machine components, the properties of forces, materials, energy, and motion, and with the application of such knowledge to devise new machines and products that improve society and people's lives. [1] Mechanical engineering combines creativity, knowledge and analytical tools to complete the difficult task of turning an idea into reality.

Mechanical engineers analyze their work using the principles of motion, energy, and force — ensuring that designs function safely, efficiently, and reliably, all at a competitive cost. [2] Mechanical engineers research, develop, design, manufacture and test tools, machines, engines, and other mechanical devices. Mechanical engineers are known for working on a wide range of products and machines. Virtually every product in modern life has probably been touched in some way by mechanical engineers. This includes solving today's problems and creating future solutions in transportation, energy, space exploration, and more.

Mechanical engineering is not all about numbers, calculations, and machines. The profession is driven by the desire to advance society through technology. Some of mechanical engineering's major achievements in the twentieth century recognized in a survey conducted by ASME are the following: [3]

1. The automobile. The development and commercialization of the automobile were judged by mechanical engineers as the profession's most significant achievement in the twentieth century. Two factors responsible for the growth of automotive technology have been high-power lightweight engines and efficient mass production processes. [4]

Henry Ford pioneered the techniques of assembly-line mass production, which enabled consumers from across the economic spectrum to purchase and own automobiles. [5] The automotive industry, in turn, has grown to become a key component of the world economy, and it has also created jobs in the machine tool and raw materials industries.

2. Power generation equipment. Mechanical engineers design, test, and manufacture power generation equipment such as internal combustion engines and gas turbines. Abundant and inexpensive electrical energy is an important factor behind economic growth and prosperity, and the distribution of electricity has improved the standard of living for billions of people across the globe. In the twentieth century, economies and societies changed significantly as electricity was produced and routed to homes, businesses, and factories. [6]

3. Agricultural mechanization. The automation of farm equipment began with tractors and with the introduction of the combine harvester (Fig. 54.1), which simplified grain harvesting. [7] Other advances have included improved high-capacity irrigation pumps, automated milking machines, and computer databases for the management of crops. With the advent of automation, more people have been able to take advantage of employment opportunities in sectors of the economy other than agriculture. That migration of human resources away from agriculture has in turn helped to promote advances across a broad range of other professions and industries.

Figure 54.1 Combine harvesters

4. The airplane. Mechanical engineers have contributed to nearly every stage of aircraft research and development. Commercial aviation has created domestic and international travel opportunities for both business and recreational purposes. The advent of air travel has enabled geographically scattered families to visit one another frequently. Business transactions are conducted in face-to-face meetings between companies located on opposite sides of a country or halfway across the world.[8]

5. Integrated circuit mass production. An amazing achievement of the electronics industry has been the miniaturization and mass production of integrated circuits, computer chips, and microprocessors.

Transistors, capacitors, resistors, and wires are built as small as possible in order to fit more memory or computing power into a given space. Wires, for instance, are as narrow as 120 nanometers in high-capacity dynamic random access memory chips. As a reference, a human hair is about 100 micrometers in diameter.[9] Mechanical engineers work on the machines that manufacture those integrated circuits.

6. Computer-aided engineering. Computer-aided engineering (CAE) encompasses the use of computers for performing calculations, preparing engineering drawings, analyzing designs, simulating performance, and controlling machine tools in a factory. Mechanical engineers use computers on a day-to-day basis.

Words and Expressions

creativity [ˌkriːeiˈtivəti] n. 创造力 (the ability to make new things or think of new ideas)
in some way 在某方面，在某种程度（意义）上，以某种方式
space exploration 太空探索，外层空间探索
power generation equipment 发电设备，电力设备
internal combustion engine and gas turbine 内燃机和燃气轮机
mechanization [ˌmekənaiˈzeiʃən] n. 机械化
distribution of electricity 配电 (在一个用电区域内向用户供电)
high-capacity 大容量，大功率
combine harvester 联合收割机，谷物联合收割机，简称 combine
milking machine 挤乳机，挤奶机 (a machine used to take milk from cows)
commercial aviation 商用航空业
integrated circuit 集成电路 (将一个电路的大量元器件集合于一个单晶片上制成的器件)
miniaturization [ˌminiətʃəraiˈzeiʃən] n. 小型化

microprocessor [ˌmaikrəu'prəusesə] n. 微处理器
transistor [træn'zistə] n. 晶体管
capacitor 电容器（a device that is used to store electrical energy），通常简称"电容"
resistor 电阻器（a device that has resistance to an electric current in a circuit），简称"电阻"
nanometer [ˈnænəˌmitə] n. 纳米（one billionth of a meter），其缩写为 nm
engineering drawing 工程图样，简称"工程图"

Notes

[1] concern with 这里指"从事于，研究的对象是"。全句可译为：机械工程领域的研究对象是机器零件、力、材料、能量和运动的特性，以及应用这些知识去设计新机器和新产品，促进社会发展和改善人们的生活。

[2] at a competitive cost 意为"在成本上具有竞争优势"。全句可译为：机械工程师们应用运动、能量和力的原理进行分析工作，以确保其设计方案的功能得以安全、有效和可靠地实现，并且在成本上具有竞争优势。

[3] 这两句可译为：通过技术推动社会进步的愿望促进了这个行业的发展。美国机械工程师学会（ASME）进行的一项调查确认了 20 世纪机械工程领域的一些主要成就如下：

[4] high-power lightweight engine 意为"功率大，重量轻的发动机"。全句可译为：功率大、重量轻的发动机和高效率的大规模生产工艺是促进汽车技术发展的两个因素。

[5] assembly-line 意为"流水装配线"，across the economic spectrum 意为"不同经济地位，各个经济阶层"。这一段可译为：亨利·福特首创了流水装配线大规模生产技术，使得人人都有能力购买和拥有汽车。现在，汽车行业也已经成为世界经济的一个重要组成部分，它也为机床行业和原材料行业创造了很多就业机会。

[6] routed to 这里指"通过线路输送到"。这两句可译为：充足和廉价的电能是促进经济增长和繁荣的重要因素，通过配电使全世界数十亿人民的生活水平得到了改善。在 20 世纪，电能的生产并通过线路进入家庭、商业企业和工厂，使得经济和社会发生了显著的变化。

[7] tractor 意为"拖拉机（a vehicle used on farms for pulling machinery）"。全句可译为：农业机械自动化是从应用拖拉机和能够简化谷物收获过程的联合收割机（见图 54.1）开始的。

[8] air travel 意为"航空旅行"，business transaction 意为"商业交易"，halfway across the world 意为"相距半个地球之遥的"。这两句可译为：航空旅行的出现使分散在不同区域居住的家庭之间可以经常地相互拜访。位于一个国家两端的公司之间或相距半个地球之遥的公司之间可以通过面对面的会议来进行商业交易。

[9] memory 意为"存储，存储器"，fit 这里指"为……提供场所（to make a place or room for）"，dynamic random access memory 意为"动态随机存取存储器，简称 DRAM"。这三句可译为：晶体管、电容、电阻和导线都做得尽可能小，以便在规定的空间内容纳更高的存储或计算能力。例如，在大容量动态随机存取存储器芯片中，导线的宽度可以窄至 120nm。作为参考，人类头发的直径大约为 100μm。

Lesson 55　Cost Estimating

In many companies cost estimating is accomplished by a professional who specializes in determining the cost of a component whether it's made in-house or purchased from an external source. This person must be as accurate as possible in his or her estimates, as major decisions about the product are based on these costs. Unfortunately, cost estimators need fairly detailed information to perform their job. It is unrealistic for the designer to give the cost estimator 20 conceptual designs in the form of rough sketches and expect any cooperation in return[1]. Thus, it is essential that the designer be able to make at least rough cost estimations until the design is refined enough to seek a cost estimator's aid. (In many small companies, all cost estimations are done by the designer.)

The first estimations need to be made early in the product design phase. These must be precise enough to be of use in making decisions about which designs to eliminate from consideration and which designs to continue refining. At this stage of the process, cost estimates within 30 percent of the final direct cost are possible. The goal is to have the accuracy of this estimate improve as the design is refined toward the final product. The more experience in estimating similar products, the more accurate the early estimates will be.

The cost estimating procedure is dependent on the source of the components in a product. There are three possible options for obtaining the components: (1) purchase finished components from a vendor; (2) have a vendor produce components designed in-house; or (3) manufacture components in-house.

There are strong incentives to buy existing components from vendors. If the quantity to be purchased is large enough, most vendors will work with[2] the product designer and modify existing components to meet the needs of the new product.

If existing components or modified components are not available off the shelf[3], then they must be produced, in which case the decision must be made as to whether they should be produced by a vendor or made in-house. This is the classic "make or buy" decision[4], a complex decision that must be based not only on the cost of the component involved, but on the capitalization of equipment, the investment in manufacturing personnel, and plans by the company to use similar manufacturing equipment in the future.

Regardless of whether the component is to be made or bought, there is a need to develop a cost estimate. We look now at cost estimate for machined components.

Machined components (see Fig. 55.1) are manufactured by removing portions of the material not wanted. Thus the costs for machining are primarily dependent on the cost and shape of the blank[5], the amount and shape of material that needs to be removed, and how accurately it must be removed. These three areas can be further decomposed into six significant factors that determine the cost of a machined component:

Figure 55.1 Machined components

1. *From what material is the component to be machined?* The material affects the cost in three ways: the cost of the raw material, the value of the scrap produced, and the ease with which the material can be machined. The first two are direct material costs and the last affects the amount of labor needed, the time and machines that will be tied up manufacturing the component.

2. *What type of machine will be used to manufacture the component?* The type of machine—lathe (see Figs. 33.1 and 33.2), milling machine (see Fig. 35.1), boring machine (see Fig. 55.2), drilling machine (see Fig. 55.3), etc. to be used in manufacture affects the cost of the component. For each type, there is not only the cost of the machine time itself, there is also the cost of the cutting tools and fixtures needed.

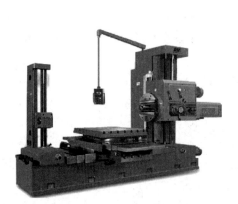

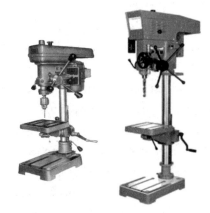

Figure 55.2 A horizontal boring machine

Figure 55.3 Drilling machines. (*Left*) Bench-type drilling machine, (*Right*) Vertical drilling machine

3. *What are the major dimensions of the component?* This factor helps determine what size machines of each type will be required to manufacture the component. Each machine in a manufacturing facility has a different cost for use, depending on the initial cost of the machine and its age.

4. *How many machined surfaces are there and how much material is to be removed?* Just knowing the number of surfaces and material removal ratio (the ratio of the final component volume to the initial volume) can aid in giving a good estimate for the amount of time required to machine the

part. More accurate estimates require knowing exactly what machining operations will be used to make each cut.

5. *How many components will be made?* The number of components to be made has a great effect on the cost. For one piece, fixturing will be minimal, though long setup and alignment times will be required. For a few pieces, some fixtures will be made. For a high volume, the manufacturing process will be automated, with extensive fixturing and numerical controlled machining.

6. *What tolerance and surface roughness are required?* The tighter the tolerance and surface roughness requirements, the more time and equipment needed in manufacture.

Words and Expressions

estimate ['estimeit]　*v.*；*n.* 估算，估价，预测
in-house　机构内部的（created, done, or existing within a company or organization）
sketch [sketʃ]　*n.* 草图（以目测估计图形与实物的比例，按一定画法要求徒手或部分使用绘图仪器绘制的图）；*v.* 画草图（to make a quick, rough drawing of sth）
rough sketch　简略草图（a preliminary sketch of a design）
vendor　卖主（one that sells sth），销售商（a company that sells a particular product）
decompose [ˌdiːkəmˈpəuz]　*v.* 分解
significant factor　重要因素
raw material　原材料（投入生产过程以制造新产品的物质）
scrap [skræp]　*n.* 切屑，边角料，废品（不能修复又不能降级使用的不合格品）
fixture ['fikstʃə]　*n.* 夹紧装置，夹具（用以装夹工件和/或引导刀具的装置）
facility [fəˈsiliti]　*n.* 设备，场所，部门
automated ['ɔːtəmeitid]　*a.* 自动化的，自动操纵的
boring machine　镗床（主要用镗刀对工件上已有的孔进行镗削加工的机床）
horizontal boring machine　卧式镗床（镗轴水平布置的镗床）
vertical drilling machine　立式钻床（主轴箱和工作台安置在立柱上，主轴竖直布置的钻床），亦为 upright drilling machine
setup　安装（在加工前对工件进行定位、夹紧和调整的作业），又称"装夹"

Notes

[1] in return 意为"作为回报"。
[2] work with 意为"与……共事，与……合作"。
[3] off the shelf 意为"现货供应的，随时可用的（from stock and readily available）"。
[4] make or buy decision 意为"自制或外购的决策"。
[5] blank 意为"毛坯（根据零件所要求的形状、工艺尺寸等制成的供进一步加工用的生产对象，a piece of material prepared to be made into something by a further operation）"。

Lesson 56 Quality and Inspection

According to the American Society for Quality (ASQ), quality is the totality of features and characteristics of a product or service that bear on its ability to satisfy given needs. The definition implies that the needs of the customer must be identified first because satisfaction of those needs is the "bottom line"[1] of achieving quality. Customer needs should then be transformed into product features and characteristics so that a design and the product specifications can be prepared.

In addition to a proper understanding of the term quality, it is important to understand the meaning of the terms quality management, quality assurance, and quality control.

Quality management includes all activites of the overall management function that determine the quality policy, objectives, and responsibilities and implements them by means such as quality assurance, quality control, and quality improvement, within the quality system. The responsibility for quality management belongs to senior management.

Quality assurance includes all the planned or systematic actions necessary to provide adequate confidence that a product or service will satisfy given needs. These actions are aimed at providing confidence that the quality system is working properly and include evaluating the adequacy of the designs and specifications or auditing the production operations for capability. Internal quality assurance aims at providing confidence to the management of a company, while external quality assurance provides assurance of product quality to those who buy from that company.

Quality control comprises the operational techniques and activities that sustain the quality of a product or service so that the outcomes will satisfy given needs. The quality control function is closest to the product in that various techniques and activities are used to monitor the process and to pursue the elimination of unsatisfactory sources of quality performance.

Many of the quality systems of the past were designed with the objective of sorting good products from bad products during the various processing steps. Those products judged to be bad had to be reworked to meet specifications. If they could not be reworked, they were scrapped. This type of system is known as a "detection correction" system. With this system, problems were not found until the products were inspected or when they were used by the customer. Because of the inherent nature of human inspectors, the effectiveness of the sorting operations was often less than 90%. Quality systems that are preventive in nature are being widely implemented. These systems prevent problems from occurring in the first place by placing emphasis on proper planning and problem prevention in all phases of the product cycle.

The final word on how well a product fulfills needs and expectations is given by the customers and users of that product and is influenced by the offerings of competitors that may also be available to those customers and users. It is important to recognize that this final word is formed over the entire life of the product, not just when it was purchased.

Being aware of customers' needs and expectations is very important, as was previously discussed. In addition, focusing the attention of all employees in an enterprise on the customers and users and their needs will result in a more effective quality system. For example, group discussions on product designs and specifications should include specific discussion of the needs to be satisfied.

A basic commitment of management should be that quality improvement must be relentlessly pursued. Actions should be ingrained in the day-to-day workings of the company that recognize that quality is a moving target in today's marketplace driven by constantly rising customer expectations. Traditional efforts that set a quality level perceived to be right for a product and direct all efforts to only maintain that level will not be successful in the long haul. Rather, management must orient the organization so that once the so-called right quality level for a product has been attained, improvement efforts continue to achieve progressively higher quality levels.

To achieve the most effective improvement efforts, management should understand that quality and cost are complementary and not conflicting objectives. Traditionally, recommendations were made to management that a choice had to be made between quality and cost, because better quality inevitably would somehow cost more and make production difficult. Experience throughout the world has shown that this is not true. Good quality fundamentally leads to good resource utilization and consequently means good productivity and low quality costs. Also significant is that higher sales and market shares result from products that are perceived by customers to have high quality and performance reliability during use.

Four basic categories of quality costs are described in the following:

1. Prevention costs incurred in planning, implementing, and maintaining a quality system that will ensure conformance to quality requirements at economical levels. An example of prevention cost is training in the use of statistical process control.

2. Appraisal costs incurred in determining the degree of conformance to quality requirements. An example of appraisal cost is inspection.

3. Internal failure costs incurred when products, components, and materials fail to meet quality requirements prior to transfer of ownership to the customer. An example of internal failure cost is scrap.

4. External failure costs incurred when products fail to meet quality requirements after transfer of ownership to the customer. An example of external failure cost is warranty claims.

Although the level of quality control is determined in large part by probability theory and statistical calculations, it is very important that the data collection processes on which these procedures depend be appropriate and accurate. The best statistical procedure is worthless if fed faulty data, and like machining processes, inspection data collection is itself a process with practical limits of accuracy, precision, resolution[2], and repeatability. All inspection and/or measurement processes can be defined in terms of their accuracy and repeatability, just as a manufacturing process is evaluated for accuracy and repeatability.

Words and Expressions

inspection 检验（通过观察和判断，结合测量、试验或估量所进行的符合性评价）
totality [təu'tæliti] n. 总和，全体，总数（the whole or entire amount of something）
specification [ˌspesifi'keiʃən] n. 技术规格，技术参数，规范
strategic [strə'ti:dʒik] a. 战略的，关键性的，对全局有重大意义的
sort [sɔ:t] v. 分类（to arrange according to kind or size），拣选（to separate from others）
rework ['ri:'wə:k] v. 重新加工，返工（为使不合格产品符合要求而对其采取的措施）
scrap [skræp] v. 报废（to get rid of something because it is damaged, no longer useful, etc.）
detection [di'tekʃən] n. 探测，发现
correction [kə'rekʃən] n. 改正，纠正（为消除已发现的不合格所采取的措施）
commitment [kə'mitmənt] n. 委托，许诺，承担义务
relentlessly [ri'lentlisli] ad. 持续地（in an extreme way that continues without stopping）
ingrained [in'greind] a. 根深蒂固的（deep-rooted），牢固的（firmly fixed）
long haul 很长一段时间（a long period of time），很长一段距离（a long distance）
irreversible [ˌiri'və:səbl] a. 不能撤回的，不能改变的
complementary [ˌkɔmplə'mentəri] a. 互补的（mutually supplying each other's lack）
quality cost 质量成本（the costs that are incurred to prevent defects in products）
prevention cost 预防成本（用于预防不合格品与故障等所需的各项费用）
statistical process control 统计过程控制（着重于用统计方法减少过程变异、增进对过程的认识，使过程以所期望的方式运行的活动），简称 SPC
appraisal cost 鉴定成本（评定产品是否满足规定的质量要求所需的费用）
internal failure cost 内部损失成本（产品出厂前因不满足规定的质量要求而支付的费用）
external failure cost 外部损失成本（产品出厂后因不满足规定的质量要求，导致索赔、修理、更换或信誉损失等而支付的费用）
repeatability [ri'pi:tə'biliti] n. 重复精度（在同一条件下，操作方法不变，进行规定次数操作所得到的连续结果的一致程度）

Notes

[1] bottom line 意为"底线，关键（the essential point），最重要的部分（the most important part of something），最应该考虑的事情（the most important thing to consider）"。
[2] resolution 意为"分辨率，是两个相邻的分散细节之间可以分辨的最小间隔。就测量系统而言，它是可以测量的最小增量；就控制系统而言，它是可以控制的最小位移增量"。

Lesson 57 Quality in the Modern Business Environment

It is essential that products meet the requirements of those who use them. Therefore, we define quality as fitness for use.[1] The term consumer applies to many different types of users. A purchaser of a product that is used as a raw material in its manufacturing operations is a consumer, and to the manufacturer fitness for use implies the ability to process this raw material with low cost and minimal scrap or rework. A retailer purchases finished goods with the expectation that they are properly packaged, labeled, and arranged for easy storage, handling, and display. You and I may purchase automobiles that we expect to be free of initial manufacturing defects, and that should provide reliable and economical transportation over time.

There are two general aspects of quality: quality of design and quality of conformance.[2] All goods and services are produced in various grades. These variations in grades are intentional, and, consequently, the appropriate technical term is quality of design. For example, all automobiles have as their basic objective providing safe transportation for the consumer. However, automobiles differ with respect to size, appointments, appearance, and performance (see Fig. 57.1). These differences are the result of intentional design differences between the types of automobiles. These design differences include the types of materials used in construction, tolerances in manufacturing, reliability obtained through engineering development of engines and drivetrains (see Fig. 57.2), and other accessories or equipment.[3]

Figure 57.1 Automobiles

Figure 57.2 Drivetrains

The quality of conformance is how well the product conforms to the specifications and tolerances required by the design. Quality of conformance is influenced by a number of factors, including the choice of manufacturing processes, the training and supervision of the work force, the type of quality assurance system (process controls, tests, inspection activities, etc.) used, the extent to which these quality assurance procedures are followed, and the motivation of the work force to achieve quality.

There is considerable confusion in our society about quality. The term is often used without making clear whether we are speaking about quality of design or quality of conformance. To achieve quality of design requires conscious decisions during the product or process design stage to ensure that certain functional requirements will be satisfactorily met. For example, the designer of a copier (see Fig. 57.3) may design a circuit component with a redundant element, because he knows that this will enhance the reliability of the product and will increase the mean time between failures.[4] This in turn will result in fewer service calls to keep the copier running, and the consumer will be far more satisfied with the performance of the product. Designing quality into the product in this fashion often results in a higher product cost. However, such cost increases are actually prevention costs, as they are intended to prevent quality problems at later stages in the life cycle of the product.

Figure 57.3　Copiers

Most organizations find it difficult (and expensive) to provide the customer with products that have flawless quality characteristics. A major reason for this difficulty is variability. There is a certain amount of variability in every product; consequently, no two products are ever identical. For example, the thicknesses of the blades on a turbojet engine impeller are not identical even on the same impeller.[5] Blade thickness will also differ between impellers. If this variation in blade thickness is small, then it may have no impact on the customer. However, if the variation is large, then the customer may perceive the unit to be undesirable and unacceptable. Sources of this variability include differences in materials, differences in the performance and operation of the manufacturing equipment, and differences in the way the operators perform their tasks. Therefore, we may define quality improvement as the reduction of variability in processes and products. Since variation can only be described in statistical terms, statistical methods are of considerable use in quality-improvement efforts.

Quality is becoming the basic consumer decision factor in many products and services. This phenomenon is widespread, regardless of whether the consumer is an individual, an industrial corporation, or a retail store. Consequently, quality is a key factor leading to business success, growth, and enhanced competitive position. There is a substantial return on investment from an effective quality-improvement program that provides increased profitability to firms that effectively employ quality as a business strategy. Consumers feel that the products of certain companies are substantially better in quality than those of their competition, and make purchasing decisions accordingly. Effective quality-improvement programs can result in increased market share, higher productivity, and lower overall costs of manufacturing and service.[6] Consequently, firms with such programs can enjoy significant competitive advantages.

All business organizations use financial controls. These financial controls involve a comparison of actual and budgeted costs, along with an associated analysis and action on the differences or variances between actual and budget. It is customary to apply these financial controls on a department level. For many years, there was no direct effort to measure or account for the costs of the quality function. However, starting in the 1950s, many organizations began to formally evaluate the cost associated with quality. There are several reasons why the cost of quality should be explicitly considered in an organization. These include:

1. The increase in the cost of quality because of the increase in the complexity of manufactured products associated with advances in technology.

2. Increasing awareness of life cycle costs, including maintenance, labor, spare parts, and the cost of field failures.

3. The need for quality engineers and managers to effectively communicate the cost of quality in the language of general management—namely, money.

Words and Expressions

fitness ['fitnis]　*n.*　适合，适当，合理
arrange for　为……做准备（to make preparation for; to prepare for）
defect [di'fekt]　*n.*　缺陷（制件与规定要求不相符的部分）
grade [greid]　*n.*　等级（同一产品或其他标准化对象按其质量水平的不同所划分的级别）
conformance [kən'fɔːməns]　*n.*　一致性，适应性，性能
quality of conformance　符合质量，符合性质量（一组固有特性满足规定要求的程度）
have as　把……当作
manufacturing process　工艺过程（改变生产对象的形状、尺寸、相对位置和性质等，使其成为成品或半成品的过程）
appointment [ə'pɔintmənt]　*n.*　指定，家具，车身内部装饰
intentional [in'tenʃənl]　*a.*　故意的，有意的
drivetrain ['draivtrein]　*n.*　动力传动系统（a mechanical system within a motor vehicle which connects the transmission to the drive axles），亦为 driveline
accessory [æk'sesəri]　*a.*　附属的，附带的；*n.*　附属品，附属装置
quality assurance　质量保证（为使人们确信某一产品、过程或服务质量能满足规定的质量要求所必需的有计划、有系统的全部活动）
quality assurance system　质量保证体系
motivation [ˌməuti'veiʃən]　*n.*　刺激，激发，诱导，动机
copier ['kɔpiə]　*n.*　复印机（a machine that makes paper copies of printed pages, pictures, etc.），亦为 copy machine
redundant [ri'dʌndənt]　*a.*　冗余（产品中具有多余一种手段执行同一种规定功能）的

service call 维修服务 (a trip made by a repairman to visit the location of sth in need of service)
prevention cost 预防成本（用于预防不合格品与故障所需的各项费用）
life cycle 生命周期，耐用周期
flawless ['flɔ:lis] a. 无瑕疵的，无缺点的，完美的
variability [,vɛəriə'biliti] n. 易变性，变化性，变异度
turbojet ['tɜ:bəudʒet] engine 涡轮喷气发动机
impeller [im'pelə] n. 叶轮，涡轮，推进器
quality improvement 质量改进
statistical [stə'tistikəl] a. 统计的，统计学的
budgeted cost 预算成本（企业按照预算期的特殊生产和经营情况所编制的预定成本）
it is customary to 通常，一般习惯于
cost of quality 质量成本（将产品质量保持在规定的质量水平上所需的有关费用）
life cycle cost 生命周期成本（指在产品经济有效使用期间所发生的与该产品有关的所有成本，包括产品设计成本、制造成本、使用成本、废弃物处置成本、环境保护成本等）

Notes

[1] fitness for use 意为"适用性（产品或服务能够成功地符合用户需要的程度）"。全句可译为：因此，我们把质量定义为适用性。

[2] quality of conformance 意为"符合质量标准的程度，符合性质量（用设计质量和制造质量，或者预期质量和实际质量的差值来评价产品制造质量的水平）"。全句可译为：一般情况下有两种质量：设计质量与符合性质量。

[3] engineering development 意为"工程技术开发"。全句可译为：这些设计差异包括结构材料的种类、制造公差、通过对发动机和动力传动系统（见图57.2）的工程技术开发获得的可靠性，以及其他附件或设备。

[4] circuit component 意为"电路元件"，redundant element 意为"冗余元件"，mean time between failures 意为"平均失效间隔时间，平均无故障时间"。全句可译为：例如，复印机的设计人员在设计某一个电路元件时，可以增加一个冗余元件，因为他知道这将增加产品的可靠性，并将增加平均失效间隔时间。

[5] turbojet engine impeller 意为"涡轮喷气发动机叶轮"。全句可译为：例如，涡轮喷气发动机叶轮上的叶片的厚度，即使对于在同一个叶轮上的叶片也不是完全一样的。

[6] 全句可译为：有效的质量改进计划可以增加市场占有率，提高生产率，降低制造和维修的总成本。

Lesson 58 Coordinate Measuring Machines

A coordinate measuring machine (CMM) is a device for measuring the geometrical characteristics of an object. It is typically used to generate 3D points from the surface of a part. It's digitizing a part in three dimensions. However, it is often used to make 2D measurements such as measuring the center and radius of a circle in a plane, or even one-dimensional measurements such as determining the distance between two points. Typically, CMMs are configured to measure in Cartesian coordinates. [1] There are also CMMs that measure in cylindrical or spherical coordinates. They can measure any part surface they can reach.

CMMs typically generate points in two ways: point-to-point mode, where the CMM touches the part and generates a single point of data every time contacting with the part; or scanning, where the CMM moves over a part, generating data as it moves. Scanning generates significantly more data than contacting.

The scanning, which includes laser scanning and white light scanning, is advancing very quickly. This method uses either laser beams or white light that are projected against the surface of the part. Many thousands of points can then be taken and used to not only check size and position, but to create a 3D image of the part as well. This "point-cloud data" can then be transferred to CAD software to create a 3D model of the part. These optical scanners often used on soft or delicate parts or to facilitate the "reverse engineering" process. Reverse engineering is the process of taking an existing part, measuring it to determine its size, and creating engineering drawings from these measurements. This is most often necessary in cases where engineering drawings may no longer exist or are unavailable for the particular part that needs replacement.

A typical stationary bridge type CMM (Fig. 58.1) is composed of three axes, an X, Y, and Z. These axes are orthogonal to each other in a typical three dimensional coordinate system. The CMM usually reads the input from the touch probe, as directed by the operator or program. The machine then uses the X, Y, Z coordinates of each of these points to determine size and position. Typical precision of a coordinate measuring machine is measured in micrometers.

A coordinate measuring machine (CMM) is also a device used in manufacturing and assembly processes to inspect a part or assembly against the design intent. By precisely recording the X, Y, and Z coordinates of the target, points are generated. These points are collected by using a probe that is positioned manually or automatically via computer control. In manual

Figure 58.1 A bridge type CMM

mode, the CMM is moved by the user. An automatic CMM is typically actuated by electric drives (using ball screws or linear motors, see Figs. 58.2 and 58.3).[2]

Figure 58.2 Ball screws

Figure 58.3 Linear motors

As well as the traditional three axis machines, CMMs are also available in a variety of other forms. Portable CMMs are different from "traditional CMMs" in that they most commonly take the form of an articulated arm. As shown in Fig. 58.4, an articulated arm CMM looks very much like a six degrees of freedom robot. Articulated arm CMMs are lightweight (typically less than 10 kilograms) and can be carried and used nearly anywhere.

The accuracy of the traditional CMM is usually better than that of a portable articulated arm CMM. But for many operations, the accuracy of articulated arm CMMs is sufficient for a variety of processes. The advantage of articulated arm CMMs is that they can reach areas that are not easy to access with traditional CMMs.

Figure 58.4 An articulated arm CMM

Thus, if the accuracy of an articulated arm CMM is sufficient for a particular application, it should be seriously considered as an alternative.

To determine the level of accuracy needed from a CMM, the tightest tolerances to be measured should be considered first. Generally, a CMM should have a level of accuracy 10 times greater than the tightest tolerance the shop needs to meet. In other words, if the tightest tolerance is 50 μm, the CMM should be accurate to 5 μm.

While the CMM hardware generates the coordinate data, the software bundled with the CMM (or in many instances sold separately) analyzes the data and presents the results to the user in a form that permits an understanding of part quality, and conformance to specified geometry.

The most important advancement in CMM technology over the past several years is that significant errors can be corrected mathematically via software. As a result, looser tolerances can be used on the system hardware, and the resulting errors (as long as they are highly repeatable) are eliminated in software.[3] This results in lower manufacturing costs, while retaining or even improving the capabilities of the CMM.

New user-friendly software that allows the CMM and probe to be accurately, quickly, and easily calibrated has also made the CMM more accurate and easier to use.[4]

A controlled environment is important for efficient CMM operation. CMMs can operate well on the shop floor if they are equipped with thermal compensation capabilities that correct for temperature changes from standard temperature (20°C).[5] In any case, the CMM should be kept in a relatively clean environment and located in a space that is isolated from vibration.

Words and Expressions

coordinate measuring machine　三坐标测量机（按空间三维坐标的测量原理设计，可用直角坐标和极坐标法测量各种复杂形状轮廓尺寸的一种精密仪器）
digitize ['didʒitaiz]　*v.* 数字化（to convert something to digital form）
configure [kən'figə]　*v.* 为某种特定目标而进行设计（to design for some specific purpose）
Cartesian [kɑː'tiːzjən]　*a.* 笛卡儿的（of or relating to the philosophy or methods of Descartes）
Cartesian coordinates　笛卡儿坐标，直角坐标（rectangular coordinates）
cylindrical [si'lindrikəl]　*a.* 圆柱的，圆柱形的（of or having the shape of a circular cylinder）
cylindrical coordinates　柱面坐标，圆柱坐标
spherical ['sferikəl] coordinates　球面坐标
point-to-point mode　点位方式，从点到点的方法
scanning ['skæniŋ]　*n.* 扫描（电子束或光点在某特定区域内以一定规律移动的过程）
white light　白光（light perceived by the eye as having the same color as sunlight at noon）
project ['prɔdʒekt]　*v.* 投影（to cause an image to appear on a surface）
point cloud　点云（在同一空间参考系下表达目标空间分布和目标表面特性的海量点集合）
point-cloud data　点云数据（指通过3D扫描仪所取得的资料形式，扫描资料以点的形式记录，每个点包含三维坐标）
delicate　脆弱的（easily broken），灵敏的，精密的（precise or sensitive in operation）
delicate part　精密零件，易损坏的零件
facilitate [fə'siliteit]　*v.* （不以人作主语的）推动，帮助，使容易，促进
reverse engineering　反求工程，逆向工程（是指用一定的测量手段对实物或模型进行测量，根据测量数据通过三维几何建模方法重构实物的CAD模型的过程，是一个从样品生成产品数字化信息模型，并在此基础上进行产品设计开发及生产的全过程）
stationary bridge type CMM　固定桥式三坐标测量机
orthogonal [ɔː'θɔgənl]　*a.* 正交的，直角的（relating to or composed of right angles），垂直的
micrometer [mai'krɔmitə]　*n.* 微米（μm），千分尺
assembly [ə'sembli]　*n.* 装配，装配体
assembly process　装配过程
design intent　设计目的，设计意图（what you want your design to do）
probe [prəub]　*n.* 测头，测量头；*v.* 探查，查明
actuate ['æktjueit]　*v.* 驱使，驱动（to put into motion or action）
electric drive　电气传动（生产过程中，以电动机作为原动机来带动生产机械，并按所给定的规律运动的电气设备）

ball screw 滚珠丝杠副（丝杠与旋合螺母之间以钢珠为滚动体的螺旋传动副。它可将旋转运动变为直线运动或将直线运动变为旋转运动）
linear motor 直线电机（是一种将电能直接转换成直线运动机械能，而不需要任何中间转换机构的传动装置），也称线性电机
portable ['pɔːtəbl] *a.* 轻便的，手提式的，便携式的（carried or moved with ease）
articulated [ɑːˈtikjulitid] *a.* 铰接的，有关节的，关节式的
articulated arm CMM 关节臂测量机
lightweight [ˈlaitweit] *n.* 重量轻的东西（something that does not weigh as much as others）
the tightest tolerance 最小公差
bundled [ˈbʌndld] *a.* 捆绑销售的（sold together, as a package, rather than separately）
in many instances 在许多情况下（in many cases），在大多数情况下
conformance [kənˈfɔːməns] *n.* 顺应，一致，符合
user-friendly 用户友好的，容易掌握使用的（easy to learn, use, understand, or deal with）
calibrate [ˈkælibreit] *v.* 校准（to check, adjust, or determine by comparison with a standard）
controlled environment 受控环境
compensation [kɔmpenˈseiʃən] *n.* 补偿（something good that acts as a balance against something bad or undesirable）
thermal compensation 热补偿（a method used to adjust a system's performance to compensate for effects caused by changes in temperature）
isolate [ˈaisəleit] *v.* 使隔离，使孤立，使绝缘（to render free of external influence）

Notes

［1］全句可译为：通常，三坐标测量机是专门为在直角坐标系内进行测量而设计的。

［2］electric drive 意为"电气传动，又称'电力拖动'。生产过程中，以电动机作为原动机来带动生产机械，并按所给定的规律运动的电气设备"。全句可译为：一个自动化的三坐标测量机通常采用电气传动的方式（通过滚珠丝杠或采用直线电动机）。

［3］looser tolerance 意为"比较大的公差"。全句可译为：结果是，系统硬件的公差可以比较大，只要可重复性很高，其所产生的误差可以通过软件予以消除。

［4］user-friendly software 意为"用户友好型软件"。全句可译为：新的用户友好型软件可以对三坐标测量机及其测量头进行精确、快速和容易的校准工作，也使三坐标测量机更为精确和更易于使用。

［5］shop floor 意为"工人，车间（the ordinary workers in a factory or the area where they work）"。全句可译为：如果三坐标测量机具有热补偿能力，能够在偏离标准温度（20℃）时，对由于温度变化带来的影响进行修正，则三坐标测量机就可以很好地在车间内工作。

Lesson 59 Reliability Requirements

Reliability requirements vary greatly depending on the type of equipment or system under consideration. They may sometimes be set by the designer or more broadly by the firm responsible for the design. In other situations the requirements may be imposed by the buyer of the product; and in some instances third outside parties[1], such as insurance companies or government agencies, may play a large role.

For any given product there are likely to be trade-offs between purchase prices and maintenance costs. The more effort put into making the product reliable, the higher will be its purchase price. At the same time, the more reliable a product is, the lower the costs for maintenance and repair are likely to be. We might be tempted to say that the solution is optimal when the total cost is minimized. In practice, however, the considerations that go into making such a trade-off vary greatly, depending on the nature of the product. The varying requirements may be understood more concretely by considering the factors that may come into play in the following three diverse examples: an air conditioner, an industrial robot (see Fig. 71.1), and an aircraft engine (see Fig. 61.1).

The reliability requirements for an air conditioner may depend to a great extent on consumer psychology. To what extent will the buyer be willing to pay a higher price in order to save later repair or replacement costs? The price is obvious, but how will the reliability be impressed on the general public: by information and advertising about the air conditioner's features, or by the willingness to offer a longer warranty than the competition? Moreover, what will the sales consequences be if the cost is too high because excessive reliability is built into the air conditioner? On the other hand, if the public buys the item at a lower price and then comes to resent the inconvenience and cost of breakdowns, will the firm's other products acquire a lingering reputation for poor design or shoddy workmanship? This, in turn, may depend significantly on the extent and efficiency of the service organization responsible for maintenance. These are complex issues. The decisions are most likely made by the manufacturer with the help of market surveys or other input to ascertain the public's preferences with regard to price and reliability.

The situation is quite different for an industrial robot or other equipment that is designed primarily for sale to large enterprises. For if the equipment is expensive, the buyer is likely to have a staff of engineers or outside consultants who are able to assess the cost trade-offs and determine their impact on the buyer's operations. With the robot, for example, a trade-off must be made between purchase price and production lost through robot breakdown. The direct cost of repair may be small in comparison to the costs of lost production. In effect, the speed at which the buyer's maintenance organization is able to accomplish repairs will be a central consideration. Thus, for these types of products, design may center not only on reliability, but also on maintainability[2]. Are they able to be repaired in a short period of time? Increasing the reliability at the expense of greatly increasing the repair time may in some situations cause greater losses in production.

For the aircraft engine the consequences of failure are generally so severe that a much higher standard of reliability is required, even though the cost of the engine is made much higher than it otherwise would be. Similarly, preventive maintenance[3] to enhance reliability will be the primary consideration, rather than the time required to repair a failed engine. Avoiding failure is likely to be so important that it may not be left entirely in the hands of the design or manufacturing organization, or to the buyer, to set reliability specifications. Insurance companies and government agencies are likely to be much more closely involved. In the design more emphasis also will be placed on early warning of incipient failures so that the engine can be taken out of service before an accident occurs.

In setting reliability requirements and in designing to ensure that they are met, we should allocate reliability among major subsystems, and then on down to lower levels of components and parts. For example, in designing an aircraft we might decide that the reliabilities of the propulsion, structural, control, and guidance systems should be approximately comparable. Then if the required reliability for the total system is to be equal to R, we would have $R = R_0^4$, where R_0 is the reliability requirement for each of the subsystems. Each subsystem reliability requirement would then be $R_0 = R^{1/4}$. Such reliability allocations are important particularly when the subsystems or components are to be designed or manufactured, or both, by different groups or by different corporations.

Thus far we have been interested only in the reliability, in the probability that the system will function properly. In reality not all failures can be treated alike, for there may be a wide range of consequences, depending on the mode of failure. In most system failures we may hope that the result is simply inconvenience and some economic loss. However, the designer must also consider carefully the possibility that a particular failure mode may endanger individuals or possibly the health or safety of a larger human population.

We return to our three diverse examples of an air conditioner, an industrial robot, and an aircraft engine to contrast safety analysis to more conventional reliability considerations. In most failures of an air conditioner its operation simply ceases. However, an electric short in an air conditioner may cause overheating and fire or even electrocute the user. The possibility of such accidents must be eliminated or reduced to a very small probability through careful design and safety analysis. If a faulty product causes such accidents, it may have to be withdrawn from the market, and the company may be brought to court in product liability lawsuits.

The failure of the industrial robot is somewhat analogous, except that accidents are likely to be occupational risks to the workers involved, rather than risks to members of the general public. In contrast, when an aircraft engine fails there is little distinction between reliability and safety analysis; for if reliability is defined in terms of engine function, and failure will significantly increase the probability of a crash.

Words and Expressions

play a large role 发挥很大的作用(to have an extremely important or fundamental role)
trade-off 平衡,权衡取舍(a giving up of one thing in return for another),综合考虑
maintenance ['meintinəns] n. 维修(为使产品保持或恢复到规定状态所进行的全部活动)
concretely ['kɔnkri:tli] ad. 清晰明确地(in a clear and definite way),具体地
diverse [dai'və:s] a. 不同的(unlike),多种多样的(made up of distinct characteristics)
warranty ['wɔrənti] n. 质保期(assurance of quality within a given time period),质保单
breakdown ['breikdaun] n. 损坏(the act or process of failing to function),故障
resent [ri'zent] v. 怨恨,愤怒(to be angry about something that you think is unfair)
lingering ['lingəriŋ] a. 延迟的,拖延的,长期的,经久不消的
shoddy ['ʃɔdi] a. 劣质的,工艺质量差的(poorly done or made; carelessly or badly made)
workmanship ['wə:kmənʃip] n. 手艺,工艺质量(the quality of something made)
market survey 市场调查(a collection of first-hand data from customers, vendors, etc.)
ascertain [,æsə'tein] v. 确定,查明(to discover with certainty)
incipient [in'sipiənt] failure 早期故障,初发故障
propulsion [prə'pʌlʃən] n. 推进(the process of driving or propelling),推进力
guidance system 导航系统
reliability allocation 可靠性分配(产品设计阶段,将产品的可靠性定量要求按给定的准则分配给各组成部分的过程),亦为 reliability apportionment
center on 集中于,以……为中心
electrocute [i'lektrəkju:t] v. 触电致死(to kill a person or animal with electricity)
in effect 实际上,事实上(in reality or fact)
liability [,laiə'biliti] n. 责任(an obligation, a responsibility, or a debt),不利条件
failure ['feiljə] n. 失效(产品丧失规定的功能,对可修复产品通常也称为故障)
faulty ['fɔ:lti] a. 有缺陷的,不完美的(containing a fault or defect; imperfect or defective)
analogous [ə'næləgəs] a. 类似的(similar in some way),相似的
occupational [,ɔkju'peiʃənəl] a. 职业的(of or relating to a person's job)

Notes

[1] third outside party 意为"……以外的第三方,外部第三方",亦可写为 outside third party。
[2] maintainability 意为"维修性(在规定条件下使用的产品,在规定条件下并按规定的程序和手段实施维修时,保持或恢复能执行规定功能状态的能力)"。
[3] preventive maintenance 意为"预防性维修(为降低产品失效的概率或防止功能退化,按预定的时间间隔或按规定准则实施的维修)"。

Lesson 60 Product Reliability

Reliability is sometimes simply defined as the probability of a product or process performing its intended or specified function. Inherent in this definition is the implication of performing under certain stated conditions or environments as well as performing for a specified length of time. [1]

Quality has a little different definition, and it is sometimes defined as "the totality of features which determine a product's acceptability". Some may choose to define this as conformance to specifications, fitness for use, or meeting requirements the first time. Reliability, as a concept, goes beyond the basic ideas in quality because it adds the concept of time to the basic definition. That is, reliability may also be viewed as meeting requirements, the first time and every time. [2] Products may have quality yet not have reliability.

Many companies have recognized that quality by itself is not sufficient in a competitive world. Products must be designed and built to last a long time in their intended function. This additional requirement goes beyond basic quality and adds a new dimension to marketing for the company. The most successful companies, in terms of product acceptance, have recognized the importance of total life cycle cost and customer satisfaction. [3] A good understanding of reliability is necessary to achieve both with minimum cost in terms of time and resources.

There are three elements that provide the basis for a reliable product—design, manufacturing, and component parts. Design is that series of operations involved in taking a product from a conceptual stage to a form that meets both company goals and customer expectations. In addition, this step should include some demonstration that the goals have indeed been met. This is often referred to as the design validation step. [4] Typically, a limited number of key features of the product are demonstrated to meet the goals over the expected life of the product. Reasons for success as well as failure can then be identified early in the product's history and corrected or improved before cost and schedule become big constraints.

Reliability techniques are needed to determine the most appropriate and effective tests as well as reduce the test time, while preserving test conditions. This is known as accelerated test. Teamwork becomes important here. The object is to finish the design process successfully in as short a time as possible. Bringing other groups into the design validation process will help minimize the number of tests needed later.

The next step, manufacturing, involves turning the design into reality without affecting it adversely. Consistency, via such techniques as statistical process control (SPC), is the key to creating products that can be tested and demonstrated as being reliable. [5] More than one reliable design has been affected by a process that damaged or weakened one or more of the components or subassemblies. This often occurs without anyone's knowledge, until a later test or a user of the product identifies the situation.

The third important element involved in producing a reliable design includes the parts and subassemblies that go into the final product. High-quality and consistent parts are needed to preserve the design and be compatible with the established manufacturing processes. This means selecting the key parts suppliers ahead of time, qualifying each of the part types, then not changing during the life of the product without some requalification. Supplier-initiated changes in purchased items should be discouraged. In order to prevent such a situation, a clear note—"No change of process without prior notification"—is needed on the procurement documents.[6]

Individual attention to the three primary elements of reliability should be coordinated through a reliability plan that makes sense for the type of product and the nature of the industry. This means understanding the marketing goals, such as expected life, and knowing the customers' expectations and end-use environment.

The reliability function includes estimating conformance to an anticipated life, as well as warranty and failure analysis information.

When developing a test plan or conducting failure analysis, teamwork becomes important. Typically, development of the original plan to test the final product involves marketing, design, manufacturing, and product management. A meeting of representatives from all of these functions to discuss a proposed test outline will almost always produce a test plan that meets the design intent and identified customer needs.

One valuable model used by reliability engineers is the so-called bathtub curve (Fig. 60.1), which is named for its particular shape.[7] This curve often describes the reliability of a complex system. The descending slope near the start represents the early history of an item and indicates a decreasing failure rate. It indicates that the first few hours or days of a product's life are the worst, followed by a period of improvement. This is sometimes referred to as the infant mortality period.

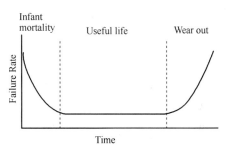

Figure 60.1 The bathtub curve

Over time, the failure rate decreases to a low point where it remains essentially constant for a long period.[8] This portion of the curve is called the useful life period. It is the constant failure rate period of a product's life. The last portion of the curve depicts an increasing failure rate and is attributed to wear out. Not surprisingly, the highest number of failures occurs during this period. Properly used, the bathtub curve can also shed valuable light on what to expect when a product or system is ultimately put to the test in the field.

Reliability and quality go hand in hand in a variety of ways. Both are vital in maintaining a competitive position in the world marketplace. It is important to understand the basic concepts and methods for improving product reliability and the ground rules for applying them at all stages of product design and manufacturing.[9]

Words and Expressions

reliability [riˌlaiəˈbiliti] n. 可靠性（产品在规定的条件下和规定的时间区间内完成规定功能的能力）

probability [ˌprɔbəˈbiliti] n. 概率，可能性，可能发生的事

totality [təuˈtæliti] n. 全体，总数，总额，完全

conformance [kənˈfɔːməns] n. 一致性，适应性，性能

specification [ˌspesifiˈkeiʃən] n. 规范（详细阐述产品、过程或服务应遵循哪些要求的文件），技术规格（机器的最大加工能力、功率等的技术参数）

fitness for use 适用性（a product fits the customer's defined purpose for that product）

attribute [ˈætribjuːt] n. 属性，特征，标志，象征

life cycle 寿命周期，耐用周期

component part 组成零件，部件，组成部分

conceptual [kənˈseptjuəl] a. 概念上的

conceptual stage 总体设计阶段，方案设计阶段

customer expectation 顾客期望（what customers predict will happen if they use a product）

validation [ˌvæliˈdeiʃən] n. 确认，验证

accelerated test 加速试验（为缩短观察产品应力响应所需持续时间或放大给定持续时间内的响应，在不改变基本的故障模式和失效机理或它们的相对主次关系的前提下，施加的应力水平选取超过规定的基本条件的一种试验）

duplication [ˌdjuːpliˈkeiʃən] n. 复制，副本

adversely [ˈædvəːsli] ad. 有害地，不利地，相反地

consistency [kənˈsistənsi] n. 一致性，连续性，稳定性

via [ˈvaiə] prep. 经过，取道

statistical process control 统计过程控制（a method of quality control which employs statistical methods to measure, monitor, and control a process），简称 SPC

subassembly [sʌbəˈsembli] n. 部件，组件

without anyone's knowledge 没有任何人知道

identify [aiˈdentifai] v. 识别，确定，发现

key parts supplier 关键零件供货厂商

ahead of 在……前头，超前于，比……提前

initiate [iˈniʃieit] v. 开始，开创，着手，引起，发动，发现

notification [ˌnəutifiˈkeiʃən] n. 通知，告示，布告

procurement [prəˈkjuəmənt] n. 采购（the purchasing of materials, services or equipment）

coordinate [kəuˈɔːdineit] v. 使同等；n. 同等物，同等者；a. 同等的，并列的

end-use 最终用途（the ultimate application for which a product has been designed）

end-use environment 使用环境

warranty [ˈwɔrənti] n. 质保单（a written guarantee of the maker's responsibility for the repair or replacement of defective products），质保期

bathtub [ˈbɑːtʌb] n. 澡盆，浴缸

bathtub curve 浴盆曲线（在产品整个使用寿命期间，典型的失效率变化曲线，形似浴盆）

descending [di'sendiŋ] n.；a. 下降（的），下行（的），递降（的）
mortality [mɔː'tæliti] a. 致命的；n. 失败率，死亡率
infant mortality period 早期失效期（失效率曲线为递减型）
depict [di'pikt] v. 描述（to describe sth using words），描绘（to show sth in a picture）
attribute M to N 认为 M 是由 N 引起的
failure rate 失效率
useful life period 正常使用期（失效率曲线为恒定型），也称 normal life period
wear out 耗损，消耗，耗尽，用坏，用完
wear out period 耗损失效期（失效率曲线为递增型）
go hand in hand 紧密相关，相随相生，并驾齐驱
shed light on sth. 使某事清楚明白地显示出来，阐明

Notes

[1] under certain stated conditions 意为"在一些规定的条件下"。全句可译为：这个定义的内在含义是在一些规定的条件下或环境中，以及在一个规定的时间长度内来实现其功能。

[2] view as 意为"把……看作"。全句可译为：也就是说，可以把可靠性看作在第一次和每一次都能满足各项要求。

[3] life cycle 意为"使用寿命，寿命周期"。全句可译为：在产品的接受程度方面，很多最成功的企业已经认识到全寿命周期成本和顾客满意的重要性。

[4] be referred to as 意为"称为，被认为"。全句可译为：这通常被称为设计验证步骤。

[5] statistical process control 意为"统计过程控制"。全句可译为：通过诸如统计过程控制这类技术所获得的一致性是制造产品的关键，产品的可靠性可以通过试验来证实。

[6] procurement document 意为"采购文件"。全句可译为：为了防止出现这种情况，在采购文件中应该有一条明确的规定"在没有预先通知的情况下，不能改变工艺"。

[7] bathtub curve 意为"浴盆曲线"。全句可译为：可靠性工程师们所使用的一个有价值的模型是浴盆曲线，它是以其特殊的形状而得名的。

[8] over time 意为"随着时间的推移（an adverb phrase that describes something which happens gradually）"。全句可译为：随着时间的推移，失效率降到了一个低点，在这个低点处，失效率将在一段较长的时间内基本保持不变。

[9] ground rules 意为"基本原则（the basic principles on which future action will be based）"。全句可译为：重要的是应该了解提高产品可靠性方面的基本概念和方法，以及将它们应用到产品设计和制造的各阶段时应遵循的基本原则。

Lesson 61 Effect of Reliability on Product Salability

Is reliability a saleable commodity? Yes and no. It all depends on the type of industry you are in. In the aerospace industry one can give an unqualified yes. Considering the technical complexity of their missions, the reliability achieved by the Apollo space missions is remarkable. One hundred percent safe returns of the Apollo astronauts must be the reliability success story of the second half of the twentieth century. Did someone say why? Simply because reliability was written into the technical specification along with other performance parameters at the conceptual design stage of the project. Tenders were therefore on the basis of achieving the standard of reliability written into the technical specification.

In the global market economy, the desire for profit provides the incentive to reach a high level of reliability. The individual firm or organization must assess the level of reliability necessary to enable the required sales targets to be achieved, along with the other engineering parameters of performance and operating characteristics. The production costs that must be maintained to enable the product to be put on the market at a competitive selling price will usually dictate the type of construction to be employed. Where this is novel, then the engineering development costs must allow for the spending associated with achieving the required level of reliability. Launching a new product which has an unsatisfactory reliability can make a complete nonsense of the best-laid marketing plans. It is good practice if the reliability is shown in the product description, for the customer is usually much less irate if he believes that the trouble has struck in spite of a sensible and feasible program of work designed to check out the reliability of the product.

Examples of the combined development and marketing strategy come from the automotive industry, the aircraft engine industry, and the fan industry. When a new model of car is announced it produces two opposite reactions in the minds of potential customers:

(1) It is new (and therefore "better" than the previous one).

(2) It is new (and therefore unproven and potentially full of teething troubles that will give unreliable operation).

The strategy is to make the customer think it is new (and therefore better than the previous one and therefore desirable) and it has been well proven by test (and therefore should have few hidden faults which will require frequent visits to the service station). To do this it is vital to include in the publicity the more readily understood tests to prove the reliability of the new car, such as "this car has been driven for 150,000 kilometers before we let the public know of its existence," or "two hundred selected drivers have been given this car to test to make sure that our new car will stand up to the rigors of everyday business[1] motoring." Both of these examples have been used in the past to tell the motoring public that extensive research and development activities have been carried out on their behalf to arrive at a desirable and reliable product.

The strategy in the aircraft engine (see Fig. 61.1) industry is similar in that there must be a convincing reply to the airline or armed-forces customer who says that he can remember the teething troubles he had with the last new engine, and what has been done to make sure that he does not have to go through a similar period all over again? The simplest way to put over the development program is to show that with the previous new engine, 10,000 hours of engine running were carried out on the test bed before entering airline or military service, but with the new engine, 15,000 hours of running will be carried out on the test stand before service operation begins. In this sort of situation it is absolutely essential to use the very best market intelligence about what the customer really wants, and is really prepared to pay for.

Figure 61.1 An aircraft engine

In the fan (see Fig. 61.2) industry the customer is concerned with price, performance, and reliability, and usually in that order. There is therefore likely to be less money in the kitty for elaborate testing, and every test must be meaningful to the engineer and convincing to the customer that he can install the product and forget it. Field testing is therefore very important for it is relatively cheap, but of course takes a fair amount of elapsed time. The customer here is particularly convinced by seeing installations, or details of installations, where the new product is running satisfactorily. Damp, steamy, or dirty installations are particularly desirable to introduce an element of "overstress" testing into the field service trials.

Figure 61.2 Fans

If the fan is to be installed in an agricultural sector, then it is important not to field test[2] the fans with the most careful user that can be found, but rather with one who cleans down in a more haphazard fashion, hosing down the equipment when he should not. The hosing down may well show that a sealing arrangement needs to be incorporated on the shaft to stop water entering the motor and ruining the reliability.

Words and Expressions

salability [ˌseiləˈbiliti]   n. 适销性（the quality of being salable or marketable），适合销售
aerospace [ˈɛərəuspeis]   n. 航空航天（飞行器在地球大气层中和太空航行活动的总称）

unqualified [ʌnˈkwɔlifaid] a. 绝对的（absolute; complete），彻底的（out-and-out）
conceptual design stage 概念设计阶段，方案设计阶段
technical specification [ˌspesifiˈkeiʃən] 技术规范，技术指标
tender [ˈtendə] n. 投标（供方应邀做出提供满足合同要求产品的报价）
competitive selling price 有竞争力的销售价格
novel [ˈnɔvəl] a. 新奇的，新颖的（new and original, not like anything seen before）
allow for 考虑到，顾及（to consider something when one makes a calculation）
launch [lɔːntʃ] n. 推出（to make a new product available for sale for the first time）
marketing strategy 营销策略
best-laid marketing plan 精心制订的营销计划（the most carefully made marketing plan）
sensible [ˈsensəbl] a. 明智的，有判断力的（having or showing good sense or judgment）
feasible [ˈfiːzəbl] a. 可行的，切实可行的
motoring [ˈməutəriŋ] n. 驾驶汽车（the act of driving an automobile）
publicity [pʌbˈlisiti] n. 宣传（the act of disseminating information to gain public interest）
rigors 严酷条件，苛刻、艰苦或极端条件（demanding, difficult, or extreme conditions）
armed forces 军队（the military forces of a nation, including the army, navy, air force, etc.）
teething troubles 初期问题（problems that happen in the early stages of doing something new）
test bed 试验台（a place equipped with instruments for testing machinery, engines, etc. under working conditions），亦为 test stand
market intelligence 市场信息（the information relevant to a company's markets）
fan 通风机（指依靠输入的机械能，提高气体压力并排送气体的机械）
elapsed time 运行时间（the amount of time that passes from the start of an event to its finish）
overstress [ˈəuvəˈstres] n.; v. 过度应力，超限应力，超载
haphazard [həpˈhæzəd] a. 偶然的，随意的（marked by lack of plan, order, or direction）
in a more haphazard fashion 以非常随意的方式
hose down 用软管水流冲洗（to wash something down with water from a hose）
sealing arrangement 密封装置

Notes

[1] everyday business 意为"日常业务，日常工作"。

[2] field test 意为"现场试验（在工作、环境、维修和测量条件均作记录的现场所进行的验证试验或测定试验）"，在此处为动词：对……进行现场试验。

Lesson 62 Computers in Manufacturing

The computer is bringing manufacturing into the Information Age. This new tool, long a familiar one in business and management operations, is moving into the factory, and its advent is changing manufacturing as certainly as the steam engine changed it more than 200 years ago.

The basic metalworking processes are not likely to change fundamentally, but their organization and control definitely will.

In one respect, manufacturing could be said to be coming full circle[1]. The first manufacturing was a cottage industry[2]: the designer was also the manufacturer, conceiving and fabricating products one at a time. Eventually, the concept of the interchangeability of parts was developed, production was separated into specialized functions, and identical parts were produced thousands at a time.

Today, although the designer and manufacturer may not become one again, the functions are being drawn close in the movement toward an integrated manufacturing system.

It is perhaps ironic that, at a time when the market demands a high degree of product diversification, the manufacturing enterprises have to increase productivity and reduce cost. Customers are demanding high quality and diversified products for less money.

The computer is the key to meet these requirements. It is the only tool that can provide the quick reflexes, the flexibility and speed, to meet a diversified market. And it is the only tool that enables the detailed analysis and the accessibility of accurate data necessary for the integration of the manufacturing system. Computers make it possible for engineers to solve a great number of challenging problems.

It may well be that, in the future, the computer may be essential to a company's survival. Many of today's businesses will fade away to be replaced by more productive combinations. Such more-productive combinations are super quality, super productivity plants. The goal is to design and operate a plant that would produce 100% satisfactory parts with good productivity.

A sophisticated, competitive world is requiring that manufacturing begin to settle for more, to become itself sophisticated. To meet competition, for example, a company will have to meet the somewhat conflicting demands for greater product diversification, higher quality, improved productivity, and low prices.

The company that seeks to meet these demands will need a sophisticated tool, one that will allow it to respond quickly to customer needs while getting the most out of its manufacturing resources.

The computer is that tool.

Becoming a "super quality, super productivity" plant requires the integration of an extremely complex system. This can be accomplished only when all elements of manufacturing—design, fabrication and assembly, quality assurance, management, materials handling—are computer integrated.

In product design, for example, interactive computer-aided design (CAD) systems allow the drawing and analysis tasks to be performed in a fraction of the time previously required and with greater accuracy. And programs for prototype testing and evaluation further speed the design process.

Additive manufacturing (AM), commonly known as 3D printing, is a computer controlled process that creates three dimensional objects by layering materials one by one based on a digital model. Many different metals and metal alloys are used in additive manufacturing, from precious metals like gold and silver to difficult-to-machine materials like stainless steels and titanium alloys.

On the shop floor, distributed intelligence[3] in the form of microprocessors controlled machines, runs automated loading and unloading equipment, and collects data on current shop conditions.

But such isolated revolutions are not enough. What is needed is a totally automated system, linked by common software[4] from front door to back.

Essentially, computer integration provides widely and instantaneously available, accurate information, improving communication between departments, permitting tighter control, and generally enhancing the overall quality and efficiency of the entire system.

Improved communication can mean, for example, designs that are more producible. The NC programmer and the tool designer have a chance to influence the product designer, and vice versa.

Engineering changes, thus, can be reduced, and those that are required can be handled more efficiently. Not only does the computer permit them to be specified more quickly, but it also alerts subsequent users of the data to the fact that a change has been made.

The instantaneous updating of production-control data permits more effective process planning and production scheduling. Expensive equipment, therefore, is used more productively, and parts move more efficiently through production, reducing work-in-process[5] costs.

Product quality, too, can be improved. Not only are more-accurate designs produced, for example, but the use of design data by the quality-assurance department helps eliminate errors due to misunderstandings.

People are enabled to do their jobs better. By eliminating tedious calculations and paperwork—not to mention time wasted searching for information—the computer not only allows workers to be more productive but also frees them to do what only human beings can do: think creatively.

Computer integration may also lure new people into manufacturing. People are attracted because they want to work in a modern, technologically sophisticated environment.

In manufacturing engineering, CAD/CAM decreases tooling design, NC-programming, and planning times while speeding the response rate, which will eventually permit in-house staff to perform work that is currently being contracted out.

Words and Expressions

metalworking ['metl,wə:kiŋ] *n.* 金属加工（the process of shaping things out of metal）
conceive [kən'si:v] *v.* 设想，想象（to think of or create something in the mind）
interchangeability [,intə,tʃeindʒə'biliti] *n.* 互换性（在同一规格的一批零件或部件中，任取其一，无须任何挑选或附加修配，如钳工修理，就能装在机器上，并达到规定的性能要求）
diversification [dai,və:sifi'keiʃən] *n.* 多样化，变化，不同，多种经营
reflex ['ri:fleks] *n.* 反射，映像，回复，复制品，反应能力
accessibility [,æksesi'biliti] *n.* 容易接近，容易利用（being easy to approach, obtain, or use）
it may well be that 表示 that 从句的内容是很有可能的（it is most likely that）
fade away 逐渐消失（to slowly disappear; to disappear gradually）
conflicting [kən'fliktiŋ] *a.* 不一致的，冲突的，矛盾的，不相容的
fabrication [,fæbri'keiʃən] *n.* 制造（the process of making goods, equipment, etc.）
materials handling 物料搬运（在场所内以改变物料存放状态和空间位置为主要目标的活动）
interactive [,intər'æktiv] *v.* 交互式的，人机对话的交互式
prototype ['prəutətaip] *n.* 原型机，样机（为验证设计或方案的合理性和正确性，或生产的可行性而制作的样品）
additive manufacturing 增材制造（采用材料逐层累加的方法制造实体零件的技术），简称 AM
shop floor 车间，工作场所，（工厂的）生产区
loading 上料（把工件送到工作位置，并实现定位和夹紧的过程）
unloading 下料（把工件从工作位置取下的过程）
process planning 工艺设计（编制各种工艺文件和设计工艺装备等过程）
scheduling ['ʃedju:liŋ] *n.* 制订计划，进度安排（setting an order and time for planned events）
optimum ['ɔptiməm] *a.*; *n.* 最佳的，最佳状态的，最适宜的
update [ʌp'deit] *v.* 使……现代化，适时修正，不断改进，革新
in-house 机构内部的（conducted within, coming from, or being within an organization）
contract out 订立合同把工作包出去（to agree by contract to pay someone outside an organization to perform a job）

Notes

[1] come full circle 意为"回到原地（to return to the original position or state of affairs）"。
[2] cottage industry 意为"家庭手工业（an industry that consists of people working at home）"。
[3] distributed intelligence 意为"分布式智能"。
[4] common software 意为"通用软件（在许多计算机系统中通用的，并有多种应用的程序）"。
[5] work-in-process 意为"在制品（在一个企业的生产过程中，正在进行加工、装配或待进一步加工、装配或待检查验收的制品）"。

Lesson 63 Computer Applications in Design and Manufacturing

Computers have been used in nearly every manufacturing job. Computers improve the efficiency, accuracy, and productivity of many manufacturing processes. Just like the other tools and machines, computers extend human capabilities and make some jobs easier. Every department in manufacturing has found a use for computers. [1]

In the management department, supervisors and managers use computers to gather information about the progress of work in all the other departments. In marketing department, researchers, advertisers, and sales people use computers to get data on potential buyers, to conduct market research, and to create advertisements. [2]

The real tool when using any computer is the software. Software is the set of coded instructions written to control the operations of the computers. Without software, a computer would be a useless machine. Many software programs are available for every department in manufacturing, but the most important uses of computers have been in the engineering and production departments. [3]

CAD. Drafters use CAD systems to make technical drawings of the product to be manufactured. In the past, drafters worked with T-squares, compasses (Figs. 25.1 and 25.2), scales, triangles, and pencils to create these drawings by hand on paper. Making drawing by hand requires a great deal of time. Each line, letter, and shape on a drawing must be created by the drafter. The drafter must have great skill to draw these features exactly the same every time. Today more and more drafters use CAD systems to perform drafting operations.

In CAD, most of these jobs are done by the computer. CAD software programs are written so that a drafter needs only to identify the type of line, letter, or shape required; the computer then draws the feature perfectly on a computer screen. Most CAD systems also include a drawing library of commonly used symbols and shapes to allow faster drawing. [4]

Another advantage of CAD is ease in revising drawings. In the past, when a product was changed or improved the drawings (on paper) had to be completely redrawn or traced. Today, CAD systems allow the drafter to redraw only those parts and features that have been revised.

CAD can also be used to simulate product testing procedures. In the past, when a product was designed by an engineer, scale models and full-sized prototypes were often made and tested for durability, strength, and performance. This process involved making dozens of different models and prototypes until the engineers found the combination of materials, parts, and design they wanted. CAD systems have simulation programs built-in so that product design models can be built, tested, and changed on a computer screen in a small portion of the time needed to make real models. [5] Designing and testing products this way takes less time, saves the company money that used to be spent on the models and prototypes, and improves the quality of manufactured products.

The use of computers in processing materials is called computer-aided manufacturing or CAM. CAM involves controlling machines with computers. In the past, all manufacturing machines were operated and controlled directly by human operators. Nowadays, CAM programmers write software programs for computers that are used to control machines.

CAD/CAM. The next step in computerizing manufacturing processes was to tie the engineering department to the production department using CAD/CAM systems. With CAD/CAM, a drafter in the engineering department makes the drawings for a product on a CAD system, and the design information is then sent directly to machines in the production department where the product is made.

Computer integrated manufacturing (CIM). After the engineering and production departments were connected with CAD/CIM, computers were used to integrate (tie together) all the departments. In CIM, the management, engineering, production, finance, marketing, and human resources departments are all linked by computer. Information about the progress in any department can be seen at once in any other department. The human resources department can prepare workers for specific jobs; the finance department can plan to buy materials to match the production schedule; marketing department can plan advertising to match the finished-product dates, engineering department and production department communicate with CAD/CAM, and management department can direct the whole company by following the progress of each department. With CIM, all separate departments can work toward the single goal of producing and selling a quality product for profit.[6]

Just in time (JIT). Once a company has started working with CIM, it is not far from JIT manufacturing. The principle of JIT is that materials and supplies are delivered to the manufacturer just in time to be used, and products are finished just in time to be delivered to the customer.[7]

For example, manufacturing companies have always ordered materials and supplies[8] in large quantities long before they were needed. These materials and supplies were then stored in nearby warehouses and delivered to the plant when needed. Renting a warehouse, storing supplies, and moving supplies from the warehouse can be expensive for a company. Since CIM helps a company know just when it needs certain supplies or materials, the company can have delivered just in time to the manufacturing plant. This saves the company money that would have been spent to rent the warehouse. Similarly, the company will finish the products and deliver them to the market immediately, again saving the cost to store finished items in a warehouse.

Words and Expressions

written instruction　书面说明，书面指示，书写指令
redraw [ri:'drɔ:]　v. 重新绘制（to draw something again, especially to improve it）
scale [skeil]　n. 比例尺（表示图上线段的长度与相应线段的实际长度之比的工具）
trace　扫描（to copy a drawing by following the lines as seen through a transparent sheet）
built-in ['bilt'in]　a. 内置的（constructed as part of a larger unit; not detachable），嵌入的

production schedule 生产进度表，生产计划
finished product 成品（the product that emerges at the end of a manufacturing process）
just in time manufacturing 准时制造（只在需要的时候，按需要的量生产所需的产品）
quality 优质的，高质量的（very good or excellent; of or having superior quality）

Notes

[1] 全段可译为：几乎制造业中的每一项工作都会用到计算机。计算机提高了许多制造过程的效率、精度和生产率。就像其他工具和机器一样，计算机扩展了人的能力，使某些工作变得更加容易。在制造业的每个部门中都能找到计算机的用武之地。

[2] 全段可译为：在管理部门中，主管和经理们采用计算机收集有关其他部门工作进度的信息。在市场营销部门，研究人员、广告制作人员和销售人员利用计算机来收集潜在客户的数据，用于市场研究和制作广告。

[3] coded instruction 意为"编码指令"，engineering department 意为"工程设计部门，负责新产品的设计和第一台样机的制造与试验"。全段可译为：使用计算机时，软件是真正的工具。软件是用于控制计算机运行的编码指令。没有软件，计算机会变得毫无用处。许多软件程序可以供制造业的每个部门使用，但计算机最重要的用途是在工程设计部门和生产部门。

[4] drawing library 意为"图库，图形库"。全句可译为：大部分CAD系统中还包括一个含有常用符号和图形的图库，来实现快速绘图。

[5] scale models and full-sized prototypes 意为"比例模型和实际尺寸样机"。这四句可译为：CAD也可以用于对产品测试过程进行仿真。在过去，当一种产品被工程师设计出来后，通常需要制作比例模型和实际尺寸样机，并对其耐久性、强度和性能进行测试。这个过程包括制造几十个不同的模型和样机，直至工程师们找到他们想要的材料、零件和设计的组合。CAD系统内部装有仿真程序，可以在计算机屏幕上构建产品设计模型，并对其进行试验和修改，所用的时间仅仅是建造实物模型的一小部分。

[6] specific job 意为"具体工作"，quality product 意为"优质产品"。这两句可译为：人力资源部门可以为具体工作准备工人；财务部门能依据生产计划制订购买物料的计划；市场营销部门能配合产品完工日期安排广告，工程设计部门和生产部门之间用CAD/CAM进行沟通，管理部门能跟踪每个部门的进展，指挥整个公司。采用计算机集成制造，公司中的各个部门都能为同一个目标而工作，即生产和销售优质产品以获取利润。

[7] 这两句可译为：一旦公司开始采用计算机集成制造（CIM），离准时制造就不远了。准时制造的原理是，只是在需要的时候才将材料和零部件送到生产厂家，只是在需要的时候才将产品生产出来，并将其交给客户。

[8] supplies 意为"用品"，这里指零部件。这段可译为：例如，制造公司通常会提前很长时间订购大量的材料和零部件。这些材料和零部件随后就存放在附近的仓库里并在需要时交付给工厂。对一个公司来说，租用仓库、存储零部件和从仓库向外运送零部件的费用是很昂贵的。因为CIM可以帮助公司确切地知道所需要的零部件或材料的数量和时间，所以使得公司能够准时地将其交付给制造厂。这就节省了公司原本用来租赁仓库的费用。同样，公司还可以将生产出来的产品立即投放到市场中，这又节省了产品的库存成本。

Lesson 64 Computer-Aided Analysis of Mechanical Systems

The major goal of the engineering profession is to design and manufacture marketable products of high quality. Today's industries are utilizing computers in every phase of the design, management, manufacture, and storage of their products. The process of design and manufacture, beginning with an idea and ending with a final product, is a closed-loop process.[1]

Factory automation is one of the major objectives of modern industry. In factory automation, all branches of the factory communicate and exchange information through a central data base. Various parts of the product are designed in the computer-aided design (CAD) branch, and then the design is sent to the computer-aided manufacturing (CAM) branch for parts manufacturing and final assembly.[2]

The computer-aided product design branch, better known as computer-aided design (CAD), may consider the design of single parts or it may concern itself with the final product as an assembly of those parts. Computerized product design requires such capabilities as computer-aided analysis, computer-aided drafting, or design optimization. The computer-aided analysis capability serves as part of the design process and is also used as a model simulator for the finished product. Analysis may be considered especially appropriate for a product whose initial design has to be modified several times during the manufacturing process. Thus computer-aided analysis can be used as a substitute for laboratory or field tests in order to reduce the cost.[3]

The computer-aided manufacturing design branch is concerned with the design of the manufacturing process. This branch considers the manufacturability of newly designed parts and employs techniques to improve the manufacturing process, in addition to on-line control of the manufacturing process.

The computer-aided analysis (CAA) process allows the engineer to simulate the behavior of a product and modify its design prior to actual production. In contrast, prior to the introduction of CAA, the manufacturer had to construct and test a series of prototypes, a process which was not only time-consuming but also costly. Most optimal design techniques require repetitive analysis processes. Although one of the major goals of an automated factory is computer-aided design, computer-aided analysis techniques must be developed first.

Computer-aided analysis techniques may be applied to the study of electrical and electronic circuits, structures, or mechanical systems. The development of algorithms for analyzing electrical circuits began in the early days of electronic computer. Similar techniques were also employed to develop computer programs for structural analysis. Today, these programs, known as finite-element techniques, have become highly advanced and are used widely in various fields of engineering.

It was not until the early 1970s that computational techniques found their way into the field of mechanical engineering. [4] One of the areas of mechanical engineering where computational techniques can be employed is the analysis of multi-body mechanical systems.

The purpose of computer-aided analysis of mechanical systems is to develop basic methods for computer formulation and solution of the equations of motion. This requires systematic techniques for formulating the equations and numerical methods for solving them. A computer program for the analysis of mechanical systems can be either a special-purpose program or a general-purpose program.

A special-purpose program is a rigidly structured computer code that deals with only one type of application. Such a program can be made computationally efficient and its storage requirement can be minimized, with the result that it will be suitable for implementation on small personal computers. The major drawback of a special-purpose program is its lack of flexibility for handling other types of applications.

A general-purpose program can be employed to analyze a variety of mechanical systems. For example, the planar motion of a four-bar linkage (see Fig. 64.1) under applied loads and the spatial motion of a vehicle driven over a rough terrain can be simulated with the same general-purpose program. [5] The input data to such a program are provided by the user and must completely describe the mechanical system under consideration. The input must contain such information as number of bodies, connections between the bodies, force elements, and geometric and physical characteristics. The program then generates all of the governing equations of motion and solves them numerically. A general-purpose program, compared with a special-purpose program, is not computationally as efficient and requires more memory space, but it is flexible in use.

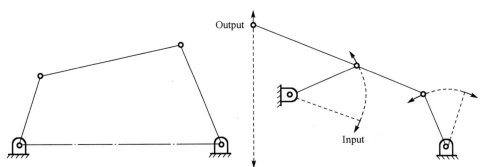

Figure 64.1 Four bar linkages

The computational efficiency of a general-purpose program depends upon several factors, two of which are the choice of coordinates and the method of numerical solution. The choice of coordinates directly influences both the number of the equations of motion and their order of nonlinearity. [6] Furthermore, depending upon the form of these equations, one method of numerical solution may be preferable to another in terms of efficiency and accuracy.

Words and Expressions

factory automation　工厂自动化
optimization [ˌɔptimaiˈzeiʃən] n. 最佳化，最佳参数选择，优化
simulator [ˈsimjuleitə] n. 模拟装置，仿真装置
field test　现场试验，实地试验
on-line control　在线控制，联机控制
analysis process　分析过程，分析程序
multi-body mechanical system　多体机械系统
general purpose program　通用程序
special purpose program　专用程序
planar [ˈplenə] a. 平面的，平的
terrain [ˈterein] n. 地形，地势，地域
governing equation　控制方程，基本方程

Notes

[1] marketable products 意为"适销产品"。全段可译为：工程技术这个职业的主要目标是设计和制造高质量的适销产品。现在，各个工业行业在设计、管理、制造和产品储存的每个阶段都应用了计算机。设计和制造过程是一个从概念开始到最终产品为止的闭环过程。

[2] 全段可译为：工厂自动化是现代工业的一个主要目标。在工厂自动化中，工厂的所有部门都通过一个中央数据库进行沟通和信息交换。在计算机辅助设计（CAD）部门中设计产品中的各种零件，然后将设计方案传送到计算机辅助制造（CAM）部门进行零件加工和最终装配。

[3] substitute for 意为"替代，取代"。这四句可译为：计算机化的产品设计需要有计算机辅助分析、计算机辅助绘图或设计优化等方面的能力。计算机辅助分析能力作为设计过程的一部分，也用来作为产成品的模型模拟器。对于在生产过程中需要多次修改初始设计的产品，这种分析可能被认为是特别合适的。因此，为了降低成本，可以用计算机辅助分析来替代在实验室中的试验和现场试验。

[4] find one's way 意为"找到道路，进入"。全句可译为：直到20世纪70年代初期，计算技术才在机械工程领域得到了应用。

[5] four-bar linkage 意为"四连杆机构，四杆运动链"，rough terrain 意为"崎岖不平地区，地形条件恶劣地区"。全句可译为：例如，一个四连杆机构在外加荷载作用下的平面运动和一辆在崎岖不平道路上行驶的汽车的空间运动都可以用同一个通用程序进行仿真。

[6] order of nonlinearity 意为"非线性的阶数"。全句可译为：坐标的选择直接影响运动方程的个数和方程非线性的阶数。

Lesson 65　Computer-Aided Process Planning

According to the *Tool & Manufacturing Engineers Handbook*, process planning[1] is the systematic determination of the methods by which a product is to be manufactured economically and competitively. It essentially involves selection, calculation, and documentation. Processes, machines, tools, operations, and sequences must be selected. Such factors as feeds, speeds, tolerances, dimensions, and costs must be calculated. Finally, documents in the form of illustrated process sheets[2], operation sheets, and process routes must be prepared. Process planning is an intermediate stage between designing and manufacturing the product. But how well does it bridge design and manufacturing?

Most manufacturing engineers would agree that, if ten different planners were asked to develop a process plan for the same part, they would probably come up with ten different plans. Obviously, all these plans cannot reflect the most efficient manufacturing methods, and, in fact, there is no guarantee that any one of them will constitute the optimum method for manufacturing the part.

What may be even more disturbing is that a process plan developed for a part during a current manufacturing program may be quite different from the plan developed for the same or similar part during a previous manufacturing program and it may never be used again for the same or similar part. That represents a lot of wasted effort and produces a great many inconsistencies in routing, tooling, labor requirements, costing, and possibly even purchase requirements.

Of course, process plans should not necessarily remain static. As lot sizes change and new technology, equipment, and processes become available, the most effective way to manufacture a particular part also changes, and those changes should be reflected in current process plans released to the shop.

A planner must manage and retrieve a great deal of data and many documents, including established standards, machinability data, machine specifications, tooling inventories, stock availability, and existing process plans. This is primarily an information-handling job, and the computer is an ideal companion.

There is another advantage to using computers to help with process planning. Because the task involves many interrelated activities, determining the optimum plan requires many iterations. Since computers can readily perform vast numbers of comparisons, many more alternative plans can be explored than would be possible manually.

A third advantage in the use of computer-aided process planning is uniformity.

Several specific benefits can be expected from the adoption of compute-aided process planning techniques:

1. Reduced clerical work in preparation of instructions.
2. Fewer calculation errors due to human error.

3. Fewer oversights in logic or instructions because of the prompting capability available with interactive computer programs.

4. Immediate access to up-to-date information from a central database.

5. Consistent information, because every planner accesses the same database.

6. Faster response to changes requested by engineers of other operating departments.

7. Automatic use of the latest revision of a part drawing.

8. More-detailed, more-uniform process-plan statements produced by word-processing techniques.

9. More-effective use of inventories of tools, gages, and fixtures and a concomitant reduction in the variety of those items.

10. Better communication with shop personnel because plans can be more specifically tailored to a particular task and presented in unambiguous, proven language.

11. Better information for production planning, including cutter life forecasting, material requirements planning, and inventory control.

Most important for computer integrated manufacturing (CIM), computer-aided process planning produces machine readable data instead of handwritten plans. Such data can readily be transferred to other systems within the CIM hierarchy for use in planning.

There are basically two approaches to computer-aided process planning: variant and generative.

In the variant approach[3], a set of standard process plans is established for all the parts families[4] that have been identified through group technology (GT)[5]. The standard plans are stored in computer memory and retrieved for new parts according to their family identification. Again, GT helps to place the new part in an appropriate family. The standard plan is then edited to suit the specific requirements of a particular job.

In the generative approach[6], an attempt is made to synthesize each individual plan using appropriate algorithms that define the various technological decisions that must be made in the course of manufacturing. In a truly generative process planning system, the sequence of operations, as well as all the manufacturing process parameters, would be automatically established without reference to prior plans. In its ultimate realization, such an approach would be universally applicable: present any plan to the system, and the computer produces the optimum process plan.

No such system exists, however. So called generative process-planning systems—and probably for the foreseeable future—are still specialized systems developed for a specific operation or a particular type of manufacturing process. The logic is based on a combination of past practice and basic technology.

Words and Expressions

computer-aided process planning 计算机辅助工艺设计（在产品制造过程中，利用计算机辅助编制工艺计划，如编制工艺路线卡和检验工序卡等的技术）

guarantee [ˌgærən'tiː] *n.*; *v.* 保证，保证书，担保，承认
retrieve [ri'triːv] *n.*; *v.* 收回，保持，更正，挽救，补救，补偿，检索，恢复
material requirements planning　物料需求计划，其缩写为 MRP
inventory ['invəntri] *n.* 清单，库存（仓库中实际储存的货物）；*v.* 编制目录，清点存货
inventory control　库存控制（对制造业或服务业生产、经营全过程的各种物品、产成品及其他资源进行管理和控制，使其储备保持在经济合理的水平上）
computer integrated manufacturing　计算机集成制造
companion [kəm'pænjən] *n.* 同伴，成对的物件之一，指南，手册，参考书
clerical work　重复性工作（typically involves repetitive tasks, such as typing or data entry）
oversight　疏忽（a mistake made because someone forgets or fails to notice something）
prompt [prɔmpt] *v.* 提醒，提示（to help someone to remember what they were going to do）
iteration [ˌitə'reiʃən] *n.* 反复，重申，迭代，逐步逼近法
up-to-date ['ʌptə'deit] *a.* 现代化的（modern），最新的（making use of what is new or recent）
concomitant [kən'kɔmitənt] *n.* 伴随物（one that occurs or exists concurrently with another）
tailor...to...　使……适合（满足）……（的要求、需要、条件等）
unambiguous [ˌʌnæm'bigjuəs] *a.* 明确的，清楚的，单值的，无歧义的
hierarchy ['haiərɑːki] *n.* 体系，系统，层次，分级结构
variant ['vɛəriənt] *a.* 不同的，各种各样的；*n.* 变型，派生，衍生，转化
generative ['dʒenərətiv] *a.* 能生产的，有生产力的，再生的，创成的
standard process　典型工艺（根据零件的结构和工艺特征进行分类、分组，对同组零件制订的统一加工方法和过程）
standard process plan　典型工艺规程
algorithm ['ælgəriðəm] *n.* 算法，规则系统
parameter [pə'ræmitə] *n.* 参数
process parameter　工艺参数（为了达到预期的技术指标，工艺过程中所需选用的技术数据）

Notes

[1] process planning 意为"工艺设计（编制各种工艺文件和设计工艺装备等的过程）"。

[2] process sheet 意为"工艺过程卡片，工艺卡片"。

[3] variant approach 意为"派生法（将相似零件归并成零件族，在进行工艺设计时检索出相应零件族的工艺规程，并根据设计对象的具体特征加以修订的 CAPP 方法）"。

[4] part family 意为"零件族（a collection of parts that are similar either because of geometric shape and size or because similar processing steps that are required in their manufacture）"。

[5] group technology 意为"成组技术（将企业的多种产品、部件和零件，按一定的相似性准则，分类编组，并以这些组为基础，组织生产的各个环节，从而实现中小批量生产的产品设计、制造和管理的合理化）"。

[6] generative approach 意为"创成法（在输入新零件的全面信息后，根据加工能力知识库和工艺数据库中的加工信息，在没有人为干预的条件下，运用某种决策逻辑与规则自动生成工艺文件。有关零件的信息可以直接从 CAD 系统中获得）"。

Lesson 66 Numerical Control

One of the most fundamental concepts in the area of advanced manufacturing technologies is numerical control (NC). Prior to the advent of NC, all machine tools were manually operated and controlled. Among the many limitations associated with manual control machine tools, perhaps none is more prominent than the limitation of operator skills. With manual control, the quality of the product is directly related to and limited to the skills of the operator. Numerical control represents the first major step away from human control of machine tools.

Numerical control means the control of machine tools and other manufacturing systems through the use of prerecorded, written symbolic instructions[1]. Rather than operating a machine tool, an NC technician writes a program that issues operational instructions[2] to the machine tool.

Numerical control was developed to overcome the limitation of human operators, and it has done so. Numerical control machine tools are more accurate than manually operated machine tools, they can produce parts more uniformly, they are faster, and the long-run tooling costs[3] are lower. The development of NC led to the development of several other innovations in manufacturing technology:

1. Electrical discharge wire cutting.
2. Laser beam cutting.
3. Abrasive waterjet cutting.

Numerical control has also made machine tools more versatile than their manually operated predecessors. An NC machine tool can automatically produce a wide variety of parts, each involving an assortment of widely varied and complex machining processes. Numerical control has allowed manufacturers to undertake the production of products that would not have been feasible from an economic perspective using manually controlled machine tools and processes.

Like so many advanced technologies, NC was born in the laboratories of the Massachusetts Institute of Technology. The concept of NC was developed in the early 1950s with funding provided by the U. S. Air Force.

The APT (Automatically Programmed Tools) language was designed at the Servomechanism laboratory of MIT in 1956. This is a special programming language for NC that uses statements similar to English language to define the part geometry, describe the cutting tool configuration, and specify the necessary motions. The development of the APT language was a major step forward in the further development of NC technology. The original NC systems were vastly different from those used today. The machine tools had hardwired logic circuits. The instructional programs were written on the punched paper tape (see Fig. 66. 1), which was later to be replaced by the magnetic plastic tape. A tape reader (see Fig. 66.2) was used to interpret the instructions written on the tape for the machine tool. Together, all of this represented a giant step forward in the control of machine tools. However, there were a number of problems with NC at this point in its development.

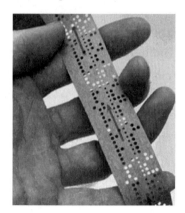

Figure 66.1 A punched paper tape

Figure 66.2 A photo-electric paper tape reader

A major problem was the fragility of the punched paper tape medium. It was common for the paper tape containing the programmed instructions to break or tear during a machining process. This problem was exacerbated by the fact that each successive time a part was produced on a machine tool, the paper tape carrying the programmed instructions had to be rerun through the reader. If it was necessary to produce 100 copies of a given part, it was also necessary to run the paper tape through the reader 100 separate times. Fragile paper tapes simply could not withstand the rigors of a shop floor environment and this kind of repeated use.

This led to the development of a special magnetic plastic tape. Whereas the paper tape carried the programmed instructions as a series of holes punched in the tape, the plastic tape carried the instructions as a series of magnetic dots. The plastic tape was much stronger than the paper tape, which solved the problem of frequent tearing and breakage. However, it still left two other problems.

The most important of these was that it was difficult or impossible to change the instructions entered on the magnetic tape. To make even the most minor adjustments in a program of instructions, it was necessary to interrupt machining operations and make a new tape. It was also still necessary to run the tape through the reader as many times as there were parts to be produced. Fortunately, computer technology became a reality and soon solved the problems of NC associated with punched paper and magnetic plastic tapes.

The development of a concept known as direct numerical control (DNC) solved the paper and plastic tape problems associated with numerical control by simply eliminating tape as the medium for carrying the programmed instructions. In direct numerical control, machine tools are tied, via a data transmission link, to a host computer[4]. Programs for operating the machine tools are stored in the host computer and fed to the machine tool as needed via the data transmission linkage. Direct numerical control represented a major step forward over punched tape and plastic tape. However, it is subject to the same limitations as all technologies that depend on a host computer. When the host computer goes down, the machine tools also experience downtime. This problem led to the development of computer numerical control.

The development of the microprocessor allowed for the development of programmable logic controllers (PLCs)[5] and microcomputers. These two technologies allowed for the development of computer numerical control (CNC). With CNC, each machine tool has a PLC (see Fig. 66.3) or a microcomputer that serves the same purpose. This allows programs to be input and stored at each individual machine tool. It also allows programs to be developed off-line[6] and downloaded at the individual machine tool. CNC solved the problems associated with downtime of the host computer, but it introduced another problem known as data management. The same program might be loaded on ten different microcomputers with no communication among them. This problem is in the process of being solved by local area networks that connect microcomputers for better data management.

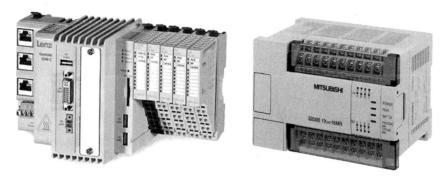

Figure 66.3　Programmable logic controllers

Words and Expressions

numerical control　数字控制（用数字化信号对机床运动及其加工过程进行控制的控制方式）
prerecord ['priːriˈkɔːd]　v. 事先录制（to record at an earlier time for later use）
numerical control machine tool　数控机床（按加工要求预先编制程序，由控制系统发出数字信息指令进行加工的机床），亦称 NC machine tool
long-run　长期的（a relatively long period of time）
tooling ['tuːlɪŋ]　n. 工艺装备（产品制造过程中所用的各种工具的总称。包括刀具、夹具、模具、量具、检具、辅具和钳工工具等），简称"工装"
innovation [ˌɪnəʊˈveɪʃən]　n. 改革，发明，革新，新方法，合理化建议
discharge [dɪsˈtʃɑːdʒ]　v. 卸下，放出，放电
electrical discharge wire cutting　电火花线切割（通过线状工具电极与工件间规定的相对运动，切割出所需工件的电火花加工），简称 EDWC，通称 wire EDM
laser beam cutting　激光切割（利用聚焦后的激光束作为主要热源的热切割方法）

waterjet ['wɔːtəˌdʒet] n. 水射流（经高压喷嘴流出的高速水流束）
abrasive waterjet cutting   磨料水射流切割，水磨料喷射切割（利用水流及微粒磨料，经高压喷嘴形成射流，对金属、非金属和复合材料等进行切割加工的方法）
predecessor ['priːdisesə] n. 被替代的事物（something that comes before something else）
a wide variety of   多种多样的
assortment [əˈsɔːtmənt] n. 种类，花色品种，分类
an assortment of   各式各样的
feasible ['fiːzəbl] a. 可行的，做得到的，合理的，可用的
hardwired   硬连线的（having permanent electronic circuits and connections），硬件实现的
interpret [inˈtəːprit] v. 解释，说明，翻译，译码
fragility [frəˈdʒiliti] n. 脆弱，脆性，易碎性
paper tape   纸带（能存储或记录信息的纸条）
magnetic tape   磁带（具有存贮数据的磁化表层的一种控制带）
punched tape   穿孔带（其上穿有代表一套具体数据的孔的控制带）
punch [pʌntʃ] n.; v. 打孔，穿孔
tape reader   读带机（能读出控制带上记录的数据的一种装置）
photo-electric paper tape reader   光电读带机
exacerbate [eksˈæsəːbeit] vt. 加重，使恶化，激怒
rerun ['riːˈrʌn] v. 再开动，重新运转，重算
successive [səkˈsesiv] a. 连续的（following each other without interruption），继承的
allow for   使……成为可能，提供机会（to make a possibility or provide opportunity for）
computer numerical control   计算机数控，简称 CNC
local area network   局域网，缩写为 LAN，指有限区域（如办公室或楼层）内的多台计算机通过共享的传输介质互连，所组成的封闭网络

Notes

[1] symbolic instruction 意为"符号指令"。

[2] operational instruction 意为"操作指令"。

[3] long-run tooling cost 意为"长期的工艺装备成本"。

[4] host computer 意为"主计算机，主机（a large, powerful computer that can handle many tasks concurrently; the main computer in a network）"。

[5] programmable logic controller 意为"可编程控制器（通过编程可实现顺序控制、定时、计数和数学运算，通过输出/输入接口实现机械产品的控制的工业控制装置）"，也称"programmable controller"，曾称"可编程逻辑控制器"。

[6] off-line 意为"离线，脱机（指设备或装置不受中央处理机直接控制的情况，not connected to a central computer or computer network）"。

Lesson 67 Numerical Control Software

Today, the product design process begins with computer-generated product concepts and designs, which are subjected to detailed analysis of feasibility, manufacturability, and even disposability. [1]

Quality criteria are determined to meet safety, environmental requirements, and conformance with company and industry standards. [2] Engineering drawings are produced as well as the bill of material. Final design decisions are made relating to styling, functional, performance, materials, tolerances, make-versus-buy, purchased parts, supplier selection, manufacturability, quality, and reliability. [3]

In the process planning stage, tooling decisions are made. The sequence of production steps is planned with actions taken at each step and controls specified, i.e. the actions taken at each step, controls to be followed, and the state of the workpiece at each workstation. In computer-aided process planning (CAPP), an application program stores prior plans and standard sequences of manufacturing operations for families of parts coded using the group technology concept, which classifies parts based on similarity of geometric shape, manufacturing processes, or some other part characteristics.

All tools required to produce the final part or assembly are specified or designed. This includes molds, forming and bending dies (see Fig. 67.1), jigs and fixtures, cutting tools, and other tooling. The tool design group typically works closely with the tool room and with suppliers to produce the necessary tools in time to meet production schedules.

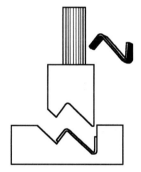

Figure 67.1 A bending die

Numerical control provides the operational control of a machine or machines by a series of computer-coded instructions comprising numbers, letters, and other symbols, which are translated into pulses of electrical current or other output signals that activate motors or other devices to run the machine. [4] With NC, machines run consistently, accurately, predictably, and essentially automatically. Quality and productivity are increased, and rework and lead times are reduced compared with manual operation.

APT (Automatic Programming of Tools) was the first NC language. APT was designed to function as an off-line, batch program used in a mainframe computer system. Because of the computer resources needed and the expense, time-sharing was employed. Eventually, interactive graphics-based NC programs used in terminals and workstations were introduced to improve visualization and provide the opportunity for immediate feedback to the user.

Interactive graphics-based NC part-programming technology increases product quality and simplifies the process by reducing setup time for lathes, milling machines, and other machine tools. Graphics software lets users easily define part geometry, obtain immediate feedback, and visualize the results while changes are made quickly and efficiently.

Today in industry nearly all production programs are generated using NC software. It's a matter of efficient use of your time and staying up with the competition. NC software brings even more to the shop than turning hours of calculating and code writing into minutes, it also programs shapes not possible any other way.

NC packages accomplish four major functions: part description, machining strategy, post-processing, and factory communications.

To describe a part, most NC programming systems provide their own geometric modeling capability as an integral component of the system. This CAD-like frontend permits users to create parts by drawing lines, circles, arcs, and splines. All NC packages support 2D geometry creation and many have optional 3D modules. The 3D module permits the creation of complex surfaces or direct machining from the solid model will become accepted in NC.

A close relationship between the model creation and machining of that model is important. Major CAD/CAM vendors provide NC software that is integrated with the design and drafting function for just this reason. Operating from the same database, the NC software can directly access a model that has been created within the design module of the CAD/CAM system. This eliminates the requirement for translation of data from one format to another.

Some NC software vendors provide stand-alone systems. In this case, a file containing the definition of the model is imported from a CAD system. An IGES or DXF translation is usually used to convert the geometry from the originating system to the NC system.[5] In other cases, the NC software can directly access the CAD database and avoid translation. In dedicated NC systems, complex NC calculations may be performed faster than for other systems and more NC-specific utilities are often available.[6]

Currently, NC programming is usually done by individuals who have actual machining experience on the shop floor. This experience and the associated knowledge gained are critical in developing the machining strategy required to cut a part. It is particularly useful in handling unusual circumstances that often arise in machining complex parts.

Knowledge-based software systems can capture the NC programmer's knowledge and establish a set of rules to create a framework that can be used to lead the NC programmer through the development of the machining strategy.

Knowledge-based systems may also be used to fully automate some NC programming tasks. This would provide for greater consistency among machining strategies and improve programming productivity.

Words and Expressions

subject to   在……条件下，根据，受到，经受
feasibility [ˌfiːzəˈbiliti]   n. 可行性，可能性，现实性
manufacturability [ˌmænjuˈfæktjurəˈbiliti]   n. 可制造性，工艺性

disposal [dis'pəuzəl] n. 废弃物清除，处置，处理
conformance [kən'fɔːməns] n. 相似，一致性，适应性
bill of material 材料表，材料清单，物料清单
tooling ['tuːliŋ] n. 工艺装备，工装
computer-aided process planning (CAPP) 计算机辅助工艺过程设计
family of part 零件族
group technology 成组技术
mold [məuld] n. 铸模，模型
forming die 成型钢模，锻模
bending die 弯曲模
jigs and fixtures 夹具（用于装夹工件和引导刀具的装置）
tool room 工具车间，工具间，工具库，也可写为 toolroom
pulse [pʌls] n. 脉冲
batch program 批处理程序
mainframe 主计算机，大型计算机
time-sharing 分时（当多用户通过终端设备同时使用一台计算机时，系统把时间分成许多极短的时间片，分配给每个联机作业，由时钟控制中断，使各作业交错使用计算机）
interactive graphics-based 基于交互式图形的
software package 软件包，程序包
dedicated ['dedikeitid] a. 专用的
terminal ['təːminl] n. 终端设备，终端机
workstation ['wəːksteiʃən] n. 工作区，工作站，〈计〉工作站（是一种高端的通用微型计算机。通常配有高分辨率的大屏幕显示器及容量很大的内存储器和外部存储器，并且具有较强的信息处理功能和高性能的图形、图像处理功能及联网功能）
visualization [,vizjuəlai'zeiʃən] n. 可视化，使看得见的
setup time 准备时间，设备调整时间，设置时间
NC package 数控程序包
strategy ['strætidʒi] n. 策略，对策，计谋
machining strategy 加工策略，加工方案
post-processing 后置处理（就是结合特定的机床把系统生成的刀具轨迹转化成机床能够识别的 G 代码指令，生成的 G 代码指令可以直接输入数控机床用于加工），后处理
frontend ['frʌnt'end] n. 前端
spline [splain] n. 仿样，样条，(pl.) 仿样函数，样条函数
specialty [,speʃi'æliti] n. 专长，专业（化），特制品，特殊产品
IGES (initial graphic exchange specification) 初始图形交换规范
DXF (data exchange format) 数据交换格式
knowledge-based 基于知识的
capture ['kæptʃə] v. 收集，吸收，吸取，记录
a set of rules 一组规则
framework ['freimwəːk] n. 构架，框架，结构

default [di'fɔːlt] *n.* 系统设定（预置）值，默认（值）； *v.* 默认

methodology [ˌmeθə'dɔlədʒi] *n.* 方法学，分类法

Notes

［1］feasibility, manufacturability and disposability 意为"可行性、工艺性和可处置性"。全句可译为：现在，产品设计过程可以从由计算机生成的产品概念和设计开始，它可以对产品的可行性、工艺性甚至可处置性进行详细的分析。

［2］company and industry standard 意为"企业和行业标准"。全句可译为：质量标准是以满足安全和环境要求，符合企业和行业标准为前提来制定的。

［3］design decision 意为"设计决策"。全句可译为：最终的设计决策是在考虑了式样、功能、性能、材料、公差、自己制作还是外购、外购件、供货厂商的选择、工艺性、质量和可靠性这些因素后做出的。

［4］computer-coded instruction 意为"计算机编码指令"。全句可译为：数字控制通过一系列的计算机编码指令对一台机器或多台机器进行操作控制。这些指令是由数字、字母和其他符号组成的，它们被转化成电流脉冲或其他输出信号来驱动电动机或其他装置，使机器运转。

［5］IGES or DXF 意为"初始图形交换规范（initial graphic exchange specification）或数据交换格式（data exchange format）"。全句可译为：通常采用 IGES 或 DXF 将几何形状从原来的系统中转换到数控系统中。

［6］dedicated NC system 意为"专用型数控系统"，utility 这里指"实用程序"。全句可译为：在专用型数控系统中，可以比在其他系统中更快地进行复杂的数控计算，通常还会有一些专门的实用程序。

Lesson 68 Computer Numerical Control

Today, computer numerical control (CNC) machine tools are widely used in manufacturing enterprises. Computer numerical control is the automated control of machine tools by a computer and computer program.

The CNC machine tools still perform essentially the same functions as manually operated machine tools, but movements of the machine tool are controlled electronically rather than by hand. CNC machine tools can produce the same parts over and over again with very little variation. They can run day and night, week after week, without getting tired. These are obvious advantages over manually operated machine tools, which need a great deal of human interaction in order to do anything.

A CNC machine tool differs from a manually operated machine tool only in respect to the specialized components that make up the CNC system. The CNC system can be further divided into three subsystems: control, drive, and feedback. All of these subsystems must work together to form a complete CNC system.

1. Control System

The centerpiece of the CNC system is the control. Technically the control is called the machine control unit (MCU), but the most common names used in recent years are controller, control unit, or just plain control. This is the computer that stores and reads the program and tells the other components what to do.

2. Drive System

The drive system is comprised of screws and motors that will finally turn the part program into motion.[1] The first component of the typical drive system is a high-precision lead screw called a ball screw (see also Fig. 58.2). A ball screw is a mechanical device that translates rotational motion to linear motion with little friction. Eliminating backlash in a ball screw is very important for two reasons. First, high-precision positioning can not be achieved if the table is free to move slightly when it is supposed to be stationary. Second, material can be climb cut safely if the backlash has been eliminated. Climb cutting is usually the most desirable method for machining on a CNC machine tool.

Drive motors are the second specialized component in the drive system. The turning of the motor will turn the ball screw to directly cause the machine table to move.[2] Several types of electric motors are used on CNC control systems, and hydraulic motors (see Fig. 68.1) are also occasionally used.

The simplest type of electric motor used in CNC positioning systems is the stepper motor (sometimes called a stepping motor). A stepper motor rotates a fixed number of degrees when it receives an electrical pulse and then stops until another pulse is received.[3] The stepping characteristic makes stepper motors easy to control.

It is more common to use servomotors in CNC systems today. Servomotors operate in a smooth, continuous motion—not like the discrete movements of the stepper motors. This smooth motion leads to highly desirable machining characteristics, but they are also difficult to control. Specialized controls and feedback systems are needed to control and drive these motors. Alternating current (AC) servomotors are currently the standard choice for industrial CNC machine tools.[4]

Figure 68.1 Hydraulic motors

3. Feedback System

The function of a feedback system is to provide the control with information about the status of the motion control system, which is described in Fig. 68.2.

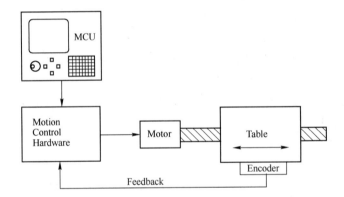

Figure 68.2 Typical motion control system of a CNC machine tool

The control can compare the desired condition to the actual condition and make corrections. The most obvious information to be fed back[5] to the control on a CNC machine tool is the position of the table and the velocity of the motors. Other information may also be fed back that is not directly related to motion control, such as the temperature of the motor and the load on the spindle—this information protects the machine from damage.

There are two main types of control systems: open-loop and closed-loop.[6] An open-loop system does not have any device to determine if the instructions were carried out. For example, in an open-loop system, the control could give instructions to turn the motor 10 revolutions.[7] However, no information can come back to the control to tell it if it actually turned. All the control knows is that it delivered the instructions. Open-loop control is not used for critical systems, but it is a good choice for inexpensive motion control systems in which accuracy and reliability are not critical.

Closed-loop feedback uses sensors to verify that certain conditions have been met. Of course, position and velocity feedback is of primary importance to an accurate CNC system. Feedback is the only way to ensure that the machine is behaving the way the control intended it to behave.

Words and Expressions

subsystem ['sʌbˌsistim] n. 子系统（a system that is part of a larger system）
centerpiece ['sentəpiːs] n. 最重要的部分（the most important part of something）
machine control unit 机床控制装置，简称 MCU
drive system 传动系统，驱动系统
backlash ['bæklæʃ] n. （机械部件间的）间隙
climb cut 顺铣（铣刀的旋转方向和工件进给方向相同），亦为 climb milling 或 down milling
drive motor 驱动电动机
lead screw 丝杠（a threaded rod that converts rotational motion into linear motion）
ball screw 滚珠丝杠副（丝杠与旋合螺母之间以钢珠为滚动体的螺旋传动副。它可将旋转运动变为直线运动或将直线运动变为旋转运动）
hydraulic motor 液压马达（做连续回转运动并输出转矩的液压执行元件）
positioning [pəˈziʃəniŋ] n. 位置控制，定位
stepper motor 步进电动机（是将电脉冲激励信号转换成相应的角位移或线位移的离散值控制电动机），亦可写为 stepping motor
servomotor [ˈsəːvəuˌməutə] n. 伺服电动机（转子或动子的机械运动受输入电信号控制做快速反应的电动机）
discrete [disˈkriːt] movement 离散的运动，不连续的运动
feedback [ˈfiːdbæk] n. 反馈（将输出量通过恰当的检测装置返回到输入端并与输入量进行比较的过程）
open-loop 开环（系统的输出量对系统的控制没有影响）
closed-loop 闭环（系统的输出量对系统的控制作用有直接的影响）
encoder [inˈkəudə] n. 编码器（把角位移或直线位移转换成电信号的装置）

Notes

[1] comprised of 意为"由……组成"，screw 这里指"丝杠"。全句可译为：驱动系统由丝杠和电动机组成，它最终会把零件加工程序转化为运动。

[2] machine table 意为"机床工作台"。全句可译为：电动机的转动会带动滚珠丝杠副转动，并使机床工作台产生运动。

[3] electrical pulse 意为"电脉冲"。全句可译为：步进电动机每接收到一个电脉冲就会转动一个固定的角度，然后停止转动，直至接收到另一个脉冲。

[4] controls 这里指"控制器，即 controllers 或 control units"，alternating current 意为"交流"。这两句可译为：需要专用的控制器和反馈系统来控制和驱动这些电动机。目前，在工业生产中使用的计算机数控机床中通常选用交流伺服电动机。

[5] feed back (*fed back*, *fed back*) 意为"反馈"，它是一个短语动词（phrasal verb）。

[6] 全句可译为：主要有两种类型的控制系统：开环系统和闭环系统。

[7] revolution 这里指"转（a single complete turn）"。全句可译为：例如，在开环系统中，控制装置可以发出让电动机转动 10 转的指令。

Lesson 69 Training Programmers

Skillful part-programmers are a vital requirement for effective utilization of NC machine tools. Upon their efforts depend the operational efficiency[1] of those machines and the financial payback of the significant investment in the machines themselves, the plant's NC-support facilities, and the overhead costs involved.

Skillful NC part-programmers are scarce. This reflects not only the general shortage of experienced people in the metalworking industries but also the increasing demand for programmers as industry turns more and more to the use of numerically controlled machines to increase the capability, versatility, and productivity of manufacturing.

On an industry-wide basis, the obvious answer is to create new programmers by training them—and there are a number of sources for such training. But first, what qualifications should programmers have, and what must programming trainees learn?

According to the National Machine Tool Builders' Assn booklet "*Selecting an Appropriate NC programming Method*", the principal qualifications for manual programmers are as follows:

Manufacturing Experience Programmers must have a thorough understanding of the capabilities of the NC machines (see Fig. 69.1) being programmed, as well as an understanding of the basic capabilities of the other machines in the shop. They must have an extensive knowledge of metal cutting principles and practices, cutting capabilities of the tools, and workholding fixtures[2] and techniques. Programmers properly trained in these manufacturing engineering techniques can significantly reduce production costs.

Figure 69.1 NC machines

Spatial Visualization Programmers must be able to visualize parts in three dimensions, the cutting motions of the machine, and potential interferences between the cutting tool, workpiece, fixture, or the machine itself.

Mathematics A working knowledge of arithmetic, algebraic, trigonometric, and geometric operations is extremely important. The knowledge of higher mathematics, such as advanced algebra, calculus, etc, is not normally required.

Attention to Details   It is essential that programmers be acutely observant and meticulously accurate individuals. Programming errors discovered during machine setup[3] can be very expensive and time-consuming to correct.

"Manual programming,"[4] the booklet notes elsewhere, "requires the programmer to have more detailed knowledge of the machine and control, machining practices, and methods of computation than does computer-aided programming. Computer-aided programming, on the other hand, requires a knowledge of the computer programming language and the computer system in order to process that language. In general, manual programming is more tedious and demanding because of the detail involved. In a computer-aided programming system, this detail knowledge is embodied in the computer system (processor, post processor, etc.)."

Experts in the NC and training fields typically agree on these qualifications and requirements — adding such subsidiary details as a knowledge of blueprint reading, machinability of different metals, use of shop measuring instruments, tolerancing methods, and safety practices.

Where should you look for candidates? First of all, in your own plant — out on the shop floor. Edward F. Schloss, a Cincinnati Milacron sales vice president, puts it this way: "We've had excellent success with good lathe operators and good milling machine operators. They don't know it, but they've been programming most of their working lives, and they know basic shop math and trigonometry. You can teach them programming rather handily. Conversely, though, it's fairly hard to make NC part-programmers out of high-powered mathematicians. The tool path programming is easy. But what to do with it — the feeds, speeds, etc. — that may take even more-extensive training."

With more-powerful computer-aided part programming, the need for metal cutting knowledge on the part of programmers is reduced. Through the use of this software, Cincinnati Milacron has been very successful in hiring new college graduates, including some with nontechnical degrees, and training them to be NC part-programmers. The trainees are given hands-on machine tool experience in the plant before they are advanced to programming.

All suppliers of NC machine tools, of course, provide some sort of training in the programming of their products, and most offer formalized training programs, Milacron's sales department, for example, has 20 full-time customer-training instructors. The company's prerequisites for programmer training include the following:

"Participants must have knowledge of general machine shop safety procedures and be able to read detail drawings and sectional views."

"Knowledge of plane geometry, right-angle trigonometry, and fundamentals of tolerancing is required."

"Knowledge of NC manual part-programming, NC machine tool setup and operating procedures[5], part processing, metal cutting technology, tooling, and fixturing is also needed."

Sending people with that kind of background to school will ensure that users of the NC machine will get the maximum benefit for their training dollar—the cost of a week of the trainee's time, travel, and living expenses, even though the training fee is waived with the basic purchase of the machine tool.

Words and Expressions

programmer [ˈprəugræmə]　*n*.　程序设计员，程序编制员，制订计划者
assn= association　协会，学会（an organization of persons having a common interest）
booklet [ˈbuklit]　*n*.　小册子（a little book usually having paper covers and few pages）
workhold [ˈwəːkhəuld]　*v*.　工件夹持
fixture [ˈfikstʃə]　*n*.　夹具（用于装夹工件和引导刀具的装置）
spatial [ˈspeiʃəl]　*a*.　空间的，立体的
visualization [ˌvizjuəlaiˈzeiʃən]　*n*.　形象化，想象（formation of mental visual images）
spatial visualization　空间想象力（ability to think in three dimensions）
algebraic [ˌældʒiˈbreiik]　*a*.　代数的
trigonometric [ˌtrigənəˈmetrik]　*a*.　三角学的，三角的
calculus [ˈkælkjuləs]　*n*.　微积分
observant [əbˈzəːvənt]　*a*.　严格遵守……的，留心的，观察力敏锐的
meticulously [miˈtikjuləsli]　*ad*.　认真地（in a way that shows great care and attention to detail）
demanding　苛求的（hard to satisfy），要求高的（requiring a lot of skill, effort, etc.）
post processor　后置处理程序（software that converts the toolpaths from a CAM system into NC code that is designed for a specific CNC machine）
subsidiary [səbˈsidjəri]　*a*.　辅助的，次要的，附属的；*n*.　附属机械
blueprint [ˈbluːˈprint]　*n*.；*v*.　（晒）蓝图，设计图，（制订）计划
handily [ˈhændili]　*ad*.　容易地（very easily），方便地（in a convenient manner）
high-powered　精力充沛的，能力很强的（extremely energetic, dynamic, and capable）
hands-on　实践的（gained by actually doing something rather than learning about it from books）
instructor [inˈstrʌktə]　*n*.　任课教师（a person who teaches a subject or skill; a teacher）
prerequisite [ˈpriːˈrekwizit]　*a*.　先决条件的，必要的；*n*.　前提，先决条件
waive [weiv]　*v*.　放弃，免除

Notes

[1] operational efficiency 意为"运行效率，经营效率"。
[2] workholding fixture 意为"工件夹具"，其主要功能为：定位（locating or positioning）、支承（supporting）和夹紧（holding or clamping）。
[3] setup 这里指"调试（the preparation and adjustment of machines for an assigned task）"。
[4] manual programming 意为"手工编程，手动编程，人工编程"。
[5] operating procedure 意为"操作程序"。

Lesson 70　History of Numerical Control
(*Reading Material*)

　　Man has been described as a tool using animal. Among the characteristics that distinguish him from other species is an ability to make and use complex devices that magnify or extend his own capabilities. These devices which we call machines have governed the rate of man's material progress throughout history. The evolution of the machine can be attributed to its inherent propagating power. Existing tools make possible the manufacture of more advanced tools which in turn serve to accelerate the evolutionary process.

　　The first machine tools are believed to have been developed more than 2,500 years ago. These early rotary devices allowed the artisan to produce intricate circular forms from wood and other hard materials. Although the early machines extended man's ability to produce relatively complex shapes, it was not until the fourteenth century that the first elementary precision machines were developed. The mechanical weight driven clock (see Fig. 70.1) became the impetus for the development of the first true machine tools, such as the screw cutting lathe. The advent of the industrial revolution greatly accelerated the evolution of the machine tool, and the development of the steam engine by James Watt in the latter half of the eighteenth century precipitated requirements for new devices and precision in metal cutting tools.

　　In 1798 Eli Whitney signed a contract with the U.S. Government to produce 12,000 muskets and promised that the parts of each musket would be interchangeable. The commitment required manufacturing control which had never before been attempted. Whitney and his associates designed water powered machinery to perform the forging, boring, grinding, and polishing operations at his factory in New Haven, Connecticut. Although Whitney had contracted to produce the muskets in two years, only 500 were completed in September 1801, and it took him eight years to satisfy the contract. Nevertheless, Whitney had developed precision manufacturing methods which were a hallmark for his time.

　　During the latter half of the nineteenth century, important refinements were made to metal cutting machine tools. Early machines were continually updated to improve production rates.

　　By 1900, the American machine shop contained basic machine tools that were not very different in function and form from those in use today. The manufacture of automobiles in England and then in the United States served as a driving force for the development of better machine tools. By 1914, the Ford Motor Company was producing over one-million "Model-Ts" (see Fig. 70.2) each year. In little more than fifty years mass production had become commonplace.

　　The twentieth century has seen machinery became more automated, thereby eliminating machine-operator intervention in the manufacturing process. With the advent of new difficult-to-machine materials and requirements for tight tolerances, the best human operators have reached the limit of their ability. The method of numerical control (NC) is used to solve these problems, and in future manufacturing is expected to be increasingly dependent on numerical control.

Figure 70.1   A weight driven clock

Figure 70.2   Model-Ts

The history of NC began in the late 1940s, when John T. Parsons proposed that a method of automatic machine control be developed which would guide a milling cutter to generate a smooth curve. In 1949 the U.S. Air Force commissioned the Servomechanisms Laboratory at the Massachusetts Institute of Technology to develop a workable NC system based on Parsons' concept.

Scientists and engineers at MIT selected punched paper tape (see Fig. 66.1) as the communication medium and initially built a two-axis point-to-point control system which positioned the drilling head over the coordinate. Later, a more sophisticated continuous path milling machine was produced. Independent machine tool builders have subsequently developed the systems currently available.

By 1957, the first successful NC installations were being used in production; however, many users were experiencing difficulty in generating part programs for input to the machine controller. To remedy this situation, MIT began the development of a computer-based part programming language called APT—automatically programmed tools. The objective was to devise a symbolic language which would enable the part programmer to specify mathematical relationships in a straightforward manner. In 1962 the first APT programming system was released for general industrial applications.

The development of numerical control technology has taken place on two major fronts. Hardware development concentrated on improved control systems and machine tools. Software development concentrated on improvements to the APT language as well as the origination of other NC programming systems.

A change in overall philosophy began in the 1970s, and numerical control was then viewed as part of a larger concept—computer-aided manufacturing (CAM). CAM encompasses not only NC but production control and monitoring, materials management, and scheduling. The emphasis on the use of computers in the manufacturing process has spawned new forms of numerical control: CNC (computer numerical control) and DNC (direct numerical control). Numerical control continues to develop, and possibilities that were once considered science fiction are now seen as attainable goals.

NC is a system that can interpret a set of prerecorded instructions in some symbolic format; it can cause the controlled machine to execute the instructions, and then can monitor the results so that the required precision and function are maintained.

Numerical control is not a kind of machine tool but a technique for controlling a wide variety of machines. For this reason NC has been applied to assembly machines, inspection equipment, wood working machines (to name only a few applications) as well as metal cutting machine tools.

The numerical control system forms a communication link which has many similarities to conventional processes. Symbolic instructions are input to an electronic control unit which decodes them, performs any logical operations required, and outputs precise instructions that control the operation of the machine. Many NC systems contain sensing devices that transmit machine status back to the control unit. It is this feedback that enables the controller to verify that the machine operation conforms to the symbolic input instructions.

Words and Expressions

material progress　物质文明
propagate ['prɔpəgeit]　v. 繁殖，传播，普及，增殖
evolutionary process　进化过程，演化过程
weight driven clock　重力驱动的时钟（a mechanical clock powered by weights）
impetus ['impitəs]　n. 推动力，促进
precipitate [pri'sipiteit]　v. 促进，加速……来临
musket ['mʌskit]　n. （旧时的）步枪，滑膛枪（枪管内无膛线的枪械）
commitment [kə'mitmənt]　n. 承诺，承诺事项（a promise to do or give sth; sth pledged）
polishing ['pɔliʃiŋ]　n. 抛光（降低工件表面粗糙度，以获得光亮、平整表面的加工方法）
hallmark ['hɔːlmɑːk]　n. 特点（a distinguishing characteristic or feature），标记，标志
driving force　驱动力（the main factor that causes something to happen）
Ford Motor Company　福特汽车公司
Model-T　T型车（世界上第一种以大量通用零部件和大规模流水线装配方式生产的汽车）
intervention [ˌintə'venʃən]　n. 干涉，干预，介入
with the advent of　随……的到来，随……的出现
generic name　总名称，属名，类名，一般名称
difficult-to-machine material　难加工材料，亦为 difficult-to-cut material
abbreviate [ə'briːvieit]　v. 缩写，简化
John Parsons　约翰·帕森斯（1913—2007），数控加工技术的创始人
commission [kə'miʃən]　n. 委托，授权
servomechanism ['sɜːvəu'mekənizəm]　n. 伺服机构
point-to-point control　点位控制
on two major fronts　在两个主要方面，在两个重要领域
spawn [spɔːn]　v. 造成，引起（to give rise to），使产生（to produce or create something）
decode [ˌdiː'kəud]　n. 解码，译码
logical operation　逻辑操作，逻辑运算
sensing device　灵敏元件，传感器

Lesson 71   Industrial Robots

There are a variety of definitions of the term industrial robot. Depending on the definition used, the number of industrial robot installations worldwide varies widely. Numerous single-purpose machines are used in manufacturing plants that might appear to be robots. These machines can only perform a single function and cannot be reprogrammed to perform a different function. Such single-purpose machines do not fit the definition for industrial robots that is becoming widely accepted.

An industrial robot is defined by the International Organization for Standardization (ISO) as an automatically controlled, reprogrammable, multipurpose manipulator, programmable in three or more axes, which may be either fixed in place or mobile for use in industrial automation applications.

There exist several other definitions too, given by other societies, e. g., by the Robot Institute of America (RIA), the Japan Industrial Robot Association (JIRA), British Robot Association (BRA), and others. The definition developed by RIA is:

A robot is a reprogrammable multifunctional manipulator designed to move material, parts, tools, or specialized devices through variable programmed motions for the performance of a variety of tasks.

All definitions have two points in common. They all contain the words reprogrammable and multifunctional. It is these two characteristics that separate the true industrial robot from the various single-purpose machines used in modern manufacturing firms.

The term "reprogrammable" implies two things: The robot operates according to a written program, and this program can be rewritten to accommodate a variety of manufacturing tasks.

The term "multifunctional" means that the robot can, through reprogramming and the use of different end-effectors, perform a number of different manufacturing tasks. Definitions written around these two critical characteristics have become the accepted definitions among manufacturing professionals.

The first articulated arm came about in 1951 and was used by the U.S. Atomic Energy Commission. In 1954, the first industrial robot was designed by George C. Devol. It was an unsophisticated programmable material handling machine.

The first commercially produced robot was developed in 1959. In 1962, the first industrial robot to be used on a production line was installed in the General Motors Corporation. It was used to lift red-hot door handles and other such car parts from die casting machines[1] in an automobile factory in New Jersey, USA. Its most distinctive feature was a gripper that eliminated the need for man to touch car parts just made from molten metal. It had five degrees of freedom (DOF). This robot was produced by Unimation.

A major step forward in robot control occurred in 1973 with the development of the T³ industrial robot by Cincinnati Milacron.[2] The T³ robot was the first commercially produced industrial robot controlled by a minicomputer. Figure 71.1 shows a T³ robot with all the motions indicated.

Since then robotics has evolved in a multitude of directions, starting from using them in welding, painting, in assembly, machine tool loading and unloading, to inspection.

Over the last three decades automobile factories have become dominated by robots. A typical factory contains hundreds of industrial robots working on fully automated production lines. For example, on an automated production line, a vehicle chassis on a conveyor is welded, painted and finally assembled at a sequence of robot stations.

Mass-produced printed circuit boards (PCBs) are almost exclusively assembled by pick-and-place robots, typically with SCARA robots,[3] which pick tiny electronic components, and place them on to PCBs (see Fig. 71.2) with great accuracy. Such robots can place tens of thousands of components per hour, far surpassing a human in speed, accuracy, and reliability.

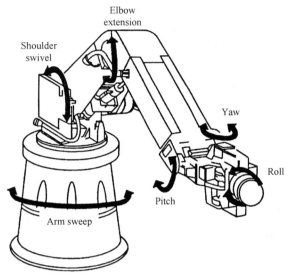

Figure 71.1 An industrial robot

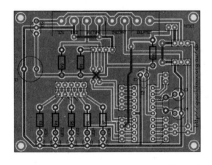

Figure 71.2 A printed circuit board

A major reason for the growth in the use of industrial robots is their declining cost. Since 1970s, the rapid inflation of wages has tremendously increased the personnel costs of manufacturing firms. In order to survive, manufacturers were forced to consider any technological developments that could help improve productivity. It became imperative to produce better products at lower costs in order to be competitive in the global market economy. Other factors such as the need to find better ways of performing dangerous manufacturing tasks contributed to the development of industrial robots. However, the fundamental reason has always been, and is still, improved productivity.

One of the principal advantages of robots is that they can be used in settings that are dangerous to humans. Welding and parting are examples of applications where robots can be used more safely than humans. Most industrial robots of today are designed to work in environments which are not safe and very difficult for human workers. For example, a robot can be designed to handle a very hot or very cold object that the human hand cannot handle safely.

Even though robots are closely associated with safety in the workplace, they can, in themselves, be dangerous. Robot workspaces should be accurately calculated and a danger zone surrounding the workspace clearly marked off. Barriers can be used to keep human workers out of a robot's workspace. Even with such precautions it is still a good idea to have an automatic shutdown system in situations where robots are used. Such a system should have the capacity to sense the need for an automatic shutdown of operations.

Words and Expressions

installation [ˌɪnstəˈleɪʃən] *n.* 整套装置，设备，结构，安装
reprogrammable 可重复编程（无须更换机械系统即可更改已编程的运动或辅助功能）
multipurpose [ˌmʌltiˈpɜːpəs] *a.* 多用途（更换机械系统后，有能力适用不同用途的性能）
manipulator [məˈnɪpjuleɪtə] *n.* 操作机（用来抓取和/或移动物体的多自由度机器）
axis [ˈæksɪs]（*pl.* axes [ˈæksiːz]） *n.* 轴（机器人以直线或回转方式运动的方向线）
multifunctional [ˌmʌltiˈfʌŋkʃənl] *a.* 多功能的
end effector 末端执行器（为使机器人完成其任务而专门设计并安装在机械接口处的装置。示例：夹持器、扳手、焊枪、喷枪等）
articulated [ɑːˈtɪkjulitid] *a.* 关节式的（connected by a joint or joints）
General Motors Corporation （美国）通用汽车公司
gripper [ˈɡrɪpə] *n.* 夹持器（供抓取和握持用的末端执行器），手爪
molten metal 熔融金属，其中 molten 是 melt 的过去分词
Unimation 万能自动化公司（Universal Automation），尤尼梅逊公司
swivel [ˈswɪvl] *n.*; *v.* 旋转
yaw [jɔː] *n.* 偏转，使左右摇转，偏摆（rotation around the vertical axis）
pitch [pɪtʃ] *n.* 俯仰（rotation around the side-to-side axis）
roll [rəʊl] *n.* 侧滚，回转（rotation around the front-to-back axis）
chassis [ˈʃæsi] *n.* 底盘（a frame upon which the main parts of an automobile are built）
production line 生产线（产品生产过程所经过的路线）
mass-produced 大量生产的（produced cheaply and in large numbers using machines in a factory）
printed circuit board 印制电路板，印刷电路板
pick-and-place robot 取放机器人，抓放机器人
setting [ˈsetɪŋ] *n.* 位置，安装，环境
welding 焊接（通过加热和/或加压，使工件达到原子结合且不可拆卸连接的一种加工方法）
parting [ˈpɑːtɪŋ] *n.* 切断（把坯料切成两段或数段的加工方法）

Notes

[1] die casting machine 意为"压铸机（将熔融金属以高压、高速压射到铸型中，并在压力下凝固获得铸件的机器）"。
[2] T^3 为 The Tomorrow Tool 的缩写，也可以写为 T3 或 T-3。
[3] SCARA robot 意为"SCARA 机器人（具有两个平行的回转关节，以便在所选择的平面内提供柔顺性的机器人），平面关节型机器人"。

Lesson 72 Robotics

The field of robotics may be defined as the science and practice of designing, manufacturing, and applying robots.[1]

Robots may be classified by coordinate system, as follows:

1. Cartesian In Cartesian coordinate robots, motion takes place along three linear perpendicular axes, that is, the arm can move up or down and in or out, in addition to a transverse motion perpendicular to the plane created by the previous two motions (see Fig. 72.1a).[2]

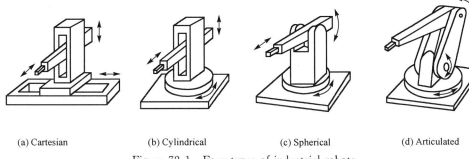

(a) Cartesian　　　　(b) Cylindrical　　　　(c) Spherical　　　　(d) Articulated

Figure 72.1　Four types of industrial robots

2. Cylindrical The motions of the arm of a cylindrical coordinate robot, like illustrated in Fig. 72.1b, are quite similar to those of the Cartesian coordinate robot, except that the motion of the base[3] is rotary. In other words, the robot's arm can move up or down, in or out, and can swing around the vertical axis.

3. Spherical The spherical coordinate robot, sometimes called polar coordinate robot, is shown in Fig. 72.1c. It has three axes of motion; two of them are rotary, and the third is linear.

4. Articulated When a robot arm consists of links connected by revolute joints only, i. e. the prismatic joint in spherical type is also replaced by another revolute joint (see Fig. 72.2), the robot is called an articulated robot (see Fig. 72.1d).[4]

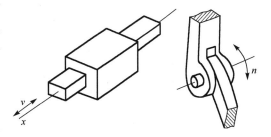

Figure 72.2　Prismatic and revolute joints

Robots may be attached permanently to the floor of a manufacturing plant, or they may move along overhead rails (gantry robots, see Fig. 72.3), or they may be equipped with wheels to move along the factory floor (mobile robots, such as AGVs used in factories for material handling purpose, see Fig. 53.1).

A classification of robots by control method is described below.

1. Pick-and-Place Robot The pick-and-place robot is programmed for a specific sequence of operations. Its movements are from point to point. Replacing unskilled human labor often in hazardous jobs, these robots had to be robust and low in initial and maintenance costs.[5]

Figure 72.3   Gantry robots

2. Playback Robot   An operator leads the playback robot and its end effector through the desired path; in other words, the operator teaches the robot by showing it what to do. The robot records the path and sequence of motions and can repeat them continually without any further action or guidance by the operator.

3. Numerically Controlled (NC) Robot   The numerically controlled robot is programmed and operated much like a NC machine. The robot is servo-controlled by digital data, and its sequence of movements can be changed with relative ease.[6] As in NC machines, there are two basic types of controls: point-to-point, and continuous-path.

4. Intelligent Robot   The intelligent robot is capable of performing some of the functions and tasks carried out by human beings. It is equipped with a variety of sensors with visual and tactile capabilities. Much like humans, the robot can determine what action to make based on information acquired through its own sensors and decision-making ability. Significant developments are taking place in intelligent robots so that they will:

(1) Behave more and more like humans, performing tasks such as moving among a variety of machines and equipment on the shop floor and avoiding collisions;

(2) Recognize, select, and properly grip the correct raw material or workpiece;

(3) Transport the part to a machine for further processing or inspection; and

(4) Assemble the components into subassemblies or a final product.

Major applications of industrial robots include the following:

(1) Material handling consists of the loading, unloading, and transferring of workpieces in manufacturing facilities. These operations can be performed reliably and repeatedly with robots. Here are some examples: (a) casting operations, in which molten metal, raw materials, and parts in various stages of manufacture are handled automatically; (b) heat treating, in which parts are loaded and unloaded from furnaces and quench baths; (c) forming operations, in which parts are loaded and unloaded from presses and various other types of metalworking machinery.

(2) Spot welding (see Fig. 72.4) produce welds of good quality in the manufacture of automobile and truck bodies.[7]

(3) Operations such as deburring, grinding, and polishing can be done by using appropriate tools attached to the end effectors.

(4) Spray painting (particularly of complex shapes) and cleaning operations are frequent applications because the motions for one piece repeat so accurately for the next.

(5) Automated assembly is again very repetitive.

(6) Inspection and gauging in various stages of manufacture make possible speeds much higher than those humans can achieve.[8]

Figure 72.4　A spot welding robot

In addition to the technical factors, cost and benefit considerations are also significant aspects of robot selection and applications. The increasing reliability and the reduced costs of intelligent robots are having a major economic impact on manufacturing operations, and such robots are gradually displacing human labor.[9]

Words and Expressions

robotics [rəu'bɔtiks]　*n.* 机器人学，机器人技术
Cartesian coordinate robot　笛卡儿坐标机器人（Cartesian robot），直角坐标机器人
perpendicular [ˌpəːpən'dikjulə]　*a.* 垂直的（at an angle of 90° to a given line, plane, or surface）
transverse motion　横向运动
cylindrical coordinate robot　柱面坐标机器人，圆柱坐标机器人（cylindrical robot）
spherical coordinate robot　球面坐标机器人，球坐标机器人（spherical robot）
polar coordinate robot　极坐标机器人，也可写为 polar robot
prismatic [priz'mætik] joint　棱柱关节（两杆件间的组件，能使其中一杆件相对于另一杆件做直线运动），又称"sliding joint（滑动关节）"
revolute joint　旋转关节（连接两杆件的组件，能使其中一杆件相对于另一杆件绕固定轴转动），又称"rotary joint（回转关节）"
articulated robot　关节机器人（手臂具有三个或更多个回转关节的机器人）
overhead rail　横梁，高架轨道
gantry robot　桁架式机器人，龙门式机器人
mobile robot　移动机器人（基于自身控制、可移动的机器人）
AGV (automated guided vehicle)　自动导引车
material handling　物料输送，物料搬运
robust [rə'bʌst]　*a.* 坚固的，耐用的
playback robot　示教再现机器人（一种将任务程序通过示教编程输入并能自动复现该程序的机器人）
point-to-point control　点位控制
continuous-path control　连续路径控制
intelligent robot　智能机器人（具有依靠感知其环境和/或与外部资源交互、调整自身行为来执行任务的能力的机器人）

decision making ability 决策能力
tactile ['tæktail] *a.* 触觉的，有触觉的
recognize ['rekəgnaiz] *v.* 辨认，识别（to identify from past experience or knowledge）
manufacturing facility 生产设备，制造工厂
quench bath 淬火槽
forming 成形加工（用与工件的最终表面轮廓相匹配的加工工具使工件成形的方法）
press 压力机（一种能使滑块做往复运动，并按所需方向给模具施加一定压力的机器）
polishing 抛光（利用机械、化学或电化学的作用，使工件获得光亮、平整表面的加工方法）
deburring [di'bə:riŋ] *n.* 去毛刺（清除工件已加工部位周围所形成的刺状物或飞边）
spray painting 喷漆，喷涂（用喷或涂的方法把涂料覆盖在制品上的过程）
cleaning ['kli:niŋ] *n.* 清洗（除去金属工件表面污物，如油脂、机械杂质、锈等的过程）
automated assembly 自动装配（以自动化机械代替人工劳动的一种装配技术）
displace [dis'pleis] *v.* 取代，置换

Notes

[1] 全句可译为：机器人学是关于机器人设计、制造和应用的一门学科。

[2] three linear perpendicular axes 意为"三个相互垂直的直线坐标轴"。全句可译为：在直角坐标机器人中，运动沿着三个相互垂直的直线坐标轴进行，即手臂可以向上或向下运动、向内或向外运动，以及与上述两种运动所形成的平面垂直的横向运动（如图72.1a所示）。

[3] base 这里指"机座（一平台或构架，操作机第一个杆件的原点置于其上）"。

[4] link 在机器人中通常译为"杆件（用于连接相邻关节的刚体）"。全句可译为：当一个机器人的手臂仅由旋转关节连接的杆件组成，即球面坐标机器人中的移动关节由另一个旋转关节（见图72.2）所取代时，这个机器人就被称为关节机器人（如图72.1d所示）。

[5] initial and maintenance costs 意为"初始成本和维护费用"。全句可译为：这些机器人通常被用来替代从事危险性工作的非技术工人，它们必须坚固耐用而且需要较低的初始成本和维护费用。

[6] servo-controlled 意为"伺服控制的"。此句与后面一句可译为：这种机器人是由数字数据伺服控制的，可以较为容易地改变其动作顺序。正如与在数控机床中一样，机器人中存在着点位控制和连续路径控制这两种基本方式。

[7] spot welding 意为"点焊，是指焊件在接头处接触面的个别点上被焊接起来"。全句可译为：点焊（见图72.4）在制造汽车和卡车车体时能生成高质量的焊缝。

[8] gauging 意为"测量"。全句可译为：在制造的各个阶段中，可以取得比人工快得多的检验与测量速度。

[9] impact 意为"影响"。全段可译为：除了技术因素，成本和效益等因素也是在选择和应用机器人时应该着重考虑的。智能机器人可靠性的提高和成本的降低正在对生产制造过程产生重大的经济影响，这种机器人正在逐渐取代人类劳动者。

Lesson 73　Basic Components of an Industrial Robot

To appreciate the functions of robot components and their capabilities, we might simultaneously observe the flexibility and capability of diverse movements of our arm, wrist, hand, and fingers in reaching for and grabbing an object from a shelf, or in using a hand tool, or in operating a machine. Described next are the basic components of an industrial robot.

Manipulator　The manipulator is a mechanical unit that provides motions similar to those of a human arm and wrist. A manipulator is formed of links (their functionality being similar to the bones of the human body), and joints (also called "kinematic pairs")[1] normally connected in series, as for the robots shown in Fig. 72.1. For a typical six degrees of freedom robot (Fig. 71.1), the first three links and joints form the arm, and the last three joints make the wrist. The function of an arm is to place an object in certain location in the three-dimensional space, where the wrist orients it.

End Effector　The end effector is a device mounted on the end of the manipulator of a robot. It is equivalent to the human hand. Depending on the type of operation, conventional end effectors may be any of the following:

(1) Grippers, electromagnets, and vacuum chucks, for material handling;

(2) Spray guns, for painting;

(3) Welding devices, for spot and arc welding;

(4) Power tools, such as electric drills; and

(5) Measuring instruments.

End effectors are generally custom-made to meet special requirements. Mechanical grippers are the most commonly used and are equipped with two or more fingers. A two-fingered gripper (Fig. 73.1) can only hold simple objects, whereas a multi-fingered gripper can perform more complex tasks. The selection of an appropriate end effector for a specific application depends on such factors as the rated load, environment, reliability, and cost.

Actuator　Actuators are like the "muscles" of a robot, they provide motion to the manipulator and the end effector. They are classified as pneumatic, hydraulic or electric, based on their principle of operation.

Pneumatic actuators utilize compressed air provided by a compressor and transform it into mechanical energy by means of pistons or air motors. Pneumatic actuators have few moving parts making them inherently reliable and reducing maintenance costs. It is the cheapest form of all actuators. But pneumatic actuators are not suitable for moving heavy loads under precise control due to

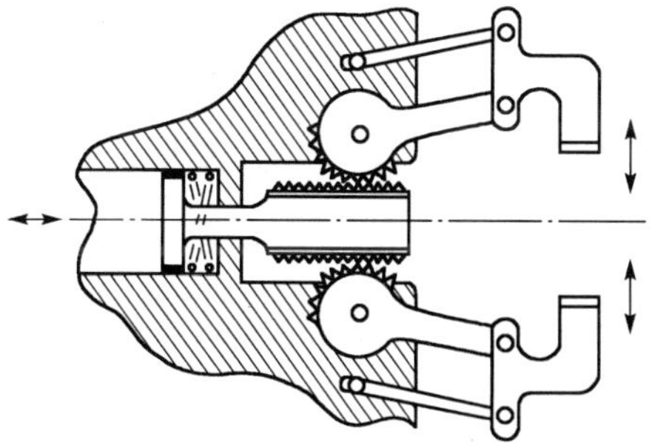

Figure 73.1　A simple gripper

the compressibility of air.

Hydraulic actuators utilize high-pressure fluid such as oil to transmit forces to the point of application desired. A hydraulic actuator is very similar in appearance to that of pneumatic actuator. Hydraulic actuators are designed to operate at much higher pressure (typically between 7 and 17 MPa). They are suitable for high power applications. Hydraulic robots are more capable of withstanding shock loads than electric robots.

Electric motors are the most popular actuator for manipulators. Direct current (DC) motors can achieve very high torque-to-volume ratios. They are also capable of high precision, fast acceleration, and high reliability. Although they don't have the power-to-weight ratio of hydraulic actuators or pneumatic actuators, their controllability makes them attractive for small and medium-sized manipulators.

Alternating current (AC) motors and stepper motors have been used infrequently in industrial robots. Difficulty of control of the former and low torque ability of the latter have limited their use.

Sensor Sensors convert one form of signal to another. For example, the human eye converts light pattern into electrical signals. Sensors fall into one of the several types: vision, touch, position, force, velocity, acceleration etc. Some of them will be explained in Lesson 74.

Digital Controller The digital controller is a special electronic device that has a CPU, memory, and sometimes a hard disk. In robot systems, these components are kept inside a sealed box referred to as a controller. It is used for controlling the movements of the manipulator and the end effector. Since a computer has the same characteristics as those of a digital controller, it is also used as a robot controller.

Analog-to-Digital Converter An analog-to-digital converter[2] (abbreviated ADC) is an electronic device that converts analog signals to digital signals. This electronic device interfaces with the sensors and the robot's controller. Typically, an ADC converts an input analog voltage (or current) to a digital number proportional to the magnitude of the voltage or current. For example, the ADC converts the voltage signal developed due to the strain in a strain gage (see Fig. 74.2) to a digital signal, so that the digital robot controller can processes this information.

Digital-to-Analog Converter (DAC) A DAC converts the digital signal from the robot controller into an analog signal to activate the actuators. In order to drive the actuators, e.g. a DC electric motor, the digital controller is also coupled with a DAC to convert its signal back to an equivalent analog signal, i.e., the electric voltage for the DC motors.

Amplifier Generally, an amplifier is any device that changes, usually increases, the amplitude of a signal. Since the control commands from the digital controller converted to analogue signals by the DAC are very weak, they need amplification to really drive the electric motors of the robot manipulator.

Words and Expressions

reach for 伸手去拿 (to move one's hand in order to get or touch something)

hand tool　手工工具（用手握持，以人力或以人控制的其他动力作用于物体的小型工具）
functionality [ˌfʌŋkəʃəˈnæliti]　n. 功能（the state or quality of being functional; usefulness）
in series　串联的
orient [ˈɔːriənt]　v. 确定位置或方向（to place something in a particular position or direction）
vacuum [ˈvækjuəm]　n. 真空（明显低于大气压力或大气质量密度的稀薄气体状态）
vacuum chuck　真空吸盘（靠空气压力差吸取重物的取物装置）
power tool　电动工具，动力工具
electric drill　电钻（钻孔用的电动工具）
custom-made　定制的（made according to the specifications of an individual purchaser）
rated load　额定负载（正常操作条件下作用于机械接口且不会使机器人性能降低的最大负载）
actuator [ˈæktjueitə]　n. 驱动器，执行机构，致动器（用于实现机器人运动的动力机构。示例：把电能、液压能、气动能转换成使机器人运动的马达）
pneumatic [njuːˈmætik]　a. 气动的（run by compressed air; operated by the pressure of air）
pneumatic actuator　气动驱动器，气动执行机构（利用有压气体作为动力源的执行机构）
air motor　气动马达，气马达（将气动能转化成转动机械能的气动元件）
inherently [inˈhiərəntli]　ad. 本能地，自然地，本质上地
compressibility [kəmˌpresiˈbiliti]　n. 可压缩性
hydraulic actuator　液压驱动器，液压致动器，液压执行元件（把液体压力能转换成机械能的装置，包括液压马达和液压缸）
direct current (DC)　直流电（方向和量值不随时间变化的电流）
alternating current (AC)　交流电（量值和方向作周期性变化且平均值为零的时变电流）
torque　转矩（作用在物体上，使它转动或具有转动趋势，同时引起物体扭转变形的力矩）
power-to-weight ratio　功率重量比（或称比功率，功率质量比，简称功重比，pwr）
controllability [kənˌtrəuləˈbiliti]　n. 可控性，可控制性
light pattern　光图像
sensor [ˈsensə]　n. 传感器（能感受规定的被测量并按照一定的规律将其转换成可用输出信号的器件或装置）
robot system　机器人系统（由机器人、末端执行器和为机器人完成其任务所需的任何机械、设备、装置或传感器构成的系统）
abbreviate [əˈbriːvieit]　v. 缩略，缩写（to reduce a word or phrase to a shorter form）
analogue signal　模拟信号（信息参数在给定范围内表现为连续的信号）
interface [ˈintə(ː)ˈfeis]　n. 界面，接口；v. 使连接，使接合：通过界面连接
strain gage　应变片（一种将被测件上的应变变化转换成电信号的敏感器件）
activate [ˈæktiveit]　v. 启动，驱动（to set in motion; to make active or more active）
digital-to-analog converter　数模转换器（把数字量转变成模拟量的器件）
amplifier [ˈæmpliˌfaiə]　n. 放大器（能把输入信号的电压或功率放大的装置）

Notes

[1] kinematic pair 意为"运动副（由两个构件直接接触而组成的可动的连接称为运动副）"。
[2] analog-to-digital converter 意为"模数转换器，即 A/D 转换器，或简称 ADC，通常是指一个将模拟信号转变为数字信号的电子元件"。

Lesson 74　Robot Sensors

Sensors in robots are like our eyes, nose, ears, mouth, and skin. Based on the function of human organs, for example the eyes or skin etc., terms like vision, tactile etc., have been used for robot sensors. Robots like humans, must gather extensive information about their environment in order to function effectively. They must pick-up an object and know it has been picked up. As the robot arm moves through the 3-dimensional space, it must avoid obstacles and approach items to be handled at a controlled speed. Some objects are heavy, others are fragile, and others are too hot to handle. These characteristics of objects and the environment must be recognized, and fed into the computer that controls a robot's movements. For example, to move the end effector of a robot along a desired trajectory and to exert a desired force on an object, the end effector and sensors must work in coordination with the robot controller. Robot sensors can be classified into two categories: internal sensors and external sensors.

Internal sensors, as the name suggests, are used to measure the internal state of a robot, i. e. its position, velocity, acceleration, etc. at a particular instant.[1] Depending on the various quantities it measures, a sensor is termed as the position, velocity, acceleration, or force sensor.

Position sensors measure the position of each joint of a robot. There are several types of position sensors, e. g. encoder, LVDT, etc.[2]

Encoder is an optical-electrical device that converts motion into a sequence of digital pulses. Encoders can be either absolute or incremental type. Further, each type may be again linear or rotary.[3]

The linear variable differential transformer (LVDT) is one of the most used displacement transducers, particularly when high accuracy is needed. An LVDT consists of three coils, as shown in Fig. 74.1. AC supplied to the input coil at a specific voltage E_i generates a total output voltage across the secondary coils. The output voltage is a linear function of the displacement of a movable core inside the coils.

Electrical resistance strain gages are widely used to measure strains due to force or torque. Gages are made of electrical conductors, usually wire or foil, as shown in Fig. 74.2. They are glued on the surfaces where strains are to be measured. The strains cause changes in the resistance of the strain gages, which are measured by attaching them to the Wheatstone bridge circuit as one of the four resistances, $R_1 \ldots R_4$ of Fig. 74.3a.[4] It is a cheap and accurate method of measuring strain. But care should be taken for the temperature changes. In order to enhance the output voltage and eliminate resistance changes due to the change in temperature, two strain gages are used, as shown in Fig. 74.3b, to measure the force at the end of the cantilever beam.[5]

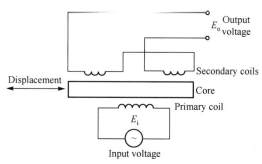
Figure 74.1 Schematic diagram of an LVDT

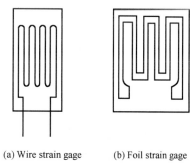
(a) Wire strain gage   (b) Foil strain gage
Figure 74.2 Strain gages

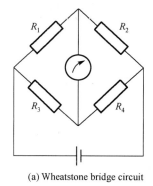
(a) Wheatstone bridge circuit

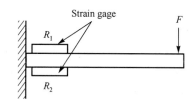
(b) Cantilever beam with strain agaes

Figure 74.3 Strain measurement

A strain gage wrist sensor is shown in Fig. 74.4. It is milled from an aluminum tube 75 mm in diameter. Eight narrow elastic beams, four running vertically and four horizontally, have each been necked down at one point, so that the strain in each beam is concentrated at the other end of the beam.[6] Strain gages cemented to each point of high strain measure the force component at that point. Foil strain gages, R_1 and R_2, are used to measure the strain. Two strain gages are mounted on each beam, so that the effects of temperature may be compensated for. This wrist sensor measures the three components of force and three components of torque in a Cartesian coordinate system.

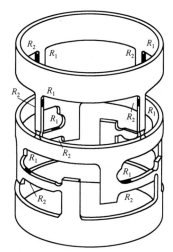
Figure 74.4 A strain gage wrist sensor

External sensors are primarily used to learn more about the robot's environment. External sensors can be divided into two categories: contact type (e. g. limit switch) and non-contact type (e. g. proximity sensor and vision sensor).

A limit switch (see Fig. 74.5) is constructed much as the ordinary light switch used at home and office. It has the same on/off characteristics. Limit switches are used in robots to detect the extreme positions of the motions, where the link reaching an extreme position switches off the corresponding actuator, thus safeguarding any possible damage to the mechanical structure of the robot arm.

A proximity sensor senses and indicates the presence of an object within a fixed space near the sensor. One of the simplest robotic proximity sensors consists of a light-emitting diode (LED) transmitter and a photodiode receiver that senses light when a reflecting surface is near. Other proximity sensors are based on capacitive and inductive principles.

Figure 74.5　Limit switches

Vision systems are used with robots to let them look around and find the parts for picking and placing them at appropriate locations.[7] Most current systems use cameras based on the charged coupled devices (CCDs) techniques. These cameras are smaller, last longer, and have less inherent image distortion than conventional robot cameras.

Words and Expressions

robot sensor　机器人传感器（用于获取机器人控制所需的内部和外部信息的传感器）
human organ　人体器官
tactile ['tæktail]　*a.* 触觉（通过直接接触，感知物体受力分布、形状和温度等信息）的
fed into　被送入，被输入
desired trajectory [trə'dʒekətəri]　预定轨迹，期望轨迹
recognize ['rekəgnaiz]　*v.* 辨认出，识别
in coordination with　与……配合
internal sensor　内部传感器（用于测量机器人内部状态的传感器），内部状态传感器
external sensor　外部传感器（用于测量机器人所处环境状态的传感器），外部状态传感器
force sensor　力传感器（能感受外力并转换成可用输出信号的传感器），force transducer
encoder [in'kəudə]　*n.* 编码器（把角位移或直线位移转换成电信号的一种装置）
LVDT (Linear Variable Differential Transformer)　线性可变差动变压器
absolute or incremental type　绝对式或增量式，绝对型或增量型
most used　用得最多的
displacement transducer　位移传感器（能感受位移量并转换成可用输出信号的传感器）
core [kɔː]　*n.* 铁芯（a soft iron rod in a coil or transformer）
secondary coil　次级线圈（变压器的输出端）
primary coil　初级线圈（变压器的输入端，一般与电源相连）
electrical resistance strain gage　电阻应变片（将机械构件上应变的变化转换为电阻变化的元件）
electrical conductor　电导体（a material that readily permits the passage of an electric current）
attach to　连接到，加入，把……放在

Wheatstone bridge　惠斯通电桥（用于测量电阻值的一种四臂电桥，被测电阻为一个臂，其余三个臂是已知电阻）
wire strain gage　电阻丝式应变片，亦为 wire strain gauge
foil strain gage　箔式应变片，亦为 foil strain gauge
wrist sensor　腕力传感器，亦为 wrist force sensor
force component　力的分量，亦为 component of force
concentrated ['kɔnsentreitid]　*a*. 集中的（existing or happening together in a small area）
limit switch　限位开关（用于限定机械设备的运动极限位置的电气开关），行程开关
light switch　照明开关，电灯开关
proximity [prɔk'simiti]　*n*. 接近（nearness in place, time, or relationship; closeness）
proximity sensor　接近传感器（以无须接触检测对象进行检测为目的的传感器的总称）
light-emitting diode　发光二极管（可以把电能转化为光能的半导体二极管），简称 LED
transmitter [trænz'mitə]　*n*. 发送器，发射器（a device that generates signals）
photodiode [ˌfəutəu'daiəud]　*n*. 光电二极管（把光信号转换成电信号的光电传感器件）
capacitive [kə'pæsitiv]　*a*. 电容的
inductive [in'dʌktiv]　*a*. 电感的，感应的
current system　现行系统
charged coupled device (CCD)　CCD 图像传感器，又称"电荷耦合器件"
image distortion　图像失真

Notes

[1] as the name suggests 意为"顾名思义"。全句可译为：内部传感器，顾名思义，是用来测量机器人的内部状态，即在某一特定时刻的位置、速度、加速度等的。

[2] joint 这里指"关节"。全段可译为：位置传感器测量机器人每个关节的位置。有几种类型的位置传感器，如编码器、线性可变差动变压器等。

[3] digital pulse 意为"数字脉冲"。全段可译为：编码器是一种光电装置，它将运动转换成一系列数字脉冲。编码器可以是绝对式或增量式的。此外，每一种又可分为直线型或旋转型。

[4] circuit 意为"电路（the complete path that an electric current travels along）"。全句可译为：应变会引起应变片电阻值的变化，将应变片作为如图 74.3a 所示的惠斯通电桥电路中 R_1 ……R_4 这四个电阻中的一个电阻，就可以测量其电阻大小。

[5] care should be taken 意为"特别要注意，值得注意"。这两句可译为：但是，应该考虑温度的变化。为了提高输出电压和消除由于温度变化对电阻变化的影响，可以采用如图 74.3b 所示的双应变片方式来测量作用在悬臂梁端部的力。

[6] neck down 意为"颈状收缩"，cemented to 意为"被黏结到某物上"。这两句可译为：它由一个直径为 75mm 的铝管铣削而成。具有八根窄长的弹性梁：四根为水平方向；四根为垂直方向。每根梁的一端呈颈状收缩，使得应变主要发生在它的另一端。在每根梁的应变较大处贴有应变片，以测量作用在该处的分力。

[7] vision system 意为"视觉系统"。全句可译为：机器人的视觉系统可使它们能够环顾四周和找到它们要抓取的零件，并将其放置到合适的位置。

Lesson 75   Mechanical Engineering in the Information Age

In the early 1980s, engineers thought that massive research would be needed to speed up product development. As it turns out, less research is actually needed because shortened product development cycles encourage engineers to use available technology. Developing a revolutionary technology for use in a new product is risky and prone to failure. Taking short steps is a safer and usually more successful approach to product development.

Shorter product development cycles are also beneficial in an engineering world in which both capital and labor are global. People who can design and manufacture various products can be found anywhere in the world, but containing a new idea is hard. Geographic distance is no longer a barrier to others finding out about your development six months into the process. If you've got a short development cycle, the situation is not catastrophic as long as you maintain your lead. But if you're in the midst of a six-year development process and a competitor gets wind of[1] your work, the project could be in more serious trouble.

The idea that engineers need to create a new design to solve every problem is quickly becoming obsolete. The first step in the modern design process is to browse the Internet or other information systems to see if someone else has already designed a transmission, or a heat exchanger that is close to what you need. Through these information systems, you may discover that someone already has manufacturing drawings, numerical control programs, and everything else required to manufacture your product. Engineers can then focus their professional competence on unsolved problems.

In tackling such problems, the availability of computers and access to computer networks dramatically enhance the capability of the engineering team and its productivity. These information age tools can give the team access to massive databases of material properties, standards, technologies, and successful designs. Such pretested designs can be downloaded for direct use or quickly modified to meet specific needs. Remote manufacturing, in which product instructions are sent out over a network, is also possible. You could end up with[2] a virtual company[3] where you don't have to see any hardware. When the product is completed, you can direct the manufacturer to drop-ship[4] it to your customer. Periodic visit to the customer can be made to ensure that the product you designed is working according to the specifications. Although all of these developments won't apply equally to every company, the potential is there.

Custom design[5] used to be left to small companies. Big companies sneered at it they hated the idea of dealing with niche markets or small-volume custom solutions. "Here is my product," one of the big companies would say. "This is the best we can make it you ought to like it. If you don't, there's a smaller company down the street that will work on your problem."

Today, nearly every market is a niche market,[6] because customers are selective. If you ignore the potential for tailoring your product to specific customers' needs, you will lose the major part of your market share perhaps all of it. Since these niche markets are transient, your company needs to be in a position to respond to them quickly.

The emergence of niche markets and design on demand has altered the way engineers conduct research. Today, research is commonly directed toward solving particular problems. Although this situation is probably temporary, much uncommitted technology, developed at government expense or written off by major corporations, is available today at very low cost. Following modest modifications, such technology can often be used directly in product development, which allows many organizations to avoid the expense of an extensive research effort. Once the technology is free of major obstacles, the research effort can focus on overcoming the barriers to commercialization rather than on pursuing new and interesting, but undefined, alternatives.

When viewed in this perspective, engineering research must focus primarily on removing the barriers to rapid commercialization of known technologies. Much of this effort must address quality and reliability concerns, which are foremost in the minds of today's consumers. Clearly, a reputation for poor quality is synonymous with bad business. Everything possible including thorough inspection at the end of the manufacturing line and automatic replacement of defective products—must be done to assure that the customer receives a properly functioning product.

Research has to focus on the cost benefit of factors such as reliability. As reliability increases, manufacturing costs and the final cost of the system will decrease. Having 30 percent junk at the end of a production line not only costs a fortune but also creates an opportunity for a competitor to take your idea and sell it to your customers.

Central to the process of improving reliability and lowering costs is the intensive and widespread use of design software, which allows engineers to speed up every stage of the design process. Shortening each stage, however, may not sufficiently reduce the time required for the entire process. Therefore, attention must also be devoted to concurrent engineering[7] software with shared databases that can be accessed by all members of the design team.

As we move more fully into the Information Age, success will require that the engineer possess some unique knowledge of and experience in both the development and the management of technology. Success will require broad knowledge and skills as well as expertise in some key technologies and disciplines; it will also require a keen awareness of the social and economic factors at work in the marketplace. Increasingly, in the future, routine problems will not justify heavy engineering expenditures, and engineers will be expected to work cooperatively in solving more challenging, more demanding problems in substantially less time. We have begun a new phase in the practice of engineering. It offers great promise and excitement as more and more problem-solving capability is placed in the hands of the computerized and wired engineer. Mechanical engineering is a great profession, and it will become even greater as we make the most of the opportunities offered by the Information Age.

Words and Expressions

as it turns out 结果表明，事实上，已经发现 (it has been found that)

geographic [dʒiə'græfik] a. 地理上的，地区性的
browse [brauz] v. 浏览（to use a special program, called a browser, to find and look at information on the Internet），翻阅（to read something superficially）
transmission [trænz'miʃən] n. 传动装置（传递运动和动力的装置），变速器（用于改变转速和转矩的机构）
pretest ['pri:test] n.; v. 事先试验，预先检验；a. 试验前的
niche market 补缺市场（小的细分市场，对于大的竞争对手而言不重要或被其忽视）
tailor... to... 使……适合（满足）……（要求，需要，条件等）
transient ['trænziənt] a. 短暂的，无常的，瞬变的
confer [kən'fə:] v. 授予，给予，使具有（性能）
evolutionary [ˌi:və'lu:ʃənəri] a. 发展的，发达的，进化的
uncommitted [ˌʌnkə'mitid] a. 自由的，不受约束的，不负义务的，独立的
write off 注销，勾销（把账面价值减低为零，to eliminate from the books of account）
undefined [ˌʌndi'faind] a. 不确定的（not clearly or precisely shown, described, or known）
viewed in this perspective 从这个角度来看
commercialization 商品化（the process of bringing new products or services to market）
junk [dʒʌŋk] n. 废品，不合格品（something of poor quality）
cost a fortune 花费一大笔钱（to cost a lot of money）
foremost ['fɔ:məust] a. 最初的（in the first place），最重要的（most important）
be synonymous [si'nɔniməs] with... 和……同义，和……的意义是一样的
expertise [ˌekspə'ti:z] n. 专门技能（the skill of an expert），专家的意见（expert opinion）
wired ['waiəd] a. 有线的，联网的（connected to computer networks）
make the most of 充分利用（make full use of; make the best of）

Notes

[1] get wind of 意为"获悉……的消息（to learn a piece of information, especially when it has been a secret; to heard about something through an unofficial source）"。

[2] end up with 意为"以……结束，结束于"。

[3] virtual company 意为"虚拟公司"。

[4] drop-ship 意为"直接出货（to ship goods from a supplier directly to a customer）"。

[5] custom design 意为"用户定制设计，按用户需求设计"。

[6] niche market 另外一种译法：利基市场。是指一些企业专注于市场的某些细分环节，通过专业化经营、见缝插针地占据有利的市场位置，从而最大限度地获取收益。

[7] concurrent engineering 意为"并行工程，将本来为串行的活动（如依次进行的研究、实验、设计、工艺、制造）通过团队工作的方式变成并行的活动，很多问题可以在团队工作中发现并予以解决，消除了串行工作方式中用后面的工序修正前面工序中的错误而造成的时间浪费，从而大大缩短新产品开发周期"。

Lesson 76　How to Write a Scientific Paper

Title　In preparing a title for a paper, the author would do well to remember one obvious fact: That title will be read by thousands of people. Perhaps few people, if any, will read the entire paper, but many people will read the title, either in the original journal or in one of the secondary (abstracting and indexing) services. Therefore, all words in the title should be chosen with great care, and their association with one another must be carefully managed. The terms in the title should be limited to those words that highlight the significant content of the paper in terms that are both understandable and retrievable.

Abstract　An Abstract should be viewed as a mini-version of the paper. The Abstract should provide a brief summary of each of the main sections of the paper. A well-prepared abstract enables readers to identify the basic content of a document quickly and accurately, to determine its relevance to their interests, and thus to decide whether they need to read the document in its entirety. The Abstract should not exceed 250 words and should be designed to define clearly what is dealt with in the paper. Many people will read the Abstract, either in the original journal or in *Engineering Index*, *Science Citation Index*, or one of the other secondary publications.

The Abstract should (i) state the principal objectives and scope of the investigation, (ii) describe the methodology employed, (iii) summarize the results, and (iv) state the principal conclusions.

The Abstract should never give any information or conclusion that is not stated in the paper. References to the literature must not be cited in the Abstract (except in rare instances, such as modification of a previously published method).

Introduction　The first section of the paper should, of course, be the Introduction. The purpose of the Introduction should be to supply sufficient background information to allow the reader to understand and evaluate the results of the present study without needing to refer to previous publications on the topic. Above all, you should state briefly and clearly your purpose in writing the paper. Choose references carefully to provide the most important background information.

Experimental Procedure　The main purpose of the Experimental Procedure section is to describe the experimental scheme[1] and then provide enough detail that a competent worker can repeat the experiments. If your method is new (unpublished) you must provide all of the needed detail. However, if a method has been previously published in a standard journal, only the literature reference should be given.

Careful writing of this section is critically important because the cornerstone of the scientific method requires that your results, to be of scientific merit, must be reproducible; and, for the results to be adjudged reproducible, you must provide the basis for repetition of the experiments by others. That experiments are unlikely to be reproduced is beside the point;[2] the same or similar results should be produced, or your paper does not represent good science.

When your paper is subjected to peer review,[3] a good reviewer will read the Experimental Procedure carefully. If there is serious doubt that your experiments could be repeated, the reviewer will recommend rejection of your manuscript no matter how awe-inspiring your results are.

Results So now we come to the core of the paper, the data. This part of the paper is called the Results section. There are usually two ingredients of the Results section. First, you should give some kind of overall description of the experiments, providing the "big picture", without, however, repeating the experimental details previously provided in Experimental Procedure. Second, you should present the data.

Of course, it isn't quite that easy. How do you present the data? A simple transfer of data from laboratory notebook to manuscript will hardly do. Most important, in the manuscript you should present representative data rather than endlessly repetitive data.

The Results need to be clearly and simply stated, because it is the Results that comprise the new knowledge that you are contributing to the world. The earlier parts of the paper (Introduction, Experimental Procedure) are designed to tell why and how you got the Results; the later part of the paper (Discussion) is designed to tell what they mean. Obviously, therefore, the whole paper must stand on the basis of the Results. Thus, the Results must be presented with clarity and precision.

Discussion The Discussion is harder to define than the other sections. Thus, it is usually the hardest section to write. And, whether you know it or not, many papers are rejected by journal editors because of a faulty Discussion, even though the data of the paper might be both valid and interesting. Even more likely, the true meaning of the data may be completely obscured by the interpretation presented in the Discussion, again resulting in rejection.

The essential features of a good Discussion are described below.

1. Try to present the principles and relationships shown by the Results. And bear in mind, in a good Discussion, you discuss—you do not recapitulate the Results.

2. Point out any exceptions or any lack of correlation and define unsettled points. Never take the high-risk alternative of trying to cover up or fudge data that do not quite fit.

3. Show how your results and interpretations agree (or contrast) with previously published work.

4. Discuss the theoretical implications of your work, as well as any possible practical applications.

Conclusions At the end of your paper, you should briefly restate your contributions.[4] Explain why your results are significant and useful, and summarize the experimental evidence, or theoretical proofs, which demonstrate how your ideas are an improvement over earlier work. Also, clearly state the limitations of your work.

Words and Expressions

abstract 摘要（a short form of a paper giving only the most important facts or ideas）
index [ˈindeks] n. 索引（a list of names or topics given in alphabetical order and showing where each is to be found）; v. 编入索引中
retrievable [riˈtriːvəbl] a. 可获取的，可检索的（that can be found and made available to be used）
well-prepared 做好充分准备的；妥善准备的（suitably prepared in advance）
entirety [inˈtaiəti] n. 全部，全体（the whole or total amount of something），总体
citation [saiˈteiʃən] n. 引用，引文，被引用的东西
methodology [ˌmeθəˈdɔlədʒi] n. 一套方法，方法学，方法论
reference [ˈrefrəns] n. 参考文献（a book, article, etc. that is mentioned in a piece of writing, showing you where particular information was found）
background information 背景资料（为论文提供相关背景知识或理论基础的其他论文、期刊、书籍等出版材料）
above all 最重要的是，尤其是（most important of all; especially）
cornerstone [ˈkɔːnəstəun] n. 基础（a fundamental basis），基石，要素（a basic element）
scientific merit 科学价值
manuscript [ˈmænjuskript] n. 稿件（a document submitted for publication）
awe-inspiring [ˈɔːinspairiŋ] a. 令人惊叹的，使人敬畏的（worthy of admiration or respect）
adjudge [əˈdʒʌdʒ] v. 宣判，认为，视为
obscure [əbˈskjuə] v. 遮掩，使模糊，使朦胧
recapitulate [ˌriːkəˈpitjuleit] v. 重述要点（to summarize and state again the main points of）
correlation [ˌkɔriˈleiʃən] n. 相互关系（a mutual relationship between two or more things）
unsettled [ʌnˈsetld] a. 未解决的，未决定的，混乱的
fudge [fʌdʒ] v. 粗制滥造，捏造
implication [ˌimpliˈkeiʃən] n. 推断，含意，暗示，意义
experimental evidence 实验证据
theoretical proof 理论证明

Notes

[1] experimental scheme 意为"实验方案，试验方案"。
[2] beside the point 意为"无关紧要的，没有价值的（of little value or importance）"。
[3] peer review 意为"同行评审（a process used to assess the quality of a manuscript by a group of experts in the appropriate field before it is published）"。
[4] contribution 这里指"对学术界的贡献，研究成果，技术成果"。

Lesson 77　How to Write a Technical Report

Communication of your ideas and results is a very important aspect of engineering. Many engineering students picture themselves in professional practice spending most of their time doing calculations of a nature similar to those they have done as students. Fortunately, this is seldom the case. [1] Actually, engineers spend the largest percentage of their time communicating with others, either orally or in writing.

When your design is done, it is usually necessary to present the results to your client, peers, or employer. The usual form of presentation is a formal technical report. Thus, it is very important for the engineering student to develop his or her communication skills. [2] You may be the cleverest person in the world, but no one will know that if you cannot communicate your ideas clearly and concisely. In fact, if you cannot explain what you have done, you probably don't understand it yourself.

The following suggestions are presented as a guide to technical writing and an aid in avoiding some of the most common mistakes.

Important Information　Emphasize important information, beware of the common error of burying it under a mass of details. [3]

Fact vs. Opinion　Separate fact from opinion. It is important for the reader to know what your contributions are, what ideas you obtained from others (the references should indicate that), and which are opinions not substantiated by fact. [4]

The following paragraphs provide the basics necessary for preparing each section of a technical report.

Abstract—A Summary of the Entire Report. An Abstract must be a complete, concise distillation of the full report, and, as such, should always be written last. It should include a brief (one sentence) introduction to the subject, a statement of the problem, highlights of the results (quantitative, if possible), and major conclusions. It must stand alone without citing figures or tables. A concise, clear approach is essential, since most Abstracts are less than 250 words. [5]

Introduction—Why Did You Do What You Did? An Introduction generally identifies the subject of the report, provides the necessary background information including appropriate literature review, and, in general, provides the reader with a clear rationale for the work described. The Introduction does not contain results, and generally does not contain equations. [6]

Analysis—What Does Theory Have to Say? The Analysis section is used to develop a pertinent theory based on the basic principles that explain the phenomenon you are investigating. Most experimental studies involve the interaction of a variety of complex influences. The purpose of the Analysis is to remove the mask of complexity and expose the underlying facts.

The Analysis is usually interspersed with equations. It is not simply a series of equations devoid of explanatory material. Symbols should be defined when they are first introduced. [7]

Experimental Procedure—What Did You Measure and How? The Experimental Procedure section is intended to describe how the experimental results were obtained and to describe any nonstandard types of apparatus or techniques that were employed. You should provide sufficient details so that the experiment can be conducted by someone else. [8]

Results and Discussion—So What Did You Find? Results are the facts. They are the data you collected and the data you calculated. Means, standard deviations, confidence intervals and errors are all results. [9]

Results of your work must be presented, as well as discussed, in this section of the report. When presenting your results remember that even though you are usually writing to an experienced technical audience, what may be clear to you may not be obvious to the reader. Often the most important vehicles for the clear presentation of results are figures and tables. Each of the figures and tables should be numbered and have a short and descriptive title. [10]

Conclusions—What Do I Know Now? The Conclusions section is where you should concisely restate your answer to the question: "What do I know now?" It is not a place to offer new facts, nor should it contain another rendition of experimental results or rationale. [11] Conclusions should be clear and concise statements of the important findings of a particular study.

It is interesting that, given the same results, two people can draw two different conclusions, and neither conclusion is necessarily incorrect. That is not to say that any conclusion is correct but that a conclusion is personal; it is your interpretation of the results and is subjective. However, the conclusion should relate to the objective of the report.

Appendixes Appendixes are the final elements in formal reports that contain supplemental information or information that is too detailed and technical to fit well into the body of the report or that some readers need and others do not. The recent trend in formal reports has been to place highly technical or statistical information in the Appendixes for those readers who are interested in such material. [12]

Words and Expressions

communication　交流（the exchange of thoughts, messages, or information）
peer [piə]　n. 同行（a person who belongs to the social group as someone else），同事
technical writing　科技写作（关于科技论文、技术报告等的写作）
beware of　注意（to be very careful about），警惕（to be cautious and watchful about）
substantiate [sʌbsˈtænʃieit]　v. 证明，证实
basic [ˈbeisik]　n. 基本要素，基础；a. 基本的
concise [kənˈsais]　a. 简明的（expressing much in few words），简洁的，简明扼要的
distillation [ˌdistiˈleiʃən]　n. 精华（the main meaning or the most important parts of something）
highlight [ˈhailait]　n. 最显著（重要）部分；v. 突出，使……显得重要，强调

quantitative [ˈkwɔntitətiv]   *a.* 数量的，定量的
literature review   文献综述 (a comprehensive summary of previous research on a topic)
rationale [ˌræʃəˈnɑːli]   *n.* 基本原理，根本原因 (fundamental reasons; the basis)
equation [iˈkweiʃən]   *n.* 方程式，公式，等式
pertinent [ˈpəːtinənt]   *a.* 有关的，相关的 (relating directly to the subject being considered)
nonstandard [ˌnɔnˈstændəd]   *a.* 不标准的，非标准的
experimental study   实验研究，试验研究
complexity [kəmˈpleksiti]   *n.* 复杂的事物，复杂性
underlying [ˌʌndəˈlaiiŋ]   *a.* 根本的，在下面的，基本的
intersperse [ˌintəˈspəːs]   *v.* (与 with 连用) 点缀，散布
appendix [əˈpendiks]   *n.* 附录 (a section of extra information at the end of a book or document)
mean [miːn]   *n.* 平均数，平均值 (the mathematical average of a set of two or more numbers)
standard deviation   标准差 (方差的算数平方根，能反映一个数据集的离散程度)，均方差
error [ˈerə]   *n.* 误差 (计算值或测量值与理论上正确值之差)
vehicle   媒介，手段 (a means by which something is expressed, achieved, or shown)
short and descriptive title   简明的标题
finding   发现，研究结果 (information or a fact that is discovered by studying something)
subjective [sʌbˈdʒektiv]   *a.* 主观的 (influenced by personal opinions)，个人的 (personal)
supplemental [ˌsʌpliˈmentl]   *a.* 补充的 (available to supply something extra when needed)
statistical [stəˈtistikəl]   *a.* 统计的，统计学的

Notes

[1] picture 这里是动词，意为"想象 (to think of something in your mind)"，seldom the case 意为"并不常见"。这三句可译为：将你的想法和成果与别人交流是工程界中一个非常重要的方面。许多工科学生想象当他们将来从事专业技术工作时，也会花费大量的时间进行类似他们当学生时所做的计算工作。幸运的是，这种情况并不常见。

[2] usual form of presentation 意为"常见的表现形式"，communication skill 意为"沟通能力，交流能力，表达能力"。这三句可译为：当你的设计完成之后，通常需要将设计结果展示给你的客户、同行或雇主。常见的表现形式是一篇正式的技术报告。因此，对于工科学生来说，提高他们的表达能力是非常重要的。

[3] a mass of 意为"大量的"。这句可译为：对重要信息要进行强调，谨防出现将其淹没在大量的细节之中这类常见的错误。

[4] contribution 这里指"工作成果，技术成果"。这两句可译为：要把事实与观点分开。对于读者来说，重要的是知道你的技术成果是什么，哪些想法是你从别人那里得到的（参考文献表明了这一点），哪些是还没有得到事实证明的观点。

[5] quantitative 意为"用数量表示的，定量的"。这几句可译为：摘要应该完整、简明地将整篇报告的精华介绍出来，因此，它总是最后写成的。它应该包括对主题的简要介绍（一句话），对问题的陈述，着重介绍结果（如果可能的话，要定量）和主要的结论。它必须在不引用图或表的情况下独立存在。由于大多数摘要中的单词少于 250 个，所以采用简明、清晰的表达方法是必要的。

[6] subject of the report 意为"技术报告的主题"。这两句可译为：引言这节通常要确定技术报告的主题，提供包括文献综述在内的必要的背景资料，通常还会为读者介绍这项工作的基本原理。引言中不包含结果，一般也没有公式。

[7] a series of 意为"一系列的，一连串的"，explanatory material 意为"解释性文字材料"。这三句可译为：在分析这节中，通常穿插着许多公式。它不仅仅是一连串没有解释性文字材料的公式。公式中第一次出现的符号应给予注释。

[8] experimental procedure 意为"实验过程，实验步骤"。这两句可译为：实验过程这节用来描述实验结果是如何获得的，以及所使用的非标准设备或实验技能。你应该提供足够的细节，以便其他人也能够完成这项实验。

[9] confidence interval 意为"置信区间"。这几句可译为：结果就是事实。它们是你收集的数据和通过你的计算得到的数据。平均值、标准差、置信区间和误差都是结果。

[10] present 意为"展示，陈述"。全段可译为：在报告的这一节中，应该展示你的工作成果，并对其进行讨论。应该记住，在展示你的成果时，即使你的读者是有经验的技术人员，那些对你来说是很简单的事情，对他们来说也并非显而易见。通常图和表是清楚地介绍实验结果的最重要的工具。每个图或表都应该有序号和简明的标题。

[11] rendition 这里指"解释（an interpretation of something）"。这句可译为：它不是一个提供新事实的地方，它也不应该再一次对实验结果和基本原理进行解释。

[12] body of the report 意为"报告正文"。这两句可译为：附录是正式报告中的最后一个组成部分，它包含一些补充资料或一些不适合写到报告正文中的太详细和技术性太强的资料，或者包含某些读者需要而另一些读者不需要的资料。目前的发展趋势是将高度技术性的资料或统计信息资料放到正式报告的附录中，以满足对这些材料感兴趣的读者们的需求。

参考译文

第1课 力学基本概念

对运动、时间和力进行科学分析的分支称为力学。它由静力学和动力学两部分组成。静力学是对静止系统进行分析,也就是在其中不考虑时间这个因素,动力学则是对随时间而变化的系统进行分析。

当一些物体连接在一起形成一个组合体或系统时,任何两个相互连接的物体之间的作用力和反作用力称为约束力。这些力约束着各个物体,使其处于特定的状态。从外部施加到这个物体系统的力称为外力。

力通过接触表面传到机器中的各构件上。例如,从齿轮传到轴上或从齿轮通过与其啮合的轮齿传到另一个齿轮(见图1.1),从V带传到带轮,或者从凸轮传到从动件。出于多种原因,人们需要知道这些力的大小。例如,如果作用在滑动轴承上的力过大,它就会将油膜挤出,造成金属与金属的直接接触,产生过热和使轴承快速失效。如果啮合的轮齿之间的力过大,就会将油膜从轮齿之间挤压出去。这可能会造成金属表层的剥落,产生噪声,直至齿轮失效。在力学研究中,我们主要关心力的大小、方向和作用点。

在力学中要用到的一些术语定义如下。

力 关于力的最初概念是由于我们需要推、提或拉各种物体而产生的。因此,力是物体之间的相互作用。我们对于力的直观概念包括作用点、方向和大小,这些被称为力的特性。

力偶 作用于同一物体上的大小相等、方向相反但不共线的两个平行力不能被合成为一个合力。作用在同一物体上的任何两个这样的力构成一个力偶(见图1.2)。力偶的唯一作用是产生一个朝特定方向的转动或转动趋势。

质量 质量是物体内所包含的物质的量。物体的质量不受重力影响,因此它与物体的重量不同,但是与其成正比。尽管一块月球岩石在月亮上和在地球上的重量不同,但它的物质含量是不变的。这不变的物质含量就称为这块岩石的质量。

惯性 惯性是物体所具有的抵抗任何外力改变其本身运动状态的性质。

重量 重量是地球或其他天体对物体的作用力,它等于物体的质量与重力加速度的乘积。

质点 当一个物体的尺寸特别小,可以忽略不计时,该物体可以称为质点。

刚体 刚体在受力后,其大小和形状都不会发生变化。实际上,所有的物体,无论是弹性体还是塑性体,在力的作用下都会发生变形。当物体的变形非常小时,为了简化计算,通常假设这个物体是刚体,也就是认为它没有发生变形。刚体是实际物体的理想化模型。

变形体 在分析由作用力引起的物体内部的应力和应变时,不能采用刚体假设。这时,我们认为物体是能够变形的。这样的分析通常被称为弹性体分析,这时所用的假设为,在作用力的范围内,物体是弹性的。

牛顿运动定律 牛顿三定律为:

第一定律 如果作用在一个物体上的所有的力平衡,那么,这个物体将保持原来的静止或匀速直线运动状态不变。

第二定律 如果作用在一个物体上的那些力不平衡,那么,这个物体将产生加速度。加速度的方向与合力的方向相同;加速度的大小与合力的大小成正比,与物体的质量成反比。

第三定律 相互作用的物体之间的作用力和反作用力大小相等,方向相反,作用在同一直线上。

力学涉及两种类型的量：标量和矢量。标量是那些只有大小的量。在力学中标量的例子有时间、体积、密度、速率、能量和质量。另一方面，矢量既有大小又有方向。矢量的例子有位移、速度、加速度、力、力矩和动量。

第4课　轴、联轴器和花键

实际上，几乎所有的机器中都装有轴。最常见的轴的形状是圆形，其横截面既可以是实心的，也可以是空心的（空心轴可以减轻重量）。有时也采用矩形轴，如一字螺丝刀的扁平形头部（见图4.1）。

为了在传递扭矩时不发生过载，轴应该具有足够的抗扭强度。它还应该具有足够的抗扭刚度，以确保安装在同一个轴上的两个零部件之间的相对转角不会过大。一般来说，当长度等于轴的直径的20倍时，轴的扭转角不应该超过1°。

轴安装在轴承中，通过齿轮、带轮、凸轮和离合器等装置传递动力。通过这些装置传来的力可能会使轴产生弯曲变形。因此，轴应该有足够的刚度以防止轴承受力过大。总而言之，在两个轴承之间，轴在每米长度上的弯曲变形不应该超过0.5mm。

此外，轴还必须能够承受弯曲载荷和扭转载荷的组合作用。因此，要考虑扭矩与弯矩（见图4.2和图4.3）的当量载荷。因为扭矩和弯矩会产生交变应力，所以在许用应力中也应该有一个考虑疲劳现象的安全系数。

直径小于75mm的轴可以采用含碳量大约为0.4%的冷轧钢，直径在75～125mm之间的轴可以采用冷轧钢或锻造毛坯。当轴的直径尺寸大于125mm时，则采用锻造毛坯，然后经过机械加工达到要求的尺寸。轻载时，广泛采用塑料轴。在电器中采用塑料轴的一个优点是安全性，这是因为塑料是电的不良导体。

齿轮和带轮等零件通过键与轴连接。在设计键和轴上与之对应的键槽（见图4.4）时，必须进行认真计算。例如，轴上的键槽会引起应力集中，键槽的存在还会使轴的横截面积减小，这会进一步减弱轴的强度。

如果轴的旋转速度为临界转速，就会发生强烈的振动，从而造成机器的严重损坏。知道这些临界转速的大小是很重要的，因为知道后就可以避开临界转速。通常，工作转速与临界转速之间至少应该相差20%。

一些轴由三个或更多的轴承支承，这意味着它是一个超静定问题。材料力学教科书介绍了求解这类问题的方法。但是，设计工作应该与特定场合的经济性相符合。例如，如果一根长轴需要由三个或更多轴承来支承，则可以对力矩做出保守的假设，按照静定轴设计，其成本可能会更低。因为，轴的尺寸变大所增加的成本可能会比进行复杂、精细的设计分析工作所多花费的成本要低一些。

轴的设计工作中的另一个重要方面是两根轴之间的连接方式。这是通过刚性或挠性联轴器等装置实现的。

联轴器是连接相邻的两个轴端的装置。在机械结构中，联轴器被用来实现相邻的两根转轴之间的半永久性连接。从某种意义上说，这种连接是永久性的，即在机器的正常使用期间内，这种连接不会被拆开。但是在紧急情况下，或者在需要更换已磨损的零件时，可以先把联轴器拆开，然后再将其连接到一起。

刚性联轴器能将两根轴紧固地连接到一起（见图3.1和图4.5），使其不能产生相对运动。刚性联轴器可用于那些需要而且能够实现两根轴的轴线精确对中的设备中。如果制造工厂中或船舶的螺旋桨需要一根特别长的轴，则可以采用分段的方式将其制造出来，然后采用刚性联轴器将各段连接起来。

在把属于不同的设备（如一个电动机和一个变速箱）中的轴连接到一起时，要实现这两轴轴线的精确对中是比较困难的，此时可以采用挠性联轴器（见图4.6和图4.7）。这种联轴器以一种能够把由于轴线相对偏移所造成的有害影响降至最低的方式将两根轴连接到一起。挠性联轴器也允许被连接的轴在它们各自的载荷系统作用下产生偏斜或在轴线方向自由移动（浮动）而不至于产生相互干扰。挠性联轴器还可以降低从一根轴传到另一根轴上的冲击载荷和振动的强度。

当轴和毂之间仅需要有轴向的相对运动时，可以通过在轴上和毂孔上加工出的花键（见图4.8）来防止它们之间的相对转动。花键分为两种：矩形花键和渐开线花键。前者形状简单，应用于某些机床中。后者的轮廓为渐开线，渐开线广泛应用于齿轮中。人们更习惯于选择渐开线花键，这是因为它能够对配合零件进行自动定心，而且可以利用加工齿轮的标准滚刀（见图4.9）对它进行加工。

第7课　紧固件和弹簧

紧固件是将一个零件与另一个零件连接到一起的装置。因此，几乎在所有的设计中都要用到紧固件。人们对于任何产品的满意程度不仅取决于其组成部件，还取决于其连接方式。紧固件为产品设计提供了以下特性：

（1）为检查和维修提供拆卸的方便；

（2）为由许多部件组成的组合式设计提供方便。采用组合式设计可以便于生产制造和运输。

紧固件可以分为以下三类：

（1）可拆卸式。采用这种紧固方式连接的零件很容易被拆卸，而且不会对紧固件造成损伤。例如，普通的螺栓螺母连接（见图7.1）。

（2）半永久式。采用此类紧固件连接的零件虽然能被拆开，但通常会对紧固件造成一些损伤。开口销（见图7.2）就是这样一个例子。

（3）永久式。采用这种紧固件就表明所连接的零件不会被分开，例如，铆钉连接和焊接接头。

对于一个特定的应用，在选择紧固件时应考虑以下因素：

（1）基本功能；

（2）外观；

（3）是采用大量的小型紧固件还是采用少量的大型紧固件（以螺栓为例）；

（4）如载荷、振动和温度等工作条件；

（5）拆卸频率；

（6）零件位置的可调性；

（7）被连接零件的材料种类；

（8）紧固件失效或者松脱造成的后果。

通过任何一个复杂的产品，都可以认识到紧固件在其中的重要性。以汽车为例，它是由成千上万个零件连接在一起而成为一辆整车的。一个紧固件的失效或松脱可能会带来像车门嘎嘎响这类小麻烦，也可能造成像车轮脱落这种严重的后果。因此，在为一个特定的用途选择紧固件时，应该考虑上述各种可能性。

弹簧是一种能够在外载荷作用下，产生相当大的弹性变形的机械零件。描述变形量与载荷成正比的胡克定律表明了弹簧的基本性能。然而，也有一些弹簧在其设计时所确定的载荷与变形量之间的关系就是非线性的。弹簧的主要用途如下所述：

（1）控制机构运动。这类应用是弹簧用途的主要部分，例如，离合器和制动器中的操纵力。此外，弹簧也用于保持两个部件之间的接触，如凸轮和它的从动件。

（2）缓冲和减振。这类应用包括汽车悬架系统弹簧和橡胶弹簧。

（3）存储能量。这类弹簧应用于钟、表和割草机中。

（4）力的测量。用来称体重的秤是这类应用中最常见的一种。

弹簧主要可以分为压缩、拉伸和扭转弹簧这三种类型（见图7.3）。压缩弹簧和拉伸弹簧是最常用的弹簧。这些弹簧的变形是线性的。扭转弹簧的变形是角位移而不是线性位移。板弹簧既可以是简支型，也可以是悬臂型。在工业中，橡胶弹簧和橡胶缓冲减震装置的适用范围日益增加。橡胶并不遵从胡克定律，其刚度随变形的增大而增加。

大部分弹簧是用钢制造的，但也有一些弹簧用硅青铜、黄铜和铍铜制造。弹簧通常由专业生产弹簧的厂家制造。圆柱螺旋弹簧是最常用的弹簧，扭杆弹簧和板弹簧也有广泛的应用。对于圆柱螺旋弹簧，如果弹簧钢丝的直径小于8mm，则通常采用冷拔钢丝或油淬火-回火钢丝通过冷卷法制成。如果弹簧钢丝的直径较大，则采用热卷法制造弹簧。弹簧制成后，需对其进行淬火和回火处理，以使其具有所需要的物理性能。

选择弹簧，尤其是在遇到重载、高温、需要承受交变应力或需要具有抗腐蚀性的时候，应该向弹簧制造厂家进行咨询。为了正确地选择弹簧，应该对弹簧的各种使用要求，包括空间限制进行全面的研究。现在有许多不同种类的专用弹簧可以满足一些特殊的要求或用途。

第8课 滚动轴承

滚动轴承可以承受径向载荷、轴向载荷或同时承受这两种载荷。因此，大部分滚动轴承按其承载方向可以被划分为三类：向心轴承，主要用于承受径向载荷；推力轴承，主要用于承受轴向载荷；角接触轴承或圆锥滚子轴承，能够同时承受径向和轴向载荷。滚动轴承的三种基本类型是：球轴承、滚子轴承和滚针轴承。一个普通的单列深沟球轴承的结构如图8.1a所示，它由内圈、外圈、球和保持架四部分组成。为了增加接触面积以承受更大的载荷，球在内、外圈上的被称为滚道的弧形的沟内滚动。这种类型的轴承既能承受径向载荷，也能承受一定的轴向载荷。几种其他类型的滚动轴承如图8.1b～图8.1d，图9.1和图10.1所示。

对于球轴承和滚子轴承，一个机器设计人员应该考虑下面五个方面：（a）寿命与载荷的关系；（b）刚度，也就是在载荷作用下的变形；（c）摩擦；（d）磨损；（e）噪声。对于中等载荷和转速，根据额定载荷选择一个标准轴承，通常都可以保证其具有令人满意的工作性能。当载荷较大时，轴承零件的变形将会变得重要起来，尽管它通常小于轴和其他与轴承一起工作的零部件的变形。在转速高的场合需要有专门的冷却装置，而这可能会增大摩擦阻力。磨损主要是由于异物入侵造成的，必须选用密封装置以防止周围环境的不良影响。

因为大批量生产方式决定了球轴承和滚子轴承不仅质量高，而且价格低，所以机器设计人员的任务是选择而不是设计轴承。滚动轴承通常是采用硬度约为900HV、整体淬火的钢来制造的。在运转过程中，轴承接触表面会承受循环应力，金属疲劳就成为其主要失效形式。目前，人们正在进行大量的研究工作以求提高这种轴承的可靠性。轴承设计是基于公认的寿命值进行的。在轴承行业中，通常将轴承的承载能力定义为这样的值，即所受的载荷小于这个值时，一批轴承中将会有90％的轴承具有超过一百万转的寿命。

尽管球轴承和滚子轴承的设计工作是由轴承制造厂家承担的，但是机器设计人员必须对轴承所要完成的任务做出正确的评价，不仅要考虑轴承的选择，而且还要考虑正确的安装条件。

轴承套圈与轴或轴承座的配合非常重要，因为它们之间的配合不仅应该保证所需要的过盈量，而且也应该保证轴承的内部间隙。内圈通常通过靠紧在轴肩上进行轴向定位。轴肩根部的圆角半径对避免产生应力集中是必要的。轴承内圈的圆角半径或倒角提供了容纳轴肩圆角半径的空间（见图8.2）。

当使用寿命不是设计中的决定因素时，通常根据轴承承受载荷时产生的变形量来确定其最大载荷。因此，"静态承载能力"这个概念可以理解为对处于静止状态的或进行缓慢转动的轴承所能够施加的载荷。这个载荷对轴承在其随后进行旋转运动时的质量没有不利影响。根据实践经验确定，静态承载能力是这样一个载荷，当它作用在轴承上时，滚动体与滚道在任何一个接触点处的总变形量不超过滚动体直径的 0.01%。这相当于一个直径为 25mm 的球的永久变形为 0.0025mm。

只有将轴承与周围环境适当地隔离开，许多轴承才能成功地实现它们的功用。在某些情况下，必须保护环境，使其不受润滑剂和轴承表面磨损生成物的污染。轴承设计的一个重要组成部分是使密封圈起到应有的作用。此外，对摩擦学研究人员来说，为了任何目的而应用于运动零部件上的密封圈都是他们感兴趣的。因为密封圈是轴承的组成部分，所以只有根据适当的轴承理论才能设计出令人满意的轴承密封圈（见图 10.2）。虽然它们很重要，但是与轴承其他方面的研究工作相比，在密封圈的研究方面所做的工作还是比较少的。

第 13 课 机械零件的强度

在设计任何机器或结构时，所考虑的主要事项之一是其强度应该比它所承受的应力要大得多，以确保安全与可靠性。要保证机械零件在使用过程中不发生失效，就必须首先知道它们在某些时候会失效的原因，然后，才能将应力与强度联系起来，以确保其安全。

理想情况下，在设计任何机械零件时，工程师应该能够利用所选材料的大量强度试验数据。那些试验使用的试样应该与他所设计的零件有着相同的热处理、表面粗糙度和尺寸，而且试验应该在与零件使用过程中承受的载荷完全相同的情况下进行。这意味着，如果零件将要承受弯曲载荷，那么就应该进行弯曲载荷的试验。如果零件将要承受弯曲和扭转的组合载荷，那么就应该进行弯曲和扭转组载荷的试验。这些种类试验可以提供非常有用和精确的数据。它们可以告诉工程师应该采用的安全系数和在规定的使用寿命时的可靠性。在设计过程中，只要能够获得这些数据，工程师就可以确信他能够很好地进行工程设计工作。

如果零件的失效可能危害人的生命安全，或者零件有足够大的产量，则在设计前收集如此广泛的数据所付出的费用是值得的。例如，汽车和冰箱的零件的产量非常大，可以在生产之前对它们进行大量的试验，使其具有较高的可靠性。如果把进行这些试验的费用分摊到所生产的零件上，则分摊到每个零件上的费用是非常低的。

你可以对下列四种类型的设计做出评价：

（1）零件的失效可能危害人的生命安全，或者零件的产量非常大，因此在设计时安排一个完善的试验程序会被认为是合理的。

（2）零件的产量足够大，可以进行适当的系列试验。

（3）零件的产量非常小，以致进行试验根本不合算；或者要求很快地完成设计，以致没有足够的时间进行试验。

（4）已经完成了零件的设计、制造和试验，但其结果不能令人满意。这时需要进行分析，来找出不能令人满意的原因和应该对其进行改进的方法。

我们将主要对后三种类型进行讨论。这就是说，设计人员通常只能利用那些公开发表的屈服强度、抗拉强度和伸长率等数据资料。人们期望工程师在利用这些不是很多的数据资料的基础上，对静载荷与动载荷、二维应力状态与三维应力状态、高温与低温，以及大零件与小零件进行设计！而设计中所能利用的数据通常是从简单的拉伸试验中得到的，其载荷是逐渐加上去的，有充分的时间产生应变。到目前为止，还必须利用这些数据来设计每分钟承受数千次复杂的动载荷的作用的零件，因此机械零件有时会失效是不足为奇的。

概括地讲，设计人员所遇到的基本问题是，无论对于哪一种应力状态或载荷情况，都只能利用通过简单拉伸试验所获得的数据并将其与零件的强度联系起来。

可能会有两种具有完全相同的强度和硬度值的金属，其中的一种由于其本身的延性而具有很好的承受过载负荷的能力。一种量度材料延性的方法是计算其断后伸长率。图 13.1 所示为一个试样在标准拉伸试验前、后的状况。在测量材料的伸长时，采用的标距通常为 50mm。为了测定伸长率，需要将断裂的两部分试样仔细地配接在一起，使其轴线处于同一直线上。断后伸长率 A 的表达式为：

$$A = \frac{L_u - L_o}{L_o} \times 100\% \tag{13.1}$$

式中，L_u 为试样的断后标距；L_o 为原始标距。

通常将 5% 的伸长率定义为延性与脆性的分界线。断裂时伸长率小于 5% 的材料称为脆性材料，大于 5% 的材料称为延性材料。

拉伸试样断裂后横截面积减少的百分率是另一种量度材料延性的方法。其定义为在颈缩区内：

$$Z = \frac{S_o - S_u}{S_o} \times 100\% \tag{13.2}$$

式中，Z 为断面收缩率；S_o 为拉伸试样的原始横截面积；S_u 为断后最小横截面积。

脆性材料在断裂前不会出现明显的屈服现象或没有屈服现象（见图 13.2）。灰口铸铁就是一个例子。延性材料在断裂前能够承受较大的应变（见图 13.3）。低碳钢就是一个典型的例子。工程师们在设计中经常选用延性材料，因为这类材料能够吸收冲击或能量，如果承受了过载负荷，它们通常会在断裂前出现较大的变形。

在选用耐磨损或抗塑性变形的材料时，硬度通常是最主要的性能。有几种可供选用的硬度测试方法，采用哪一种方法取决于最需要测量的材料特性。最常用的四种硬度是布氏硬度、洛氏硬度、维氏硬度和努氏硬度。

大多数硬度测试系统是将一个标准载荷作用在与被检验材料相接触的球、棱锥或圆锥形压头上。然后将硬度表示为所产生的压痕（见图 13.4）尺寸的函数。因为硬度的测量是非破坏性试验，而且不需要专门的试样，所以这意味着硬度是一个容易测量的性能。通常可以直接在实际的机械零件上进行硬度试验。

第 17 课　材 料 选 择

近些年来，工程材料的选择显得非常重要。此外，选择过程应该是一个对材料的连续不断的重新评价过程。新材料不断出现，而一些原有的材料中可以被利用的数量可能会减少。环境污染、材料的回收利用、工人的健康及安全等方面经常会对材料选择附加新的限制条件。为了减轻重量或节约能源，可能要求使用不同的材料。来自国内和国际的竞争、对产品可维修性要求的提高和顾客的反馈等方面的压力，都会促使人们对材料进行重新评价。由于材料选用不当造成的产品责任诉讼，已经产生深刻的影响。此外，材料与材料加工之间的相互依赖关系已经被人们认识得更清楚。新的加工方法的出现，通常会促使人们对被加工材料进行重新评价。因此，为了能够在合理的成本和确保质量的前提下，获得满意的结果，设计工程师和制造工程师都必须认真仔细地选择、确定和使用材料。

制造任何产品的第一步工作都是设计。设计通常可以分为几个不同的阶段：（a）概念设计；（b）功能设计；（c）生产设计。在概念设计阶段，设计者着重考虑产品应该具有的功能。通常要设想和考虑几个方案，然后决定这种想法是否合理；如果合理，则应该对其中一个或几个方案做进一步的改进。在此阶段，关于材料选择唯一需要考虑的问题是：是否有性能符合要求的材料可供选用；如果没有的话，是否有较大的把握在成本和时间都允许的限度内研制出一种新材料。

在功能设计或工程设计阶段，要做出一个切实可行的设计。在这个阶段需要绘制相当完整的图样，选择并且确定各种零件的材料。通常要制造出样机或实物模型，并对其进行试验，评价产品的功能、可靠性、外观和可维修性等。虽然这种试验可能表明，在产品进入生产阶段之前，应该更换某些材料，但是，绝对不能将这一点作为不认真选择材料的借口。应该结合产品的功能，认真仔细地考虑产品的外观、成本和可靠性。一个非常成功的公司在制造所有样机时，所选用的材料应该和其在生产中使用的材料相同，并尽可能使用同样的制造技术。这样做，对公司是很有好处的。功能完备的样机如果不能根据预期的销售量经济地制造出来，或者样机与产品在质量和可靠性方面有很大的不同，则这种样机就几乎没有价值。设计工程师最好能在这一阶段全部完成材料的分析、选择和确定工作，而不是将其留到生产设计阶段去做。因为，在生产设计阶段材料的更换是由其他人进行的，这些人对产品的所有功能的了解可能不如设计工程师。

在生产设计阶段中，与材料有关的主要问题是应该把材料完全确定下来，使它们与现有的设备相适应，能够利用现有设备经济地进行加工，而且材料的数量能够比较容易地保证供应。

在制造过程中，不可避免地会出现对使用中的材料做一些更改的情况。经验表明，可以采用某些价格便宜的材料作为替代品。然而，在大多数情况下，在进行生产以后改换材料比在开始生产前改换材料所花费的代价要高。在生产设计阶段做好材料选择工作，可以避免大多数的这种材料更换情况。在生产制造开始后出现了可供使用的新材料，是更换材料的最常见的原因。当然，这些新材料可能降低成本、改进产品性能。但是，必须对新材料进行认真的评价，以确保其所有性能都被人们所了解。应当时刻牢记，新材料的性能和可靠性很少能像现有的材料那样为人们所了解。大部分的产品失效和产品责任事故案件是由于在选用新材料作为替代材料之前，没有真正了解它们的长期使用性能而引起的。

产品的责任诉讼迫使设计人员和公司在选择材料时，必须采用最好的程序。在材料选择过程中，最常见的五个错误是：（a）不了解或未能利用关于所选用材料的最新和最有用的信息资料；（b）未能预见和考虑产品可能的合理用途（若有可能，则设计人员还应进一步预测和考虑由于产品使用方法不当造成的后果。在近年来的许多产品责任诉讼案件中，由于错误地使用产品而受到伤害的原告控告生产厂家，并且赢得了判决）；（c）所使用的材料的数据不全或有些数据不确定，尤其当其长期性能数据是如此的时候；（d）质量控制方法不适当和未经验证；（e）由一些完全不称职的人员选择材料。

通过对上述五个问题的分析，可以得出这些问题是没有充分理由存在的结论。对这些问题的分析和研究可以为避免这些问题的出现指明方向。尽管采用最好的材料选择办法也不能避免因产品责任而产生的索赔，但是设计人员和工业界按照适当的程序进行材料选择，可以大大减少它们的数量。从上面的讨论可以看出，选择材料的人们应该对材料的性质、特点和加工方法有一个全面而基本的了解。

第 21 课　润　　滑

润滑的主要目的之一是减少摩擦力，任何可以控制两个滑动表面之间摩擦和磨损的物质——无论是液体、固体还是气体——都可以归类为润滑剂。

润滑的种类

无润滑滑动　经过精心处理的、去除了所有外来物质的金属在相互滑动时会产生咬死和冷焊。当达不到这么高的纯净度时，吸附在表面的气体、水蒸汽、氧化物和外界异物会降低摩擦力并减小黏附的趋势，但通常会产生严重的磨损，这种现象被称为"无润滑"滑动或干滑动。

流体膜润滑　在滑动面之间引入一层流体膜，把滑动表面完全隔离开，就产生了流体膜润滑（见图 21.1）。这种流体可能是有意引入的，如汽车主轴承中的润滑油；也可能是无意中引入的，如在光滑的橡胶轮胎与潮湿的路面之间的水。虽然流体通常是油、水和很多其他种类的液体，但是它也可以是气体。最常用的气体是空气。

为了把零件隔离开，润滑膜中的压力必须和作用在滑动面上的负荷保持平衡。如果润滑膜的压力是由外源供给的，则这种系统称为流体静压润滑。如果滑动表面之间的压力是由于滑动面本身的形状和运动所共同产生的，则这种系统就称为流体动压润滑。

边界润滑　处于无润滑滑动和流体膜润滑之间的状态被称为边界润滑（见图 21.2）。它可以被定义为这样一种润滑状态，在这种状态中，表面之间的摩擦力取决于表面的性质和润滑剂中除黏度外的其他性质。边界润滑包括大部分润滑现象，通常在机器启动和停止时出现。

固体润滑　当普通润滑剂不具有足够的承受载荷的能力或不能在极限温度下工作时，石墨和二硫化钼这一类固体润滑剂就得到了广泛的应用。但润滑剂不只是以润滑脂（见图 21.3）和粉末这样一些为人们所熟悉的形态出现，在一些精密的机器中，金属也常作为滑动面。

润滑剂的作用

尽管润滑剂主要是用来控制摩擦和磨损的，它们也能够而且通常也确实起到许多其他的作用，这些作用随其用途不同而不同，但通常相互之间是有关系的。

控制摩擦力　滑动面之间润滑剂的数量和性质对所产生的摩擦力有极大的影响。例如，不考虑热和磨损这些相关因素，只考虑两个油膜润滑表面间的摩擦力，它能比在两个同样表面间，但没有润滑时的摩擦力小 200 倍。在流体膜润滑状况时，摩擦力与流体的黏度成正比。一些诸如石油衍生物这类润滑剂，可以有很多种黏度，因此能够满足范围宽广的功能要求。在边界润滑状态，润滑剂黏度对摩擦力的影响不像其化学性质的影响那么显著。

控制磨损　磨料、腐蚀和固体与固体之间的接触都会造成磨损。适当的润滑剂将能够帮助减缓这些磨损现象。它们通过形成润滑膜来增加滑动表面之间的距离，从而减轻了磨料磨损和固体与固体之间接触造成的磨损，因此，也减轻了由于磨料污染物和表面微凸体（见图 21.4）造成的损伤。

控制温度　润滑剂通过减少摩擦和将产生的热量带走来控制温度。其效果取决于润滑剂的用量、环境温度和外部冷却措施。冷却剂的种类也会在较小程度上影响表面的温度。

控制腐蚀　润滑剂在表面腐蚀的控制方面有双重的作用。当机械闲置不工作时，润滑剂起到防腐剂的作用。当机械工作时，润滑剂通过在被润滑零件表面形成的保护膜来控制腐蚀，其中可以含有抗腐蚀添加剂。润滑剂控制腐蚀的能力与润滑剂保留在金属表面的润滑膜的厚度和润滑剂的化学成分直接相关。

其 他 作 用

除减少摩擦外，润滑剂还经常有其他的用途。其中的一些用途如下所述。

传递动力　润滑剂被广泛用来作为液压传动中的工作液体。

绝缘　在像诸如变压器（见图 21.5）这些特殊用途中，具有高介电常数的润滑剂起到了电气绝缘材料的作用。为了获得最好的绝缘性能，润滑剂中不能含有任何杂质和水分。

减振　在像减振器这种能量传递装置中和在承受高频率的间歇性载荷的齿轮这类机器零件的周围，润滑剂被作为减振液使用。

密封　润滑脂通常还有一个特殊作用，就是形成密封层以防止润滑剂外泄和异物侵入。

第 22 课　摩擦学概论

摩擦学是一门研究在相对运动中相互作用表面的科学与技术。它起源于希腊语 tribos，意思是摩擦。它研究工作表面的摩擦、润滑和磨损，目的是详细地理解表面间的相互作用，以便为实际应用制定改进方案。摩擦学研究人员的工作是跨学科的，包括物理、化学、力学、热力学和材料科学等学科，并且包括一个涉及有关表面间相对运动的机械设计、可靠性和工作性能的庞大复杂而且交织在一起的领域。

估计目前世界上大约有23%的能源是以某种摩擦和磨损形式消耗的。其中的20%用于克服摩擦力，3%用于磨损部件的再制造。摩擦学研究的目的是减少或消除在各种表面摩擦技术中不必要的浪费。

摩擦学研究的一个重要目标是根据我们的需要调节摩擦力的大小，可以将其调到最小（如在机器中），或者最大（如作为防滑表面）。然而，必须着重指出，只有在对温度、滑动速度、润滑、表面粗糙度和材料性能等所有条件下的摩擦过程有了基本理解之后，这个目标才能实现。

从设计角度来看，对于一个实际应用的最重要判别标准，是在滑动界面上由干状态或润滑状态起主导作用。在许多场合，如在机器中，尽管可能存在着几种不同的润滑状态，但是只有一种状态（通常是流体润滑）起主导作用。然而在少数情况下，事先无法知道界面上究竟是干还是湿，显然这对于进行任何设计都是很困难的。这种现象最常见的例子是充气轮胎。在干摩擦的情况下，光滑的轮胎外表面在光滑的路面上可以获得最大的接触面积，使摩擦的黏附分量达到最大值。然而，这个组合在湿的情况下会产生非常低的摩擦系数，因此无法获得所期望的相对运动。不过在后一种情况下，如果在轮胎表面有合理的花纹（见图22.1）和采用合理的路面结构，仍然能够达到最佳的状态，尽管在干燥的气候下这种组合会获得较低的摩擦系数。

润滑状态可以划分为流体动力润滑、边界润滑和弹性流体动力润滑。目前使用的各种轴承是完全流体动力润滑特性的最好例子，其滑动表面完全被界面间的润滑油膜隔开。边界润滑或混合润滑（见图22.2）是在相对运动表面之间既存在流体动力润滑，又有固体之间接触的一种混合状态。通常认为，在一个特定的产品设计中，当流体动力润滑失效时，才会出现这种混合状态。例如，一个滑动轴承在规定的载荷和速度下被设计成完全流体动力润滑的，但是当速度下降或载荷增加时，就会造成轴颈和轴承表面一部分是固体接触，一部分是流体动力润滑这种状态。这种边界润滑是不稳定的，通常可以恢复到完全流体动力润滑，或者变成表面之间完全咬死。当薄的润滑油膜中的压力达到可使润滑剂的边界面发生弹性变形时，则滑动界面上的润滑状态属于弹性流体动力润滑。

固体润滑剂处于干状态和润滑状态之间，即尽管接触面通常是干的，但是固体润滑材料的存在使得它像在一开始就是湿润的一样。这是在特定的载荷和滑动条件下，固体润滑剂覆盖层表面上产生的一种物理-化学相互作用的结果，这样就产生了一种相当于润滑的效应。石墨和二硫化钼是最常用的固体润滑剂材料，其结构如图22.3所示。这些固体的润滑性是由其层状结构产生的。

通过利用减少摩擦和防止磨损的新技术，可以显著减少世界各地的车辆、机械和设备中因摩擦和磨损造成的能量损失。

可以减少摩擦和磨损的现代方法和技术有：
1. 新型润滑剂解决方案，如基于纳米技术的减摩抗磨添加剂、低黏度润滑油和气相润滑；
2. 新材料解决方案，如选用新材料和表面改性技术。

第23课 产品图样

制造企业的成立是为了生产一种或多种产品。这些产品是采用被称为产品图样的文件来进行完全定义的。产品图样上的尺寸和技术要求将保证零件具有互换性和能够可靠地实现其设计性能。产品图样通常包括零件图和装配图。

零件图是一个零件的标注有尺寸的多个视图，它描述了零件的形状、尺寸、材料和表面粗糙度。这张图样本身含有制造这个零件所需要的全部信息。大部分零件需要采用三个视图对其形状做完整的描述。

装配图表明设计中的所有零件是如何被装配到一起的，以及整个装置的功能。因此，在其中完整的形状描述并不重要。应该采用必要和尽可能少的视图对装配体中各个零件之间的相互关系进行描述。

所有的零件都必须在某种程度上与其他零件相互作用，以产生设计方案所期望得到的功能。在绘制零件图之前，设计人员应该对装配图进行透彻的分析，以保证零件之间有适当的配合，所标注的公差准确无误，接触表面经过适当的机械加工和零件之间可以产生正确的运动。

英寸是英制的基本单位，实际上在美国所有的制造图样中都是用英寸来标注尺寸的。

毫米是公制的基本单位。标注尺寸时可以略去数字后面公制单位的缩写 mm，因为标题栏附近的 SI 符号表明所有的单位都是公制的。

在美国，有些图样同时用英寸和毫米两种单位标注尺寸，通常将用毫米标注的尺寸放在圆括号或方括号中。也可以首先用毫米作为单位标注尺寸，然后将尺寸单位换算成英寸，并将其放在方括号中。两种单位之间的换算会产生必须舍入的小数位误差。每张图样所用的主要单位制，必须在标题栏中加以说明。

标题栏 在实践中，标题栏中通常包括图样名称或零件名称、制图员、日期、比例、单位名称和图样代号。另外，还可以包含审核人员和材料等信息。在第一次制图之后，任何为了改进设计方案而做的变更和改进都应该在更改区中标明。标题栏一般位于图纸的右下角。

比例是图样中机件要素的线性尺寸与实际机件相应要素的线性尺寸之比。无论采用什么比例，在图样上的尺寸都是指机件的真实尺寸，而不是视图的尺寸。在一套产品图样中可以采用各种不同的比例。

根据项目复杂程度的不同，一套产品图样中图样的数量少则一张，多则一百张以上。因此，在每张图样上都必须写上这套图样的总张数和该张图样所在的张次（例如，共 6 张第 2 张，共 6 张第 3 张，等等）。

零件名称和序号 给每个零件一个名称和序号，通常采用的字母和数字的高度为 1/8 英寸（3 毫米）。零件的序号应放在该零件视图的附近，以便清楚地表明它们之间的关系。

明细栏 在明细栏中，零件的序号和名称必须与产品图样中的每一个零件相一致。此外，还要列出所需要的相同零件的数量和制造每种零件所使用的材料。

直到 1900 年，世界各国的图样通常都采用第一角投影的画法绘制。在第一角投影中，俯视图放在主视图的下面，左视图放在主视图的右面，依次类推。现在，美国、加拿大和英国通常采用第三角投影，而世界上许多国家仍然采用第一角投影。在第三角投影中，俯视图放在主视图的上面，右视图放在主视图的右面，左视图放在主视图的左面。

实际上，第一角投影与第三角投影的唯一区别就是视图的位置。当使用者阅读第一角投影图样而将其当作第三角投影图样时会产生混乱和制造误差，反之亦然。为了避免误解，可以采用投影识别符号（见图 23.1）来区分第一角投影与第三角投影。如果某些图样的投影方式可能会引起误解，那么这些符号就会出现在标题栏中或者靠近标题栏的位置。

产品图样是法律合同，它记录了在设计工程师指导下制订的设计细节和技术要求。因此，图样必须尽可能地清楚、准确和详尽。对于一个项目，在生产时进行修改或改进所付出的代价要远比在设计初始阶段进行这项工作昂贵得多。为了在经济上有竞争优势，图样必须尽可能没有差错。

图样的审核人员应该具有特殊的素质，他们能够发现图中的错误，并提出修改和改进意见，以便制造出物美价廉的产品。审核人员通过装配图和零件图来检查设计方案是否正确无误。此外，他们还负责检查图样的完整性、质量和清晰度。

除对每处修改都要做好记录外，制图员应该记录下整个项目期间所做的一切更改。随着项目的进行，制图人员应该将所有的更改、日期和有关人员的姓名记录下来。这样的记录使得任何人在将来对项目进行复核时，会很容易和很清楚地了解为了获得最后设计方案所经历的过程。

第 26 课　尺 寸 公 差

由于大多数制造行业均具有高度的竞争性，所以寻找降低成本的途径这个问题一直是人们所关心的。降低成本的最好起点是在产品的设计阶段。设计工程师在进行设计工作时需要考虑哪些是可以供他选择的方案。如果没有对可能的生产成本进行认真的分析，那么通常不可能选定最好的方案。在对产品的功能、互换性、质量和经济性进行设计时，需要对其公差、表面粗糙度、工艺过程、材料及加工设备等问题进行认真的研究。

在企业中工作的工程师总要面对这样一个事实，即任何两个机器零件都不能制造得完全相同。他知道在设计中必须考虑在重复性生产中所产生的微小尺寸差异，在图样上标注合适的公差，将尺寸的变化限制在允许的范围内。加工后的零件的外形轮廓必须位于公差规定的范围内。采用适当的公差可以保证产品在功能和使用寿命方面都能达到预期的目标。

每位设计人员都非常清楚，如果零件都以较小的公差来加工制造，则产品的成本就会迅速增加。因此，工程师们不断地得到劝告，要采用尽可能大的公差。然而，有时可能出现对功能要求所需要的各种公差之间的相互关系没有进行充分研究的情况。在这种情况下，为了保证零件在装配时不发生问题，设计人员通常不恰当地将公差规定得过于严格。相对于认真、透彻地对公差进行分析来说，这显然是一个价格昂贵的替代方式。

要使产品能够以较低的价格被生产出来并且能够满足预期要求，规定适当的加工公差是最为重要的工作。公差的大小是由设计人员确定的，它取决于许多与设计相关的条件，以及过去在设计类似产品时所获得的经验（如果有这方面经验的话）。车间中生产制造过程和机床性能方面的知识可以帮助人们以最有效的方式确定公差的数值。如果所规定的公差太小，以致采用现有的加工设备加工工件的这个尺寸时无法达到设计要求，就需要对设计进行修改。

工程图样是制造机器零件的依据。因此，从事制造业的人员都直接或间接地需要能够识读那些应用于整个生产过程的图样。

工程图样中含糊不清的地方会引起很多混乱和经济损失。在拟定公差时，设计人员必须充分认识到，要完全达到其设计目的，图样上必须包含所有必要的信息。因而，图样上必须给出全部信息，并且尽可能简单明了。图样中的每部分内容都应该能被大家所理解。图样中所表示的含义对于所有使用它的人员（设计、采购、刀具设计、生产、检验、装配和维修部门）来说都应该是唯一的。

公差在图样上可以采用不同的标注方式。在单向制中，一个极限偏差是零，另一个极限偏差就是尺寸允许的全部变动量。在双向制的尺寸标注中，采用基本尺寸和在其正、负两个方向上的变动量来表示。

当所有的尺寸都处在允许零件含有的材料量为最多的极限状态时，就称这个零件处于最大实体状态（MMC）。对于一根轴或一个外形尺寸，它的基本尺寸为最大极限尺寸，它在公差范围内变动时，只能使尺寸减小。对于一个孔或内部尺寸，它的基本尺寸为最小极限尺寸，在公差范围内的变动，只能使尺寸增大。

当所有的尺寸都处在允许零件含有的材料量为最少的极限状态时，就称这个零件处于最小实体状态（LMC）（见图26.1）。按LMC标注公差时，对于外形尺寸，它的基本尺寸为最小极限尺寸；对于内孔尺寸，它的基本尺寸为最大极限尺寸。在公差范围内的尺寸变动，会使零件包含的材料量增加。

按最大实体尺寸标注公差对生产有利。对于一个外形尺寸，工人按照其基本尺寸或最大极限尺寸进行加工，如果其去除量过小，则还可以通过重新加工，使工件尺寸在允许的范围内。一个工人按平均尺寸进行加工时，加工偏差只能在小范围内变动。无论怎样，上述概念为以不同方式在零件图样上标注公差提供了方便的表达形式。

在机械制造过程中，虽然会尽可能保持稳定的生产条件，但是加工后获得的尺寸仍然不可避免地出现误差。在完全相同的制造过程中，按某一指定尺寸加工一批零件，加工后所得到的尺寸却并不完全相同。产生这种现象的原因值得人们研究。一般将产生这种现象的原因分为两大类，即可确定的原因和随机原因。

可确定的原因。生产过程中某些因素的微小变动可以引起尺寸变化。原材料性能的微小变化也会引起尺寸变化。刀具会产生磨损并需要重新装夹。速度、润滑剂、切削温度、操作人员及其他条件都可能会发生变化。通过系统的分析研究，一般可以找出这些原因并可采取相应的步骤来消除它们。

随机原因。随机原因的出现是具有偶然性的。它们是由一些既无法确定又不能控制的力所造成的。它们是生产过程的固有误差，即使尽可能地保持所有条件完全一致，它们仍然不可避免地存在。

当依次检查由可确定的原因造成的误差，并且将其逐一排除后，即可达到理想的稳定状态或控制状态。如果随机原因对尺寸变化的影响过大，则一般来说，采用更精密的加工设备要比花费更多精力来改变生产过程更为有效。

第 28 课　公差与配合

现代技术要求零件的尺寸越来越精确。对每个尺寸都允许其有一个在规定范围内的变动量，称为公差。例如，一个零件的尺寸可以被表示成 20 ± 0.06，其公差（尺寸变动量）为 0.12mm。为了把生产成本降至最低，在不影响零件功能的情况下，应当采用尽可能大的公差。公差越小，制造成本越高。

一些公差与配合相关术语的定义如下。

公称尺寸　由图样规范确定的理想形状要素的尺寸。它过去被称为"基本尺寸"。

实际尺寸　加工后零件的实测尺寸。

极限尺寸　尺寸要素允许的尺寸的两个极限值。为了满足要求，实际尺寸应位于上、下极限尺寸之间，含极限尺寸。

上极限尺寸　尺寸要素允许的最大尺寸。

下极限尺寸　尺寸要素允许的最小尺寸。

偏差　实际尺寸减其公称尺寸所得的代数差。偏差是一个带符号的值，可以是正值、负值或零。

上极限偏差　上极限尺寸减其公称尺寸所得的代数差。它过去被称为"上偏差"。

下极限偏差　下极限尺寸减其公称尺寸所得的代数差。

基本偏差　最接近公称尺寸的那个极限偏差。基本偏差用字母表示（如 H，k）。

公差　上极限尺寸与下极限尺寸之差。公差是一个没有正、负号的绝对值。公差也可以是上极限偏差与下极限偏差之差。

标准公差等级　用常用标示符表征的线性尺寸公差组。标准公差等级标示符由 IT 及其之后的数字组成（如 IT6），字母 "IT" 为 "国际公差" 的英文缩略语。共有 20 个 IT 等级：IT01，IT0，

IT1，…，IT18。

公差带 公差极限之间（包括公差极限）的尺寸变动值。其过去的英文术语为"tolerance zone"。公差带包含在上极限尺寸与下极限尺寸之间，由公差大小和相对于公称尺寸的位置确定。

公差带代号 基本偏差和标准公差等级的组合。公差带代号由基本偏差标示符和公差等级组成（如H6，k8等）。

配合 公称尺寸相同且待装配的内尺寸要素（孔）和外尺寸要素（轴）之间的关系。

间隙配合 孔和轴装配时总是存在间隙的配合。此时，孔的下极限尺寸大于或在极端情况下等于轴的上极限尺寸。

过盈配合 孔和轴装配时总是存在过盈的配合。此时，孔的上极限尺寸小于或在极端情况下等于轴的下极限尺寸。

过渡配合 孔和轴装配时可能具有间隙或过盈的配合。

基孔制配合 孔的基本偏差为零的配合，即其下极限偏差等于零。基孔制配合的可能示例：H7/m6 和 H6/k5。人们经常使用标准的拉刀（见图28.1）、铰刀（见图28.2）和其他种类的标准刀具来加工高精度的孔，并使用标准的塞规（见图28.3）来检验孔的尺寸。另外，可以很容易地将轴加工成各种需要的尺寸。因此，一般情况下，应该优先选用基孔制配合。在这种配合制中，孔的下极限尺寸与公称尺寸相同，其原因是孔的尺寸可以通过机械加工变大，但不能减小。

基轴制配合 轴的基本偏差为零的配合，即其上极限偏差等于零。基轴制配合的可能示例：G7/h6 和 M6/h6。有些情况下需要采用基轴制配合。例如，当需要在一根轴上装配几个具有相同公称尺寸、不同配合的零件时，就应该选用基轴制配合。在这种配合制中，轴的上极限尺寸与公称尺寸相同，其原因是轴的尺寸可以通过机械加工变小，但不能增大。

选择装配法 通过手工试配来选择并装配零件的方法。通过这种方法，可以装配在较低的成本下制造出来的公差较大的零件，它可以作为高的制造精度和易于装配的零件之间的一种折中方法。

在选择装配法中，所有零件都经过测量，并根据实际尺寸的大小将其分为几个等级，使"小"轴能与"小"孔配合，"中"轴能与"中"孔配合，依次类推。

由于表面结构会影响零件的功能，因此必须精确地确定它的参数，表面结构是表面上的差异，包括表面粗糙度、表面波纹度、表面纹理和表面缺陷。

表面粗糙度 由使工件表面形成光滑表面的加工工艺所造成的最细微的表面不平度。表面粗糙度的高度以微米或微英寸为单位进行测量。

表面波纹度 是超过表面粗糙度间距界限的大间隔偏差，以毫米或英寸为单位进行测量。可以将粗糙度看作叠加在表面波纹度上的表面不平度。

表面纹理 为所采用的加工方法所产生的表面刀痕图案的方向。

表面缺陷 不经常出现或在很大区间内才会出现的表面瑕疵，其中包括裂纹、气孔、微细裂纹、划痕等。

第29课　机械设计概论

机械设计通过应用科学和技术知识进行新产品设计或对现有产品进行改进，以满足人类的需求。它是一个广阔的工程技术领域，除要构思产品的尺寸、形状和详细结构外，还要考虑产品在制造、销售和使用等方面的问题。

进行各种机械设计工作的人员通常称为设计人员或设计工程师。机械设计是一项创造性的工作。除在工作中要有创新性外，设计工程师还必须在机械制图、运动学、动力学、工程材料、材料力学和工艺过程等方面具有扎实的基础知识。

如前所述，机械设计的目的是生产满足人类需求的产品。发明、发现和科学知识本身并不一定能给人类带来益处，只有当它们被应用在产品设计中时才能产生效益。因此，应该认识到在对一个特定产品进行设计之前，必须首先确定人们是否需要这种产品。

应当把机械设计看作设计人员运用创造性的才能进行产品设计、系统分析并对产品的制造方法做出正确判断的一个机会。掌握工程基础知识要比熟记一些数据和公式更为重要。仅仅使用数据和公式不足以在一个好的设计中做出所需要的全部决策。另外，应该认真精确地进行所有的计算工作。例如，如果将一个小数点的位置放错，就会使正确的设计变成错误的。

在好的设计方案中，设计人员应该勇于尝试新思路，愿意为此承担一定的风险；并且知道，当新的方法不适用时，就恢复采用现有的方法。因此，设计人员必须要有耐心，因为所花费的时间和努力并不能保证带来成功。提出一个全新的设计方案要求摒弃许多陈旧的、为人们所熟知的方法。由于许多人易于墨守成规，故这样做并不是一件容易的事情。一位设计工程师应该不断地探索改进现有产品的办法，在此过程中应该认真选择原有的、经过验证的设计原理，将其与未经过验证的新观念结合起来。

新的设计方案中通常会有许多缺陷和未能预料的问题发生，只有在这些缺陷和问题得到解决之后，才能体现出新产品的优越性。因此，一个性能优越的产品诞生的同时，也会伴随着较高的风险。应该强调的是，如果设计本身不要求采用全新的方法，就没有必要仅仅为了达到变革的目的而采用新办法。

在设计的初始阶段，应该允许设计人员充分发挥创造力，不受很多约束。尽管可能会出现许多不切实际的想法，也通常会在设计的早期阶段，即确定产品的制造细节之前被改正。只有这样，才不至于堵塞创新的思路。经常需要提出几种设计方案，然后加以比较。很有可能在最后选定的方案中，采用了某些未被接受的方案中的一些想法。

另一个应该认识到的重要问题是，设计工程师必须能够同其他有关人员进行交流和沟通。与其他人就设计方案进行交流和沟通是设计过程的最后和关键阶段。毫无疑问，许多伟大的设计、发明或创造之所以没有为人类所利用，就是因为那些创新者不善于或不愿意向其他人介绍自己的成果。介绍方案是一项说服别人的工作。当一个工程师向行政人员、管理人员或其主管人员介绍自己的新方案时，就是希望说服他们或向他们证明自己的方案是比较好的。只有成功地完成这项工作，为得出这个方案所花费的大量时间和精力才不会被浪费掉。

基本上人们只有三种表达自己思想的方式，即文字材料、口头表述和绘图。因此，一个优秀的工程师除掌握技术外，还应该掌握这三种表达方式。如果一个技术能力很强的人在上述三种表达方式中的某一方面能力较差，他就会遇到很大的困难；如果上述三种能力都较差，那将永远没有人知道他是一个多么能干的人！

一个有能力的工程师不应该害怕在向别人介绍自己的方案时可能会遭遇失败。事实上，偶然的失败是肯定会发生的，因为每个真正有创造性的设想似乎总是有失败或批评伴随着它。从一次失败中可以学到很多东西，只有那些不怕遭受失败的人们才能获得最大的收益。总之，决定不向别人介绍自己的方案，才是真正的失败。为了进行有效的交流，必须回答以下问题：

（1）所要设计的这个产品是否真正为人们所需要？
（2）此产品与其他公司的现有同类产品相比有无竞争能力？
（3）生产这种产品是否经济？
（4）产品的维修是否方便？
（5）产品有无销路？是否可以盈利？

只有时间才能对上述问题给出正确的答案，但是，产品的设计、制造和销售只能在对上述问题的初步肯定答案的基础上进行。设计工程师还应该通过零件图和装配图，与制造部门一起对最终设计方案进行沟通。

通常，在制造过程中会出现某种问题。可能需要对某个零件的尺寸或公差做一些变更，使零件的生产变得容易。但是，工程上的变更必须经过设计人员批准，以保证不会对产品的功能产生不利的影响。有时，在产品装配时或装箱外运前的试验中才发现设计中的某种缺陷。这些事例恰好说明了设计是一个动态过程。总是存在着更好的方法来完成设计工作，设计人员应该不断努力，寻找这些更好的方法。

第 31 课　几条机械设计准则

"需求是发明之母"这句古老的谚语现在仍然是正确的。设计是从实际或者假想的需求开始的。现有的设备可能需要在耐久性、效率、重量、速度或成本等方面做一些改进工作，也可能需要有新的设备来完成以前由人来做的工作，如物料搬运或产品装配。当目标完全或部分被确定以后，设计工作的下一个步骤是对能够实现所需要功能的那些机构及其布局进行构思。对于此项工作，徒手绘制的草图很有价值，它不仅可以记录设计者的想法，而且还有助于与别人进行讨论，特别是和自己的大脑进行交流，从而促进创新想法的产生。

当一些零件的大致形状和几个尺寸被确定后，就可以开始认真的分析工作了。分析工作的目的是要在重量最轻、成本最低的情况下，获得令人满意或优良的工作性能，并且还要安全耐用。对于每个关键承载截面，应该寻求最佳的比例和尺寸。要选择材料和材料的处理方式，只有根据力学原理进行分析才能达到这些重要目的，诸如根据静力学原理分析支反力；根据动力学原理分析惯性和加速度；根据弹性力学和材料力学分析应力和变形；根据流体力学来分析润滑和液压传动。

最后，在基于功能要求和可靠性所进行的设计工作完成之后，可能需要制造出一台样机。如果样机的试验结果令人满意，而且该装置将要进行批量生产，就应该对最初提出的设计方案做一些修改，使其能够以较低的成本进行批量生产。在以后的制造和使用期间内，如果产生了新的想法，或者根据试验和经验所做的进一步分析结果表明可以有更好的替代方案，则很可能对原设计方案进行修改。产品质量、客户的满意程度和制造成本均与设计有关。

为激发创造性思维，建议设计人员遵循以下规则。

1. 创造性地利用所需要的物理性能和控制不需要的物理性能　可以利用自然法则或物质的性能（如强度、刚度、惯性、浮力、离心力；杠杆原理和斜面原理、摩擦、黏性、流体压力和热膨胀），以及许多电学、光学和化学现象来满足一台机器的设计要求。一种性能在某一种场合可能是有用的，而在另一种场合则可能是有害的。例如，离合器面片（见图 6.4 和图 31.1）需要有摩擦力，而离合器轴承却不需要摩擦力。设计时，要创造性地利用和控制所需要的物理性能，将不需要的物理性能减至最小。

2. 在重量最轻的情况下，提供合理的应力分布和刚度　对于承受交变应力的零件，应该特别注意减少应力集中和提高内圆角（见图 8.2）、螺纹、孔和配合等处的强度。可以通过改变零件的形状来降低它所承受的应力；也可以对零件施加预应力，如表面滚压以使其得到强化。空心轴和箱形截面能获得有利的应力分布，同时具有刚度高而重量最轻的特点。轴和其他零件必须具有足够的刚度，以避免产生共振。

3. 利用基本公式来计算和优化尺寸　力学和其他学科的基本公式是公认的计算依据。有时需要将这些公式进行移项而变成特殊形式，以简化尺寸的计算或对尺寸进行优化。例如，利用梁和表面应力计算公式来计算齿轮的轮齿尺寸。在不能采用解析法计算的情况下，可以在基本公式内引入系数。例如，对于薄壁钢管，考虑到腐蚀性，可将根据压力算出的厚度增加一些。当必须应用一个基本公式来确定形状、材料和使用条件，而这些被确定的量仅与在公式推导中的假设比较接近时，要采取措施使结果"安全可靠"。

4. 根据性能组合选择材料　选择材料时要考虑相关的性能组合，不仅要考虑强度、硬度和重量，而且有时还要考虑抗冲击性、耐腐蚀性和耐高温或低温的能力。成本和制造性能都是应该考虑的因素，这些因素包括焊接性、可加工性、对热处理温度变化的敏感性和所需要的涂层。

5. 在现有零件和整体零件之间进行认真的选择　若一个以前研制出的零件能够满足性能要求和可靠性要求，并适用于所设计的这台机器而无须附加研制费用，那么设计人员及其公司通常会从零件制造厂的现货中选取该零件。但是，只有充分了解其性能，才能进行认真的选择工作，因为任何一个机器零件的失效都会影响公司的信誉，并使公司承担相应的法律责任。在其他情况下，若机器设计人员自己来设计零件，则零件的强度、可靠性和成本等方面的要求都可以更好地得到满足。可以将某个零件与其他零件设计成一个整体零件，如将齿轮与轴设计为一个整体（见图31.2），其优点是结构紧凑。

6. 保证零件在装配中能够准确定位和不发生干涉　一个好的设计能够保证零件定位准确，便于装配和维修。在装配过程中轴肩不需要测量就能够提供准确的位置。零件的形状应该被设计为能够保证这个零件不会被装反或装错位置。必须能够预见和防止诸如不同的螺纹孔（见图31.3）中的螺钉之间的干涉和不同的连杆机构之间的干涉。

第32课　计算机在设计和制图中的应用

　　计算机在工程和相关领域中得到广泛的应用，而这种应用预计会得到越来越快的发展。工程技术专业的学生应该通晓计算机，了解计算机的用途和它们的优点，否则，就会在将来的职业生涯中处于非常不利的地位。

　　计算机辅助设计（CAD）应用计算机来帮助解决设计问题，生成工程图样和其他技术文件。在CAD中，传统的绘图工具，诸如丁字尺、绘图圆规（见图25.1和图25.2）和绘图板等被电子输入和输出装置所取代。当采用CAD系统时，设计人员可以比较容易地在计算机屏幕上对所要设计的物体进行构思，而且能够考虑多个替代设计方案，或者为了满足一些必要的设计要求，而对一个特定设计方案进行快速修改。然后，设计人员可以对设计方案进行各种工程分析，以找出其中可能存在的问题（如过大的载荷或变形）。这种分析工作的速度和精度远远超出了传统方法所能达到的程度。CAD用户通常利用键盘和/或鼠标输入数据，以便在计算机屏幕上生成图像。这些图像可以通过绘图仪或打印机（见图32.1和图32.2）形成纸质文本。

　　绘图效率得到了显著提高。当某个图形被绘制一次后，它再也不需要被重新绘制。它可以被从数据库中检索出来，并可以被复制、拉伸、改变尺寸，在不需要重画的情况下进行多种变化。剪切和粘贴技术可以用来帮助提高劳动效率。

　　工程师们普遍认为，计算机不会改变设计过程的性质，但它是一个能够显著提高工作效率和生产效率的工具。设计人员和CAD系统可以被视为一个设计团队：设计人员提供知识、创造力和控制；计算机生成准确且易于修改的图形，以很高的速度进行复杂的设计分析工作，以及存储和检索设计信息。有时，计算机可对工程师所使用的许多其他工具起增强或替代作用，但是它不能取代由设计人员控制的设计过程。

　　根据设计问题的本质和所使用计算机系统的先进程度的不同，对设计人员或制图人员来说，计算机可以有下列的一些或全部优点。

1. 易于绘制和修改图样　采用计算机可以比用手工更快地绘制出工程图样。在对图样进行变更和修改时，也比用手工完成的效率更高。

2. 对图样有更形象化的理解　许多系统可以显示同一物体的不同视图，而且三维图像（见

图32.3）可以在计算机屏幕上旋转。

3. 辅助制图数据库 建立和维护设计数据库（设计方案库）可以存储设计方案和符号，便于检索和应用于解决新问题。

4. 快速和方便地进行设计分析 由于计算机可使分析变得容易，设计人员可对很多可供选择的设计方案进行评价。因此，在加快设计过程的同时，可以考虑更多的可能性。

5. 对设计进行仿真和检验 一些计算机系统可对产品的工作情况进行仿真，并对在各种状况和应力作用下的设计方案进行检验。用计算机进行检验可以改进或取代模型和样机。

6. 提高精度 计算机绘图比手工更为精确。许多CAD系统都能检测错误，并通知用户。

7. 改善档案管理工作 图样可以更为方便地存档、检索和采用磁盘或磁带传送。

计算机图形学在工程技术领域的应用几乎没有任何限制，大部分可以用铅笔完成的图解方法都可以用计算机完成，而且通常效率更高。应用范围从三维建模和有限元分析到绘制平面图和进行数学计算。

许多曾经只能在大型计算机上运行的先进应用软件，现在也可以在微型计算机上运行。CAD的一个重要应用领域是制造业。计算机辅助设计/计算机辅助制造(CAD/CAM)系统可以用来设计零件或产品，生成主要的生产工艺，并且采用电子方式将这些信息传给包括机器人在内的制造设备并控制它们的运行。与传统的制造系统相比，这些系统具有许多优点，包括减少设计工作量、提高材料利用率、缩短产品开发周期、提高精度和改进库存控制。

第33课 车床和车削参数

车床在工业生产中被广泛用来加工各种类型的机械零件。一些车床是通用机床，而另一些车床则被用来完成某些专门工序的加工任务。

普通车床 普通车床（见图33.1和图34.1）是在全世界的生产车间和维修车间中广泛使用的通用机床。它的尺寸范围很广，从小型的台式车床到巨大的重型车床。

仿形车床 仿形车床被设计用来对零件进行自动化加工。这种车床的基本操作如下：在夹持装置上安装平面或立体形状的样板，然后，导向触头或指针沿着样板的外形移动，从而控制切削刀具的运动。像电动机的轴、主轴、活塞和其他很多种类的工件，都可以采用这种车床来进行切削加工。

转塔车床 在普通车床这类通用车床上加工形状复杂的工件时，需要花费很多时间来对加工时所用的一些刀具进行更换和调整。对普通车床的早期改装工作之一是采用一个可以安装多把刀具的转塔来代替尾座，使它能够更好地适应大批量生产的需要。这种机床被称为转塔车床（见图33.2）。

转塔车床能够进行车外圆、车端面、钻孔、车螺纹、切断和车孔等多种切削加工。数把刀具（通常多达6把）被安装在六角形转塔中。当每项加工工作完成后，就可以转动转塔。用于车螺纹、切断、车孔和其他加工的刀具为了完成其特定的任务而具有特殊的形状，或者采用不同形状的刀片（见图33.3）。

各种转塔车床的主要特点是能按适当顺序连续完成一系列的加工工作。一旦刀具被安装调整好后，这种机床就不需要技术水平很高的操作工人。另外还有立式转塔车床，它们更适合加工短的、重的工件，工件的直径可达到1.2米。

车削中心 当前，许多技术更为先进的车床被称为车削中心。这是因为，在其上除能够进行常规的车削加工外，还可以进行一些铣削和钻削加工。车削中心基本上可以认为是转塔车床和铣床的组合体。

车削加工的切削参数 在车削加工中，工件上多余的材料被刀具以切屑的形式切除，以获取需要的几何形状、公差和表面粗糙度。这就需要刀具的硬度比工件的硬度高。

车削加工过程是以二维表面成型法为基础的。也就是说，刀具与工件材料之间需要两种相对运动。这两种相对运动，一种被称为主运动，它是决定切削速度大小的主要因素；另一种被称为进给运动，它向切削区提供新的加工材料。工件的旋转运动是主运动；刀具的直线运动是进给运动。车削加工的基本参数如图33.4所示。

1. 切削速度 切削速度 v 是主运动中工件相对刀具（在切削刃的指定点）的瞬时速度，可以用下式表示：

$$v = \pi dn \tag{33.1}$$

式中，v 为切削速度（m/min）；d 为工件上需要切削部分的直径（m）；n 为工件或主轴的转速（r/min）。

2. 进给量 车床的轴向进给量可以被定义为：主轴每转一周时，刀具沿工件长度方向的移动距离。主运动和刀具进给运动 f 的共同作用可以重复或连续地以切屑的形式去除工件材料，从而获得所需要的已加工表面。

3. 切削深度（背吃刀量） 切削深度 a 是刀具切削刃切入工件原来的表面以下的距离。切削深度决定了工件的最终尺寸。在车削加工中采用轴向进给时，切削深度可以通过直接测量工件半径的减少量来确定；在车削加工中采用径向进给时，切削深度等于工件长度的减少量。

4. 切屑厚度 未变形时的切屑厚度 h_1，就是在垂直于切削方向的平面内垂直于切削刃测量得到的切屑厚度。切削后的切屑厚度（实际切屑厚度 h_2）大于未变形切屑厚度，这意味着削比或切屑厚度比 $r = h_1/h_2$ 总是小于1。

5. 切屑宽度 未变形时的切屑宽度 b，是在与切削方向垂直的平面内沿切削刃测得的切屑宽度。

6. 切削面积 采用单刃刀具进行切削加工时（见图33.4），切削面积 A 是未变形切屑厚度 h_1 和切屑宽度 b 的乘积（$A = h_1 b$）。切削面积也可以用进给量 f 和切削深度 a 来表示：

$$h_1 = f\sin\kappa \text{ 及 } b = a/\sin\kappa \tag{33.2}$$

式中，κ 为主偏角。

因此，可以由下式求出切削面积：

$$A = fa \tag{33.3}$$

第34课 普通车床

通常认为车床是最古老的机床。车床上主要的加工表面为外圆表面。其加工过程为：在工件旋转时，单刃刀具沿与旋转轴平行的方向移动，切除工件上不需要的材料（见图2.2）。

普通车床是最常见的车床。由于早期的车床是由蒸汽机（steam engine）驱动的，普通车床英文名称中的 engine 来源于此。它也称为卧式车床。

普通车床的基本部件有：床身、主轴箱、尾座、溜板、丝杠和光杠，如图34.1所示。

床身是车床的基础件。它通常是由经过充分正火或时效处理的灰口铸铁或球墨铸铁制成的。它是一个稳重坚固的刚性机架，所有其他基本部件都安装在床身上。通常在床身上都有内侧和外侧两组平行的纵向导轨（见图34.2）。有些制造厂对全部四条导轨都采用凸三角形导轨，而有的制造厂则在其中的一组中采用或两组中都采用一个凸三角形导轨和一个矩形导轨。为了耐磨损，大多数现代机床的导轨都是经过表面淬火的。

主轴箱安装在内侧导轨的固定位置上，通常在床身的左端。它提供动力，并可使工件具有不同的转速。它基本上由一根安装在精密轴承中的空心主轴和一系列变速齿轮——类似于卡车变速器——所组成。通过变速齿轮，主轴可以有许多种转速。大多数车床有8~18种转速，通常按等比级数排列。一种日益增长的趋势是通过电气或机械传动进行无级变速。

由于车床的精度在很大程度上取决于主轴，因此，主轴的结构尺寸较大，通常安装在预紧后的重型圆锥滚子轴承或球轴承（见图10.1b和图10.1c）中。主轴中有一个贯穿全长的通孔，长棒料可以通过该孔送入。主轴孔径是车床的一个重要尺寸，因为当工件必须通过主轴孔送入时，它确定了能够加工的棒料的最大尺寸。

卡盘是车床上最常用的夹紧装置。三爪自定心卡盘（见图34.3a）在夹紧和松开工件时，所有的卡爪都同时移动，它适用于横截面为圆形或六角形的工件。四爪单动卡盘（见图34.3b）通过每个卡爪的单独移动将工件夹紧。这种卡盘对工件的夹紧力比较大而且能够对非圆形（正方形、长方形）的工件进行精确定心。

尾座用于支承工件的另一端。它特别适用于细长工件的加工。尾座中可以安装固定顶尖（死顶尖），也可以安装能够随工件一起自由旋转的顶尖（活顶尖）。尾座可沿床身上的内侧导轨移动，以便于加工不同长度的工件。尾座套筒中还可以安装诸如中心钻（见图34.4）、麻花钻（见图36.1）、铰刀（见图28.2）等刀具，用来加工和精加工位于工件回转轴线上的孔。

三爪和四爪卡盘通常只适用于短工件。长工件要装夹在两顶尖之间进行加工（见图34.5）。在将工件放到车床顶尖间之前，需要在每个端面钻一个中心孔。安装在主轴和尾座中的顶尖通过这些中心孔确定了工件轴线的位置。然而，这些顶尖不能将主轴的运动传递给工件。为此，通常采用拨盘和鸡心夹头来带动工件旋转。在图34.5中，安装在主轴内的顶尖为固定顶尖（见图34.6a），安装在尾座内的顶尖为回转顶尖（见图34.6b）。顶尖的柄部通常被精加工成莫氏锥度，用来与主轴或尾座中的锥孔相配合。

在普通车床上，工件的最大尺寸由两个参数表示。第一个参数称为最大回转直径。它大约是两顶尖连线与导轨上最近点之间距离的两倍。第二个参数是顶尖距。最大回转直径表示在这台车床上能够车削的工件的最大直径，而顶尖距则表示在两个顶尖之间能够装夹的工件的最大长度。

在车床上我们可以进行下列切削加工：

车外圆 车外圆是通过去除工件上的多余材料，获得具有所需尺寸的外圆柱面的加工方法。

车平面 车平面是一种在车床上加工平面的方法。采用这种方法时，刀具的进给方向与旋转轴垂直。

车孔 车孔与车外圆类似。它是一种内圆车削方法，用来扩大工件中已有的孔或圆柱状空腔的内表面。

切断 这是一种采用切断刀将工件切成两段的加工方法。

滚花 用滚花刀在工件表面滚压出直纹或网纹的方法称为滚花。

在车床上还可以进行其他几种切削加工，如车锥面、车螺纹、车槽、钻孔、钻中心孔和倒角。

第35课 铣床和磨床

铣床属于用途最广泛和最有用的机床中的一种，这是因为，在铣床上可以进行很多种类的切削加工。第一台铣床是由伊莱·惠特尼在1818年制造的。目前，存在着各式各样的具有许多不同特征的铣床可以供人们使用。

升降台式铣床是最常见的铣床，可以在其上进行多种工件的铣削加工。安装铣刀的主轴可以是水平布置的（见图35.1a），用于周铣；也可以是垂直布置的（见图35.1b），用于端铣。

在床身式铣床中，工作台直接安装在床身上，这使得工作台只能做水平运动。这种铣床的刚度高，通常用于大批量生产中。只有一个主轴的床身式铣床被称为单轴床身式铣床（见图35.2a）。床身式铣床的主轴可以是水平的或垂直的，并且也可以是双主轴型或三主轴型的（也就是说，具有两个或三个主轴），可以对工件的两个或三个表面同时进行加工。还有一些其他种类的铣床，如加工重型工件的龙门铣床（见图35.2b）和各种专用铣床。

万能分度头（见图35.3）是最常用的铣床附件之一。它是一个能够容易地和精确地将工件转动预定角度的夹具。其典型用途是铣削零件的多边形表面和加工齿轮的轮齿。分度头也可以用在包括钻床和磨床在内的许多其他种类的机床上。

磨削是一种应用最广泛的零件精加工方法，可以得到非常小的公差和非常低的表面粗糙度。现在，有许多不同种类的磨床。外圆磨床和平面磨床是两种最常用的磨床。其他种类的磨床包括无心磨床、内圆磨床和螺纹磨床。

外圆磨床用来磨削圆柱形或圆锥形工件。如图35.4所示，工件通常被装夹在两个顶尖之间，采用鸡心夹头和拨盘带动其旋转。这两个顶尖分别装在磨床的头架和尾座中。旋转的砂轮与工件接触，在其外圆表面上去除金属。这种加工方法与铣削有点类似，用砂轮取代了铣刀，而其上数以千计的磨粒可以被认为是微小的铣刀刀齿。实际上，在加工过程中，它们会产生微小的切屑。

平面磨床用来磨削平面。通常用安装在磨床工作台上的磁力吸盘吸紧工件；对于非导磁材料，采用虎钳或其他夹具夹持工件。工作台在砂轮下面做纵向往复运动（见图35.5）。当每个纵向行程终了时，工作台就作一次横向进给。工作台既可以采用自动进给也可以采用手动进给。

内圆磨床被用来磨削精密的孔、汽缸孔及各种类似的、需要进行精加工的孔。螺纹磨床被用来磨削螺纹量规上的精密螺纹和用来磨削螺纹的中径与轴的同心度公差很小的精密零件上的螺纹。

零件的设计特征在很大程度上决定了需要采用的磨床的种类。当加工成本太高时，就值得对零件进行重新设计，使其能够通过采用既便宜又具有高生产率的磨削方法加工出来。例如，只要有可能，就应该对零件进行恰当的设计，尽量用无心磨削加工，以获取经济效益。

在磨削过程中，砂轮上的锋利磨粒会被磨圆变钝，丧失其具有的切削能力。通过对砂轮进行整形和修锐（见图35.6）可以恢复其切削能力。整形可以消除因磨损不均匀而产生的形状误差，恢复砂轮正确的几何形状。修锐通过去除磨钝的磨粒，使砂轮变得锋利。

对于普通砂轮，整形和修锐通常是同时进行的。对于超硬磨料砂轮，整形和修锐通常是分开进行的。经过适当修整的砂轮所需要的磨削功率小，不会产生磨削烧伤和颤振，可以加工出高质量的零件。

第38课　齿轮制造方法

齿轮上相邻的轮齿之间的空间形状是复杂的，并随着齿轮的齿数及模数的变化而变化。因此，大多数的齿轮制造方法采用展成法加工齿面而不是采用成形法加工。

插齿　插齿是用途最广泛的齿轮切削加工方法。尽管插齿主要用于直齿轮和斜齿轮的轮齿切削，但是它也适用于内齿轮、人字齿轮、椭圆齿轮（见图12.2和图19.1）和齿条的切削加工。

图38.1展示了采用插齿刀进行插齿加工的原理。在这个过程中，装有插齿刀的主轴在转动的同时做轴向往复运动。在刀具与工件啮合并进行切削时，工件主轴与刀具主轴同步地缓慢转动。刀具向下运动是主要的切削运动。在回程（向上运动）时，刀具必须向后退让约1mm，以留出空隙。否则，在回程时，刀具会因为摩擦而导致快速磨损失效。

插齿加工的优点是生产率比较高和可以将齿插到接近轴肩处。可惜的是，加工斜齿轮时，需要有一个螺旋导轨为插齿行程附加一个回转运动，如图38.2所示。由于这种螺旋导轨不易制造，或者说其制造成本比较高，而且对每种不同螺旋角的斜齿轮，均需要制造专用的插齿刀和螺旋导轨，因此这种方法仅适用于斜齿轮的大批量加工。能够加工内齿轮是插齿加工的一大优点。

滚齿 滚齿加工的原理与利用齿条进行展成法加工的原理相同，除此之外，为了避免缓慢的往复运动，每个旋转刀具上都带有很多个"齿条"。这些"齿条"轴向排布，形成了一个开槽的蜗杆（见图4.9）。滚齿加工过程如图38.3所示。

由于滚刀和工件都不做往复运动，所以滚齿加工时的金属去除率很高。对于普通滚刀可以采用40m/min的切削速度，对于硬质合金滚刀可以采用高达150m/min的切削速度。

滚齿被广泛用来加工直齿轮、斜齿轮、蜗杆和蜗轮上的轮齿。滚齿机不能用来加工内齿轮和锥齿轮。

拉齿 通常不采用拉削的方式加工斜齿轮，但拉削很适合用于内直齿轮的加工。在这种情况下，拉削主要被用来加工其他任何一种方法都不容易加工的内花键（见图4.8）。拉削加工生产效率高，表面粗糙度低和尺寸精度高。然而，因为拉刀的价格昂贵，而且一种拉刀只能加工同一尺寸的齿轮，所以这种方法主要应用于大批量生产中。

剃齿 剃齿是一种精加工方法，它采用剃齿刀从齿廓上去除厚度为2～10μm的微薄切屑层。剃齿是目前应用最为广泛的对切削加工后和淬火之前的直齿轮和斜齿轮的轮齿进行精加工的方法。剃齿的目的是降低表面粗糙度和提高齿形精度。

剃齿刀形状如齿轮，在刀齿的根部有附加的凹槽（用于排除细小切屑和切削液），在齿面上有许多小槽以形成切削刃。如图38.4所示，剃齿刀与经过粗加工的齿轮成交错轴啮合转动，这样在理论上沿着轮齿有一个具有相对速度的点接触，从而产生刮削作用。剃齿刀的刀齿具有一定的弯曲柔性，因此只有当它们在两个轮齿之间并且与这两个轮齿都接触时，才能有效地进行切削工作。加工周期可能少于半分钟，机床的价格也不昂贵，但是刀具比较精密，难于制造。

磨齿 磨齿是非常重要的，因为它是加工淬硬齿轮的主要方法。当对热处理变形的预先校正达不到齿轮所要求的高精度时，就必须采用磨削加工。

最简单的磨齿方法是成形磨削法（见图38.5）。这种方法的加工速度相当缓慢，但是可以持续地获得较高的加工精度。最快的磨齿方法采用与滚齿相同的原理，不同之处是使用蜗杆砂轮（见图38.6和图38.7）来取代滚刀（见图4.9）。尽管在磨削过程中砂轮和工件有产生不同的变形量的可能，可能需要用砂轮的形状补偿机床变形的影响，但是磨削加工的精度还是比较高的。将砂轮展成为一个蜗杆形状是一个缓慢的过程，这是因为修整砂轮的金刚石不仅要使砂轮形成齿条外形，而且还需要在砂轮旋转时进行轴向移动。一旦砂轮的整形和修锐工作完成后，就能快速地进行磨齿加工，直到砂轮需要重新修整时为止。这是一个最常用的，能够高效率地加工小齿轮的方法。

第43课 特种加工工艺

人类通过使用工具和智慧，制造出各种能使其生活变得更便利、更舒适的物品。这种方法把人类与其他种类的生命区分开来。许多世纪以来，工具和为工具提供动力的能源的种类都在不断地发展，以满足人类日益完善和越来越复杂的想法。

在远古时期，工具主要是由石器构成的。就所制造物品相对简单的形状和被加工的材料来说，石头作为工具是适用的。当铁制工具被发明出来以后，耐用的金属和更精致的物品便能够被制造出来。到了20世纪，已经出现了一些由有史以来最耐用，同时也是最难加工的材料制造的产品。为了迎接这些材料给制造业带来的挑战，刀具材料已经发展到高速钢、硬质合金、涂层硬质合金、立方氮化硼、金刚石和陶瓷。

给我们的工具提供动力的方法也发生了类似的改进。最初，是以人或动物的肌肉为工具提供动力；随后，水力、风力、蒸汽和电力得到了利用，人类通过使用新型机器，以及更高的精度和更快的加工速度来进一步提高制造能力。

每当采用新的工具、新的材料和新的能源时，制造效率和制造能力都会得到很大的提高。然而，当旧的问题解决之后，就会有新的问题和挑战出现。例如，现今制造业面临着下面一些问题：如何钻一个直径为2mm，长度为670mm的孔，而不产生锥度和偏斜？用什么办法能够有效地去除形状复杂的铸件内部通道中的毛刺，而且保证去除率达到100%？是否有一种焊接工艺，能够避免目前在产品中出现的热损伤？

从20世纪40年代以来，制造业中发生的大变革一次又一次地促使制造厂商去满足日益复杂的设计方案和那些耐用的，但是在许多情况下非常难加工的材料所带来的各种要求。这种制造业的大变革无论是现在还是过去都集中在采用新型工具和新型能源上。结果是产生了用于材料的去除、成型、连接的新型加工工艺。这些工艺目前称为特种加工工艺。

在目前所采用的传统加工工艺（如锯削、钻削和拉削加工）中，材料的去除主要是采用电动机和比工件材料硬度高的刀具材料进行的。传统的成型加工是利用电动机、液压和重力所提供的能量进行的。同样，材料连接的传统方法是采用诸如燃烧的气体和电弧等热能进行的。

与之相比，特种加工工艺采用按照以前的标准来说是非传统的能量来源。现在，材料的去除可以利用电化学反应、电火花（见图43.1）、高温等离子、高速液体和磨料射流。过去很难进行成型加工的材料，现在可以利用大功率的电火花所产生的磁场、爆炸和冲击波进行成型加工。采用超声波和电子束可以使材料的连接能力有很大的提高。

在过去的70年间，人们发明了20多种特种加工工艺，并且将其成功地应用于生产之中。这么多种特种加工工艺存在的原因与有许多种传统加工工艺存在的原因是一样的。每种工艺都有它自己的特点和局限性。因此，不存在一种对任何制造环境来说都是最好的工艺方法。

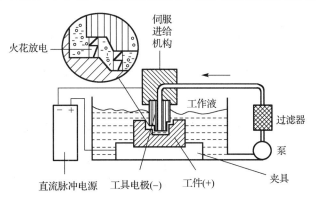

图43.1　电火花加工过程示意图

例如，有时特种加工工艺可以通过减少生产某个产品所需要的加工工序的数量，或者通过采用比以前使用的方法更快的工序来提高生产率。

在其他情况下，采用特种加工工艺可以通过提高重复精度，减少易损坏工件在加工过程中的损伤，或者通过减少对工件性能的有害影响来减少采用原来的加工工艺所产生的废品的数量。

由于前面提到的这些特点，特种加工工艺从其诞生时起就开始了稳定的发展。这些工艺在将来会有更快的增长速度，原因如下：

（1）目前，与传统加工工艺相比，除材料的体积去除率外，特种加工工艺几乎具有不受限制的能力。在过去几年中，某些特种加工工艺在提高材料去除率方面有了很大的进展，而且有理由相信这种趋势在将来也会继续下去。

（2）大约半数的特种加工工艺目前采用计算机控制加工参数。使用计算机可以使人们不熟悉的加工过程变得简单，因而加大了人们对这种技术的接受程度。此外，计算机控制可以保证可靠性和重复性，这也加大了人们对这种技术的接受程度及其应用范围。

（3）大多数特种加工工艺可以通过视觉系统、激光测量仪和其他加工过程中的测量技术来实行自适应控制。例如，如果加工过程中的测量系统表明，产品中那些正在加工的孔的尺寸在变小，那么就可以在不更换硬的加工工具（如钻头）的情况下修正孔的尺寸。

第 45 课　工程陶瓷的机械加工

工程陶瓷材料具有许多引人注目的特性，如高硬度、高耐热性、化学惰性、低导热性和导电率等。然而，这些特性使得陶瓷的机械加工非常困难，无论是采用磨削的方法还是采用非磨削的方法都是如此。当采用非磨削加工方法时，必须提高材料去除率才能在经济上划算。其他要求包括降低表面粗糙度、控制几何形状特征和降低生产设备的成本。

磨削涉及工件的材料性能、砂轮规格和机床的选择等许多变量之间复杂的相互影响。这种相互影响叫作"磨削加工性"，它可以用材料去除速度、所需要的功率或磨削力、表面粗糙度、公差和表面完整性等项指标的数量大小来表示。

通常采用粗粒度砂轮实现工件材料的快速去除。在磨削过程中，工件表面会产生一些影响表面完整性的缺陷（见图 45.1），在每个工件上所产生的缺陷的大小和多少都与在其他工件上的不同。这就造成了一批工件的断裂强度值的大小不同。机械加工中的切屑去除过程，也会在加工表面上和表面层下面产生残余应力。表面层中的缺陷的长度可能在 $10\sim100\mu m$ 之间。这些损伤通常必须采用细粒度砂轮精磨的方式除去。

对于陶瓷的常规磨削加工，磨床与砂轮的选择也会影响磨削加工性。一般来说，只能采用金刚石砂轮加工陶瓷。加工烧结后的陶瓷时，用树脂结合剂的金刚石砂轮要比用金属或陶瓷结合剂的砂轮的效果好。树脂结合剂砂轮产生的摩擦力较小，磨损较快，比较容易露出新的磨粒，可以将陶瓷工件的表面缺陷尺寸减至最小。

因为陶瓷的性能对最终的磨削结果起重要作用，所以必须深入了解它们与机械加工变量之间的相互作用，以便确定成本。对于导热性差的材料，冷却液的应用是非常关键的，它可防止产生热致裂纹。陶瓷材料的孔隙率、晶粒尺寸与微观组织结构都会影响表面粗糙度和表面质量；孔隙率高会使表面粗糙度变差。一般来说，加工陶瓷需要较高的磨削力和功率，这会缩短砂轮的使用寿命。

有几种磨削方法不需要使用昂贵的金刚石砂轮，因而不存在由金刚石砂轮带来的问题。水射流加工（见图 44.2）是采用高速液体射流来冲蚀工件材料。射流可以是仅使用水流或是在水中夹带磨料粒子。射流可以有脉冲或连续两种方式，对大多数陶瓷的加工采用连续射流方式。这种加工方式最适用于切削缝隙、沟槽或大孔的套料加工。

水射流加工既无热影响也不需要大的机械力，因此在对热冲击最敏感、最脆弱的陶瓷进行加工时，都不会造成工件的损坏。这种加工方法可以避免产生颤振、振动、工件的表面变形或亚表面的损伤。

超声加工（见图 44.1）是另一种磨削加工过程，它具有一些优于传统磨削加工的特点。超声加工有时被称为冲击磨削，它是一种使用超声换能器的机械加工过程。当电能转换为机械运动后，具有所需形状的工具会以超声频率（$20\sim25kHz$）在工件上做纵向振动，其振幅为 $15\sim30\mu m$。

磨料悬浮液在工具和工件之间不断流动。常用的磨料有金刚石、碳化硼、碳化硅和氧化铝，这些磨料颗粒悬浮在水或适当的化学溶液中。除向切削区提供磨料外，悬浮液还用来带走被粉碎下来的材料微粒。磨料的粒度决定了工件的表面粗糙度、工件与工具之间的加工间隙和加工速度。超声振动工具与磨料悬浮液的共同作用可以均匀地蚀除工件材料，将工具的形状精确地复现到工件上。

超声加工几乎可以用来加工任何一种硬脆材料，但是加工 40 HRC 以上的材料时更为有效。这些材料包括硅、玻璃、石英、光纤材料、结构陶瓷（如碳化硅）和电子基板（如氧化铝）。

非磨削加工，诸如激光束加工和电子束加工，不受陶瓷硬度的限制，因此可以用来代替传统的磨削加工。在激光束加工中，激光器将能量密度高的光束聚焦在工件上，使工件材料蒸发。由于激光束是由机械定位的，所以没有电子束加工快。与电子束加工不同的是它不需要真空室，因而工件的装入较快，并且不受尺寸限制。另一个优点是设备的价格较低。通常采用脉冲方式钻孔，采用连续的方式进行切割加工。激光束可以用来加工任何硬质材料，其中包括金刚石。

加工速度是由激光使材料熔化、汽化蒸发和去除的速度所决定的。材料的去除是热对流和光束的压力所引起的。如果采用气流吹走熔化、汽化的材料，就可以显著地提高材料的去除率。温度梯度能够引起裂纹，因此，厚陶瓷工件应该在尚未烧结的状态下加工。相反，因为加热能产生有益的残余应力，所以经过激光加工的陶瓷的强度可以高于经过金刚石加工的陶瓷。根据需要去除的材料量来确定激光器的功率，一台 150 瓦的激光器就可以用来加工薄陶瓷，而加工厚陶瓷则通常需要采用 15,000 瓦的激光器。

第 46 课　振动的定义和术语

所有的物质——固体、液体和气体——都能够产生振动。例如，在喷气发动机尾管中产生的气体振动会发出令人讨厌的噪声，而且有时还会使金属产生疲劳裂缝。一台常见的机器中有许多运动零件，每个零件都是潜在的振动源或冲击激励源。设计人员需要处理好振动和噪声的允许值与降低激励所需要的费用之间的关系。

物体的振动是它相对于静平衡位置的周期性位置变化或位移。与振动有关的几个相互关联的物理量是加速度、速度和位移。例如，一个不平衡的力在系统中产生的加速度（$a = F/m$）会因为系统的抵抗而引起振动作为响应。振动大致可以分为三类：(1) 瞬态的；(2) 连续的或稳态的；(3) 随机的。

瞬态振动　瞬态振动是逐渐衰减的，而且通常与不规则的扰动有关。例如，冲击或碰撞力、刀具的切入运动，也就是那些在确定的期间内不重复的力。尽管瞬态振动是振动的暂时性成分，它们能够产生大的初始振幅和引起高的应力。但是，在大多数情况下，它们持续的时间很短，因而人们可以将其忽略不计而只考虑稳态振动。

稳态振动　稳态振动通常和机器的连续运转有关，而且尽管这种振动是周期性的，但不一定是简谐振动或遵从正弦函数的振动。由于产生振动需要能量，所以振动会通过消耗能量来降低机器和机构的效率。能量的消耗有多种方式。例如，摩擦力和将由其产生的热量传递到周围环境、声波和噪声、通过机架与基础的应力波等。因此，稳态振动总是需要连续的能量输入才能维持其存在。

随机振动　随机振动是一个用来描述非周期性振动的术语。也就是说，这种振动不具有周期性，而且通常是不重复的。

在下面的段落中，对一些与振动有关的术语和定义加以明确，其中一些可能是工科学生们都已经知道的。

周期、频率和振幅　稳态机械振动是系统在一定时间间隔内的重复运动，该时间间隔被称为周期（见图 2.7）。单位时间内的振动次数被称为频率。系统任何部分离开其静平衡位置的最大位移就是该部分的振幅，总的行程是振幅的两倍。因此，"振幅"并不是"位移"的同义词，而是距离静平衡位置的最大位移。

自由振动　除重力外，在没有任何其他作用力时产生的振动称为自由振动（见图46.1）。通常，一个弹性系统离开它的稳定平衡位置后并被释放时，这个系统就会产生振动，也就是说，自由振动是在弹性系统固有的弹性恢复力的作用下产生的，而固有频率则是系统的一个特性。

受迫振动　受迫振动是在外力的激励下产生的。这个激励通常是时间的函数。例如，不平衡的转动零件，或者齿轮在制造过程中产生的缺陷均会引起这种振动。受迫振动的频率就是激振力或外加力的频率，也就是说，其频率是一个与系统固有频率无关的量。

自激振动　自激振动通常被称为颤振。它是由切屑去除过程与机床结构之间的相互作用而产生的。自激振动通常是由切削区的扰动开始的。

一种最重要的自激振动是再生颤振，它是由刀具在前一次切削后产生的表面粗糙度和振纹的表面上进行再次切削时引起的。其结果是，切削深度的变化造成了切削力的变化，使得刀具产生振动。这个过程重复地继续进行（因此就有了"再生"这个名词）。在起伏不平的道路上开车就很容易观察到这种振动。自激振动通常有很大的振幅。

共振　当作用力的频率与系统的固有频率相同或相近时就会产生共振。在这种临界条件下，机械系统中会出现具有危险性的大振幅和高应力。但是，在电学中，收音机和电视机的接收器则被设计成在共振频率时工作。因此，在所有各种振动系统中，计算或估计出系统的固有频率是非常重要的。当转轴或主轴发生共振时，这时的转动速度称为临界转速。在没有阻尼或其他振幅限制装置的情况下，共振就是在有限激振下产生无穷大响应的状态。因此，预测和消除机械装置产生共振的条件是非常重要的。

阻尼　阻尼是振动系统能量的耗散，从而防止过度的响应。可以观察到，自由振动的振幅会随时间的推移而减小，因此，振动最终将由于某些限制或阻尼的影响而停止。所以，如果要使振动持续下去，一定要有外部的能源对因阻尼而耗散的能量进行补充。

能量耗散在某种程度上与系统的部件或元件之间的相对运动有关，它是由于某种类型的摩擦引起的。例如，在结构中，材料内部的摩擦和由空气或液体阻力等造成外部摩擦称为"黏性"阻尼（见图46.1b），这时假定阻力与运动部件之间的相对速度成正比。一种能够提供黏性阻尼的装置称为阻尼器。阻尼器不能存储能量，仅能消耗能量。

第49课　工程师在制造业中的作用

许多工程师的职责是进行产品设计，设计方案将会通过对材料的加工制造而变成现实产品。设计工程师在选择材料、制造方法等方面起着关键的作用。设计工程师应该比其他人更清楚地知道他或她的设计应该达到什么目的。他知道对使用荷载和使用要求所做的假设、产品的使用环境、产品应该具有的外观形貌。为了满足这些要求，他必须选择和规定要使用的材料。通常，为了利用材料且使产品具有所期望的形状，设计工程师知道应该采用哪些制造方法。在许多情况下，选择某种特定材料，可能意味着已经确定了某种必须采用的加工方法。同时，当决定采用某种加工方法后，很可能需要对设计方案进行修改，以便能够高效率、低成本地应用这种加工方法。某些尺寸公差可以决定产品的加工方法。

总之，在将设计转变为产品的过程中，必须有人做出这些决定。在大多数情况下，如果设计人员在材料和加工方法方面具有足够的知识，他会在设计阶段做出最为合理的决定。否则，做出的决定可能会降低产品的性能，或者使产品变得过于昂贵。显然，设计工程师是制造过程中的关键人物，如果他们能够进行面向生产（可以进行高效率生产）的设计，就会给公司带来效益。

制造工程师选择和调整所采用的加工方法和设备，或者监督和管理这些加工方法和设备的使用。其中一些工程师进行专用工艺装备的设计，将其安装在通用机床上进行特定产品的生产。这些工程师在机床、工序能力和材料方面必须具有广博的知识，以使机床在没有过载和损坏，而且对被加工材料没有不良影响的情况下，有效且高效率地完成所需要的加工工作。这些制造工程师在制造业中也起到重要作用。

少数工程师设计在制造业中使用的机床和设备。显然，他们是设计工程师。而且对于他们的产品而言，他们同样关心设计、材料和制造方法之间的相互关系。然而，他们更多地关心所设计的机床将要加工的材料的性能和机床与材料之间的相互作用。

还有一些工程师——材料工程师，他们致力于研制新型的、更好的材料，他们也应该关心这些材料的加工方法和加工对材料性能的影响。

尽管工程师所起的作用可能会有很大差别，但很明显，大部分工程师都必须考虑材料与制造工艺之间的相互关系。低成本制造不是凭空产生的。在产品设计、材料选择、加工方法和设备的选择、工艺装备选择和设计之间都有非常密切的相互依赖关系。这些步骤中的每一个，都必须在开始制造前仔细地加以考虑、规划和协调。从产品设计到实际生产所需要的时间，特别是对于复杂产品，可能需要数月甚至数年，并且可能花费很多钱。典型的例子有，对于一种全新的汽车，从设计到投产所需要的时间大约为两年，而一种现代化飞机则可能需要四年。

随着计算机和由计算机控制的机器的出现，我们进入了一个生产规划的新时代。采用计算机将产品的设计功能与制造功能集成（见图 49.1 和图 49.2），称为 CAD/CAM（计算机辅助设计/计算机辅助制造）。这种设计用来制订加工工艺规程和提供加工过程本身的编程信息。可以根据供设计与制造使用的中央数据库内的信息绘制零件图，需要时可以生成加工这些零件的程序。此外，对加工后零件的计算机辅助试验与检测也得到了广泛的应用。随着计算机价格的降低和性能的提高，这种趋势将毫无疑问地不断加速发展。

第 55 课 成 本 估 算

许多公司内部的成本估算是由一个专业人员来完成的，他专门从事确定零件成本这项工作，无论这个零件是在公司内部生产的，还是从其他公司购买的。因为对于产品的决定主要是在这些估算的基础上做出的，所以这个人必须尽可能精确地进行估算工作。可惜的是，成本估算人员需要相当详细的信息才能进行工作。期望成本估算人员从设计人员那里获得 20 个概念设计方案的简略草图，给出合理的估算结果是不现实的。因此，设计人员至少应当能够在设计方案基本完成，可以交给成本估算人员之前，对成本做出粗略的估算（在许多小公司，所有的成本估算工作都由设计人员完成）。

应该在产品设计阶段的早期就进行最初的估算工作。这些估算工作应该足够精确，以便确定哪些设计方案应该被排除，哪些设计方案需要进一步完善。在这个阶段，成本估算值与最终的直接成本差值在 30% 以内是可能的。其工作目标是在进行设计改进和将其完善到最终的产品的过程中，提高成本估算的精度。对类似产品成本估算的经验越丰富，早期成本估算的精度就越高。

成本估算过程依赖于产品中各个零件的来源。可以通过三种可能的选择来获得这些零件：(1) 从供应商处购买成品零件；(2) 让供应商生产本公司所设计的零件；(3) 在本公司内部制造这些零件。

应该积极鼓励从供应商处购买现有的零件。如果需要购买的数量足够大，则大部分供应商将会与产品设计人员共同工作，对现有的零件进行改进，以满足新产品的需要。

如果现有的零件或经过改进的零件目前已经脱销，没有存货，那么就应该生产这些零件。在这种情况下，就应该做出是由供应厂商生产这些零件，还是在本公司内部生产的决定。这是一个典型的"自制或外购"决策。这个复杂的决定不仅取决于此零件的成本，还取决于对设备的投资，对从事生产制造的人员的投资，以及公司是否有将来还会使用类似制造设备的计划。

无论这个零件是在公司内部制造的，还是从公司外部购买的，都应该对它进行成本估算。我们现在着眼于对经过机械加工后的零件的成本估算。

零件的机械加工（见图 55.1）是通过去除其毛坯上不需要的部分而实现的。因此，机械加工费用主要取决于零件毛坯的成本和形状、需要去除的材料的数量和形状，以及零件的精度。这三个方面可以进一步分解为六个主要因素，这些因素决定了经过机械加工后零件的成本。

1. 零件是由什么材料制成的？材料对成本的影响有三种方式：原材料的成本、被切除材料的价值和材料加工的难易程度。前两项是直接的材料成本，后一项决定了生产这种零件所需要的工作量、加工时间和机床。

2. 采用哪一种机床来生产这种零件？生产这种零件所使用的机床种类，如车床（见图 33.1 和图 33.2）、铣床（见图 35.1）、镗床（见图 55.2）和钻床（见图 55.3）等会影响零件的成本。对于每种机床来说，不仅应该计算机床使用时间的成本，还应该计算所需要的刀具和夹具的成本。

3. 零件的主要尺寸是多大？这个因素决定了制造这种零件所需要的每种类型机床的尺寸。在制造零件过程中，每台机床由于其购买时的价格和使用时间的不同，具有不同的使用成本。

4. 有多少需要加工的表面和有多少材料应该被去除？知道了需要加工表面的数量和材料去除率（零件加工后体积与其加工前体积的比值），可以帮助人们较好地估算出加工这个零件所需要的时间。如果要进行更精确的估算，则应该确切地知道实现每次切削所用的加工方式。

5. 需要生产的零件的数量是多少？零件的生产数量对成本有很大的影响。如果生产的零件仅为一件，则应该利用尽可能少的夹具，尽管这样做需要较长的安装、调整和找正时间。如果需要生产少数几件，则需要制造一些夹具。在进行大批量生产时，将采用有大量夹具和数控加工的自动化的生产过程。

6. 需要何种公差和表面粗糙度？对公差和表面粗糙度的要求越严格，则在零件的制造中就需要更多的加工时间和设备。

第 56 课　质量与检验

根据美国质量学会的定义，质量是产品或服务能够满足规定需求而具有的特性和特征的总和。这个定义表明，应该首先确定顾客的需求，因为满足这些需求是实现质量目标的关键。然后应该把顾客的需求转化为产品的特性和特征，据此进行设计工作和制订产品的技术要求。

除正确理解质量这个术语外，对质量管理、质量保证和质量控制等术语的理解也很重要。

质量管理包括确定质量方针、目标和职责，并在质量体系中通过诸如质量保证、质量控制和质量改进等方法实施的整个管理职能的全部活动。质量管理是高层管理人员的职责。

质量保证包括了为使人们确信某种产品或服务质量能够满足规定的要求所必需的有计划、有系统的全部活动。这些活动的目标是给人们提供对质量体系能够正常工作的信心，而且包括对设计方案和技术要求妥善性的评价或对生产作业能力的审核。内部质量保证的目标在于使公司的管理人员对产品质量具有信心，而外部质量保证的目的在于向公司产品的用户在产品质量方面做出担保。

质量控制包括了为保持某种产品或服务质量满足规定的要求所采取的作业技术和活动。质量控制功能与产品的关系最为密切，它采用各种技术和活动来监控加工过程，并且消除生成不合格产品的根源。

许多过去的质量体系的设计目标是，在各个加工阶段，将合格的产品与不合格的产品区分开来。那些被判定为不合格的产品必须经过重新加工以满足技术要求。如果它们不能被重新加工，就应该当成废品处理。这种体系称为"检测—改正"体系。在这种体系中，只有在对产品进行检验时，或者当用户使用产品时，才会发现其中存在的问题。源于检查人员本身固有的因素所决定，这种分类拣选工作的有效性经常低于90%。目前，预防型质量体系得到了越来越广泛的应用。在这种体系中，为了避免发生质量问题，首先强调对产品生产的各个阶段进行合理的规划和制定预防发生问题的措施。

对于一种产品满足需求或期望值的好坏程度的最终评价是由顾客和用户做出的，而且会受到可以向这些顾客和用户出售同类产品的竞争对手的影响。这个最终评价是基于产品在整个使用寿命周期内的表现做出的，而不是只根据其在购买时的性能做出的。认识到这一点是很重要的。

如前所述，了解顾客的需求和期望是很重要的。此外，将一个企业中全体员工的注意力集中在顾客和用户及他们的需求上，会产生一个更有效的质量体系。例如，在对产品设计方案和技术要求进行小组讨论时，要对顾客应该得到满足的需求进行专门的讨论。

管理人员的一个基本职责是持续进行质量改进。应该认识到，在当今的市场中，质量是一个由顾客对产品日益增高的期望所驱动的动态目标，质量行动必须落实到公司每天的工作中。传统的做法是对某种产品确定一个适当的质量等级，然后集中全部力量使该产品满足这个质量等级。这种方法不适合在长期的生产中使用。与之相反，管理人员应该给企业确定这样的方向：某种产品达到适当的质量等级之后，还应对其进行不断的改进，使其达到更高的质量等级。

为了获得最有效的改进成果，管理人员应该明白质量与成本是互补的，而不是相互矛盾的。传统的观念认为，管理人员应该在质量与成本之间做出选择。这是因为，质量好的产品，其成本必然较高，而且其生产难度也较大。目前，世界各地的经验均已表明，这种观念是不正确的。高质量的产品从根本上实现了资源的优化利用，从而意味着高生产率和低质量成本。此外，被顾客认为质量高和在使用过程中性能可靠的产品，其销售量和市场占有率都会明显地高于同类产品。

质量成本的四种基本类型如下所述：

1. 预防成本是规划、实施和维持一个质量体系，使其能在经济的水平上保证产品性能满足质量要求所需的费用。进行统计过程控制方面知识的培训可以作为预防成本的一个例子。

2. 鉴定成本是指确定产品性能与质量要求之间符合程度所需的费用。检验可以作为鉴定成本的一个例子。

3. 内部损失成本是指在将产品、元件和材料的所有权转移给顾客之前因出现不满足质量要求而产生的费用。废品可以作为内部损失成本的一个例子。

4. 外部损失成本是指在将所有权转移给顾客之后，因产品不能满足质量要求而产生的费用。根据产品保修单进行索赔可以作为外部损失成本的一个例子。

尽管质量控制水平的确定工作大部分是采用概率论和统计计算进行的，但是采用适当的和精确的数据采集过程为这项工作提供依据也是非常重要的。提供错误的数据会使最好的统计工作程序变得毫无意义。与机械加工过程类似，检验数据采集本身也是一个受到准确度、精密度、分辨率和重复精度等因素限制的过程。所有的检验和（或）测量过程都可以依据它们的准确度和重复精度来确定范围，这与采用精确度和重复精度来评价制造过程是一样的。

第59课　可靠性要求

对于不同种类的设备或系统，可靠性要求可能会有很大的差异。它们有时由设计人员制订，但更多的时候是由负责设计工作的公司制订的。在有些场合，这种要求可能是由产品购买者方提出的，有时第三方，如保险公司或政府机构，也可能起很大作用。

对于任何一种产品，很可能需要在购买费用和维修费用之间做出权衡取舍。在提高产品可靠性方面付出的努力越多,则产品的购买费用就越高。与此同时,产品的可靠性越高其维修费用就越低。我们可能会说,当总费用为最低时,所对应的是最好的解决方案。然而,在实践中,产品性质不同,做出权衡取舍时需要考虑的事项就会有很大的差异。通过对空调器、工业机器人(见图71.1)和航空发动机(见图61.1)这三种产品中起作用的各种因素的分析,可以更清楚地理解不同产品之间可靠性要求的差异。

对空调器的可靠性要求,在很大程度上取决于消费者的心理因素。购买者在多大程度上愿意付较高的价钱以节省其后的修理或更换费用?产品价格是显而易见的,但是如何使可靠性给公众留下深刻的印象呢?具体的做法有:可以提供关于空调器特点的信息和广告,或者愿意提供比竞争对手更长的质保期。如果由于具有太高的可靠性而使空调器的价格过高,那么销售结果会是什么样呢?另一方面,如果公众以较低的价格购买了产品,然后他们对使用过程中出现的不方便或故障修理费用表示不满,是否会长期地给这家公司的其他产品带来设计质量不好和工艺质量低劣的声誉呢?这在很大程度上又取决于负责维修的服务机构的效率和工作范围。这些问题错综复杂,很可能需要制造厂商借助市场调查结果或其他信息来源,确定公众对价格和可靠性的偏好,然后做出决定。

对于工业机器人或其他设备,情况就完全不同了。这些产品主要是为销售给大型企业而设计的。如果设备很昂贵,购买方可能会派自己的工程师或雇用咨询顾问对费用进行综合分析,并确定它们对购买方工作的影响。以机器人为例,必须综合考虑购买费用和因机器人故障而造成的生产损失。与生产损失费用相比,机器人的直接修理费用可能是很少的。实际上,购买方的维修部门能够完成修理工作的速度将是值得重点考虑的问题。因此,对于这些种类的产品,设计方案不仅要注意可靠性,也要注意可维修性。它们能否在短期内得到修复?通过增加修理时间作为代价来提高可靠性,在某些情况下将带来较大的生产损失。

对于航空发动机而言,失效事故的后果通常都非常严重,因此就需要有更高的可靠性标准,这就使得航空发动机的制造费用要比其他发动机高很多。同样地,主要考虑的因素将是为了提高可靠性而进行的定期检修,而不是修复一台出现故障的发动机所需要的时间。防止出现失效事故是非常重要的,以致不能把制订可靠性技术规范的工作完全交给设计或制造企业或购买方。保险公司和政府机构可能会更密切地介入。在设计中更应该强调对初期故障的早期报警,以便使发动机能够在发生事故之前退出使用。

在制订可靠性要求,并在设计中保证这些要求能够得到满足时,我们应该在主要的子系统之间分配可靠性指标,然后再将其分到下一个层次的部件和零件中去。例如,在飞机设计中要使推进、结构、控制和导航系统都具有大体相同的可靠性。如果整个系统要求的可靠性为R,则我们有$R=R_0^4$,式中R_0是每个子系统的可靠性要求。因此,每个子系统的可靠性要求为$R_0=R^{1/4}$。这种可靠性的分配是非常重要的,特别是当各子系统或部件是由不同的部门或公司设计与制造时尤为重要。

到目前为止,我们只关心可靠性,即系统能够正常工作的概率。实际上,并不是所有的失效事故都应同样对待,因为不同的失效模式所产生的后果可能会相差很多。对于大多数系统来说,我们希望出现失效事故的后果只是造成使用上的不方便和一些经济上的损失。但是,设计人员必须慎重考虑那些可能危及人身安全或危及较大范围内人群的健康与安全的特殊失效模式。

我们回到空调器、工业机器人和航空发动机这三个例子上来,以便在安全性分析和传统的可靠性分析之间进行对比。对空调器来说,大部分失效事故造成的后果只是停止工作。然而空调器中的电线短路可能造成过热或火灾,甚至使用户触电身亡。必须进行认真的设计和安全性分析工作,消除发生这种事故的可能性或将其降低到很小的概率。如果一个有缺陷的产品导致了这样的事故,它可能不得不退出市场,而且,该公司可能会因产品责任诉讼而被告上法庭。

工业机器人的失效事故多少与之有些类似，不同之处是这些失效事故只会对与从事这项工作有关的工人造成危险，而不会危及一般公众。与之相比，当一台航空发动机出现失效事故时，在可靠性和安全性分析之间没有什么区别，这是因为可靠性是用发动机的功能定义的，出现失效事故会显著增大飞机坠毁的概率。

第61课　可靠性对产品适销性的影响

可靠性是一种可销售的商品吗？答案为既是又不是。这完全取决于你所处的行业类型。在航空航天业，人们可以毫无保留地说"是"。考虑到阿波罗太空飞行任务的技术复杂性，它所达到的可靠性是卓越非凡的。阿波罗号宇航员百分之百的安全返回地球，是20世纪后半叶可靠性技术的成功事例。有人可能会问，这种成功是怎么取得的？这完全是因为在这个项目的方案设计阶段，就已经把可靠性连同其他性能参数一起写入技术规范中。因此，投标是建立在能够满足技术规范所要求的可靠性标准的基础上的。

在全球性市场经济的环境中，对利润的追求提供了达到高可靠性的动力。为了完成要求的销售目标，每个公司或机构都应该在规定其他工程性能参数和运行特性的同时，确定必须达到的可靠性水平。制造费用通常可以决定所采用的结构形式，产品的制造费用应该维持在能使产品的售价在市场上具有竞争力的水平上。对于一种新产品，在工程研制开发费用中必须包括使产品达到要求的可靠性水平的费用。将一种可靠性不能令人满意的新产品投放到市场中，会使精心制订的营销计划变得毫无意义。在产品说明书中标明可靠性是一种很好的做法，如果顾客相信产品故障是在进行了为验证产品可靠性而设计的切合实际的试验工作之后发生的，他的不满意度通常会显著降低。

汽车行业、航空发动机行业和通风机行业可以作为将产品开发与市场营销策略相结合的实例。当一种新型汽车面市后，在可能会购买这种汽车的顾客们的头脑中会产生两种完全不同的想法：

(1) 它是新型的（因此比前一种型号"更好"）。

(2) 它是新型的（因此没有经过验证，在初期可能会出现许多故障，因而不能可靠地工作）。

应该采取的策略就是要使顾客相信，它既是新型的（所以比前一种型号更好和更能令人称心如意），又是经过试验验证的（所以几乎没有那些需要经常去维修站的潜在故障）。为了能够做到这一点，在宣传时用一些通俗易懂的试验来证明新型汽车的可靠性是非常重要的。例如，"这种汽车在我们让公众知道它的存在之前已经行驶了15万千米"，或者"我们挑选了200名司机来测试这种车，以确保我们的新车能够经得起日常商业驾驶的严酷考验"。在过去，这两个例子被用来告诉广大汽车驾驶人员，为了他们的利益，厂方已经进行了大量的开发工作，生产出来的产品是令人满意的和性能可靠的。

飞机发动机（见图61.1）行业的策略与此类似之处在于，必须对航空公司或军队方面客户所提出的问题做出令人信服的答复。他们会说，他还记得前一种发动机在刚开始使用时出现的那些问题，你们采取了哪些措施来确保这种发动机不再出现类似的问题？使产品开发工作能够被接受的最简单方法就是去说明，前一种型号的发动机在交给航空公司或军队使用前在试验台上进行了10000小时的运行试验，但是新型发动机在交付使用前将在试验台上进行15000小时的试验。在这种情况下，要最大限度地利用市场信息来了解：顾客真正需要什么？顾客真正准备购买什么？

通风机（见图61.2）行业的顾客所关心的问题，按通常的主次顺序为：价格、性能和可靠性。因此，资金中用于进行复杂试验的费用就比较少。对工程师来说，每一项试验都必须是有意义的和有说服力的，并且能使顾客相信，他安装这台设备之后就不需要费心维护。现场试验由于比较便宜，因而是非常重要的，但花费的时间比较长。在这种试验中，顾客可以看到设备的安装过程，安装细节和新产品令人满意的工作情况，这会使他对这种产品的性能产生信心。在现场试验中，潮湿、有蒸汽或脏的安装环境特别适合于进行"超载"试验。

如果某种通风机是用于农业环境的，在进行通风机的现场试验时，重要的是不要找一位最细心的用户，而是要找一位能够做出用水冲洗设备的方式来进行清洁工作的这种以非常随意的方式办事的用户。用水冲洗这件事能够充分表明，需要在轴上安装密封装置，以阻止水进入电动机并破坏其可靠性。

第62课　计算机在制造业中的应用

计算机正在将制造业带入信息时代。计算机长期以来在商业和管理方面得到广泛的应用，它正在作为一种新的工具进入工厂，而且它如同蒸汽机在200多年前使制造业发生改变那样，正在使制造业发生变革。

尽管基本的金属切削过程不太可能发生根本性的改变，但是它们的组织形式和控制方式必将发生改变。

从某一方面可以说，制造业正在完成一个循环。最初的制造业是家庭手工业：设计者本身也是制造者，每次构思和制造一件产品。后来，形成了零件的互换性这个概念，生产被按照专业功能进行划分，每次可以生产数以千计的相同零件。

今天，尽管设计者与制造者不再可能是同一个人，但在向集成制造系统前进的途中，这两种功能已经越来越靠近了。

可能具有讽刺意味的是，在市场需求高度多样化产品的同时，制造业必须提高生产率和降低成本。消费者要求用较少的钱去购买高质量和多样化的产品。

计算机是满足这些要求的关键因素。它是能够提供快速反应能力、柔性和速度来满足多样化市场的唯一工具。而且，它是实现制造系统集成所需要的、能够进行详细分析和利用精确数据的唯一工具。计算机使工程师们有可能得出大量富有挑战性的问题的解决方案。

将来，计算机很可能是一个企业生存的基本条件。许多现今的企业将会被生产能力更高的企业组合逐渐取代。这些生产能力更高的企业组合是一些具有非常高的质量、非常高的生产率的工厂。目标是设计和运行一个能以高生产率的方式生产100%合格产品的工厂。

一个采用先进技术、充满竞争的世界正在要求制造业满足更多的需求，并且促使其自身采用先进的技术。为了适应竞争，一个公司会满足某些多少有点相互矛盾的需求，比如说，既要生产多样化和高质量的产品，又要提高生产率和降低价格。

在努力满足这些需求的过程中，公司需要一个采用先进技术的工具，一个能够对顾客的需求做出快速反应，而且能够从制造资源中获得最大收益的工具。

计算机就是这个工具。

成为一个具有"非常高的质量、非常高的生产率"的工厂，需要对一个非常复杂的系统进行集成。这只有通过采用计算机对机械制造的所有组成部分——设计、加工、装配、质量保证、管理和物料搬运进行集成才能完成。

例如，在产品设计期间，交互式计算机辅助设计系统使得完成绘图和分析工作所需要的时间比以前大大减少，并且具有更高的准确性。此外，样机的试验与评价程序进一步加快了设计过程。

增材制造（AM），俗称3D打印，是一种由计算机控制的工艺过程，它采用逐层堆积材料的方法制造出基于数字模型的三维物体。许多不同种类的金属和金属合金，从金、银等贵金属到不锈钢、钛合金等难加工材料都可用于增材制造。

在车间里，分布式智能以微处理器这种形式来控制机床、操纵自动上下料设备和收集关于当前车间状态的信息。

但是这些各自独立的改革还远远不够。我们所需要的是由一个通用软件从始端到终端进行控制的全部自动化的系统。

一般来说，制造系统经过计算机集成后具有如下优点：可以利用更为广泛、及时而准确的信息，可以改进各部门之间的交流与沟通，实施更严格的控制，而且通常能提高整个系统的全面质量和效率。

例如，改进交流和沟通意味着会使设计具有更好的可制造性。数控编程人员和工艺装备设计人员有机会向产品设计人员提出意见，反之亦然。

因而可以减少技术方面的变更，而对于那些必要的变更则可以更有效地进行处理。计算机不仅能够更快地对变更之处做出详细说明，而且还能够把变更之后的数据告诉随后的使用者。

利用及时更新的生产控制数据可以更有效地进行工艺设计和制订生产计划。因此，可以使昂贵的设备得到更好的利用，提高零件在生产过程中的运送效率，降低在制品的成本。

产品质量也可以得到改进。例如，不仅可以提高设计精度，还可以使质量保证部门利用设计数据，避免由于误解而产生错误。

可使人们更好地完成工作。通过避免冗长的计算和书写工作——这还不算查找资料所浪费的时间——计算机不仅使人们更有效地工作，而且还能把他们解放出来去做只有人类才能做的工作：创造性地思考。

计算机集成制造还会吸引新的人才进入制造业。人才被吸引过来的原因是他们希望到一个现代化的、技术先进的环境中工作。

在制造工程中，CAD/CAM 减少了工艺装备设计、数控编程和工艺设计所需要的时间。同时加快了响应速度，这最终将会使目前外委加工的那些工作由公司内部人员来完成。

第 65 课　计算机辅助工艺设计

根据《工具与制造工程师手册》，工艺设计是系统地确定一套能够经济地、有竞争力地将产品制造出来的方法。它主要由选择、计算和编写工艺文件组成。对加工方法、机床、刀具、工序和顺序必须进行选择。对于一些参数（如进给量、速度、公差、尺寸和成本等）应该进行计算。最后，应该编写带工序简图的工艺卡片、工序卡片和工艺路线等方面的工艺文件。工艺设计是产品设计和制造的中间环节。那么，它是如何将设计与制造连接起来的呢？

大部分制造工程师都会同意，如果 10 个工艺人员对同一个零件进行工艺设计，他们可能会得出 10 种不同的方案。显然，并不是所有这些方案都是由最有效的制造方法组成的，事实上，也不能保证它们中的任何一个方案都是由加工这个零件的最佳方法组成的。

目前制造过程中一个更为混乱的现象是，对于一个零件来说，现在所编制的工艺规程可能与以前在制造过程中所编制的同一个零件或相似零件的工艺规程相差很多，而且这个工艺规程可能再也不会应用于同一个零件或相似零件。这说明很多工作成果都被浪费了，而且在工艺路线、工艺装备、对工人的要求和成本等方面都不一致，甚至对外购件的要求也不一样。

当然，工艺规程不应该是一成不变的。随着产品批量的变化和新技术、新设备、新的加工方法的出现，加工制造某一特定零件最适当的方法也会发生变化，而且这些变化应该在车间目前使用的加工工艺规程中反映出来。

工艺人员应该管理和检索大量的数据和很多文件，其中包括已经建立了的标准、可加工性数据、机器规格、工艺装备清单、原材料库存量和一些目前正在应用的工艺文件。这主要是一些信息处理工作，而计算机是完成这项工作的一个理想助手。

应用计算机辅助工艺设计时还有一个优点。由于这项工作涉及许多相互关联的事情，在确定最优方案时，需要进行许多次迭代。因为计算机可以很容易地进行大量的比较工作，所以它比人工所能分析的可供选择的方案要多得多。

采用计算机辅助工艺设计的第三个优点是所设计的工艺过程具有一致性。

采用计算机辅助工艺设计可以获得以下几点好处：

1. 在准备工艺文件时，减少了重复性工作。
2. 减少了由于人为过失导致的计算错误。
3. 由于交互式计算机程序的提示功能而减少了在逻辑和说明方面的疏漏。
4. 通过中央数据库可以直接利用最新信息。
5. 由于每个工艺设计人员都利用相同的数据库，因此可以保证信息的一致性。
6. 对由其他业务部门的工程技术人员所提出的修改意见做出快速反应。
7. 自动地利用最新版本的零件图纸。
8. 采用文字处理技术，产生更详细、更一致的工艺文件。
9. 更有效地利用库存的刀具、量具和夹具，并能够减少这些物品的种类。
10. 由于能够使工艺规程适合某项特定的工作，而且能用清楚的、有理有据的语言表达出来，因此可以与车间的人员进行更好的交流。
11. 可以更好地获得编制生产计划所需的信息，其中包括刀具寿命预测、物料需求计划和库存控制。

对计算机集成制造最为重要的是，计算机辅助工艺设计可以生成机器可以阅读的数据，而不是手写的规程。这种数据可以传递到计算机集成制造体系的另一个系统中，用于进行工艺设计。

计算机辅助工艺设计通常有两种类型：派生式和创成式。

在派生式中，对采用成组技术确定的一个零件族中的所有零件编制一套典型的加工工艺规程。这个典型工艺规程存储在计算机的存储器中，根据新零件的零件族编码进行检索。成组技术可以帮助把新零件归类于适当的零件族中。通过对典型工艺规程的编辑，可以满足特定工作的专门要求。

在创成式中，通过采用确定加工制造过程中各种工艺决策的适当算法，将各个单独的工艺规程综合起来。在一个真正的创成式计算机辅助工艺设计系统中，工序的排列和所有制造过程参数都可以在不必参考以前的工艺规程的情况下自动生成。在它最终实现之后，这种方式将是普遍适用的：将任何一个零件提交给这个系统，计算机都会生成最优的工艺规程。

然而，这种系统目前还不存在。所谓的创成式计算机辅助工艺设计系统——大概在可以预料到的将来——仍然是应用于一个特定工序或特定加工过程的专用系统，其逻辑原理是以过去的经验与基本理论的组合为基础的。

第 66 课　数 字 控 制

先进制造技术中一个最基本的概念是数字控制（NC）。在数控技术出现之前，所有的机床都是由人工操纵和控制的。在与人工控制的机床有关的很多局限性中，操作者的技能大概是最突出的问题。采用人工控制时，产品的质量直接与操作者的技能有关。数字控制代表了从人工控制机床走出来的第一步。

数字控制意味着采用预先录制的、存储的符号指令，控制机床和其他制造系统。一个数控技师的工作不是去操纵机床，而是编写能够发出机床操作指令的程序。

发展数控技术是为了克服人类操作者的局限性，而且它确实完成了这项工作。数控机床比人工操纵的机床的精度更高，生产出来的零件的尺寸一致性更好，生产速度更快，而且长期的工艺装备成本更低。数控技术的发展导致了制造工艺中其他几项新发明的产生：

1. 电火花线切割技术。
2. 激光切割技术。
3. 磨料水射流切割技术。

数字控制还使得机床比采用人工操纵的"前辈"的用途更为广泛。一台数控机床可以自动生产多种多样的零件，每个零件都可以有许多不同种类的、复杂的加工过程。数控可使生产厂家承担那些对于采用人工控制的机床和工艺来说，在经济上不划算的产品的生产任务。

与许多先进技术一样，数控诞生于麻省理工学院的实验室中。数控这个概念是20世纪50年代初在美国空军的资助下提出来的。

APT（自动编程工具）语言是1956年在麻省理工学院的伺服机构实验室中被设计出来的。这是一个专门适用于数控的编程语言，它使用类似英语的语句来定义零件的几何形状，描述切削刀具的形状和规定必要的运动。APT语言的研究和发展是数控技术进一步发展过程中的一大进步。最初的数控系统与今天应用的数控系统是有很大差别的。在那时的机床中，只有硬连线逻辑电路。指令程序写在穿孔纸带上（见图66.1），它后来被塑料磁带所取代，采用读带机（见图66.2）将写在纸带或磁带上的指令为机床翻译出来。所有这些共同构成了机床数控方面的巨大进步。然而，在数控发展的这个阶段还存在着许多问题。

一个主要问题是穿孔纸带的易损坏性。在机械加工过程中，载有编程指令信息的纸带断裂和被撕坏是常见的事情。在机床上每加工一个零件，都需要将载有编程指令的纸带放入读带机中重新运行一次。因此，这个问题变得很严重。如果需要制造100个某种零件，则应该将纸带分别通过读带机100次。易损坏的纸带显然不能承受严酷的车间环境和这种重复使用。

这就导致了一种专用的塑料磁带的研制。与在纸带上通过采用一系列的小孔来载有编程指令不同，在塑料带上通过采用一系列的磁点来载有编程指令。塑料带的强度比纸带的强度要高很多，这就可以解决常见的撕坏和断裂问题。然而，它仍然存在着两个问题。

其中最重要的是，很难或不可能更改输入到磁带上的指令。即使对指令程序进行最微小的调整，也必须中断加工并制作新的磁带。而且磁带通过读带机的次数还必须与需要加工的零件的个数相同。幸运的是，计算机技术的实际应用很快解决了数控技术中与穿孔纸带和塑料磁带相关的问题。

在形成直接数字控制（DNC）这个概念之后，可以不再采用纸带或塑料带作为编程指令的载体，这样就解决了与之有关的问题。在直接数字控制中，几台机床通过数据传输线路连接到一台主计算机上。操纵这些机床所需要的程序都存储在这台主计算机中。当需要时，通过数据传输线路提供给每台机床。直接数字控制是在穿孔纸带和塑料带基础上的一大进步。然而，它也有着与其他依赖于主计算机的技术一样的局限性。当主计算机出现故障时，由其控制的所有机床都将停止工作。这个问题促使了计算机数字控制技术的产生。

微处理器的发展使可编程控制器和微型计算机的发展成为可能。这两种技术为计算机数控（CNC）的发展提供了条件。采用CNC技术后，每台机床上都有一个可编程控制器或微型计算机对其进行数字控制。这可以使程序能够被输入和存储在每台机床的内部。采用CNC还可以实现离线编程，并可将该程序下载到每台机床中。计算机数控解决了主计算机发生故障所带来的问题，但是它产生了另一个被称为数据管理的问题。同一个程序可能要被分别装入十台相互之间没有通信联系的微型计算机中。这个问题目前正在解决之中，可以采用局域网将各台微型计算机连接起来，以实现更好的数据管理。

第 69 课　培训编程人员

熟练的零件编程人员是能够有效地利用数控机床的一个重要条件。他们的工作决定了这些机床的运行效率和在机床本身、工厂的数控辅助设备及管理费用等方面的投资所能得到的经济回报。

目前，熟练的零件加工数控编程人员非常短缺。这不仅表明了在机械加工业普遍缺少有经验的人员，而且也表明随着在这个行业中越来越多地通过应用数控机床来增加生产能力、通用性和生产率，对编程人员的需求也日益增多。

从行业的层面来看，显而易见的答案是通过培训来培养新的编程人员，而且可以通过多种途径进行这种培训。但是，首先应该确定：编程人员应该具备什么条件和参加培训的编程人员应该学习什么？

根据全国机床制造厂商协会编写的《选择适当的数控编程方法》这本小册子，手工编程人员主要应该具备下列各项条件：

机械制造经验　编程人员对要进行编程的数控机床（见图 69.1）的性能应该有透彻的了解，还要了解车间中其他机床的基本性能。他们还应该在金属切削原理和实践、刀具的切削能力、工件夹具和夹持技术等方面具有广博的知识。编程人员还应当在能够显著降低生产成本的制造工程技术方面得到适当的培训。

空间想象力　编程人员应该能够想象出零件的三维形状，机床的切削运动，在刀具、工件、夹具或机床本身之间可能产生的干涉。

数学　算术、代数、三角学、几何等方面的知识是非常重要的。高等数学，诸如高等代数、微积分等，通常是不需要的。

对细节的关注　编程人员应该是具有非常敏锐的观察能力并且是非常认真、仔细的人。要改正在机床调试过程中发现的编程错误，可能会花费很多金钱和时间。

在这本小册子的另一个地方提到："与计算机辅助编程相比，手工编程要求编程人员在机床及其控制系统、加工过程、计算方法等方面掌握更多的知识。另一方面，采用计算机辅助编程，应该掌握计算机编程语言和运用这种语言所需要的计算机系统知识。一般来说，由于涉及许多细节，所以手工编程速度更慢并对编程人员的要求更高。在计算机辅助编程中，这些细节知识都包含在计算机系统（处理程序、后置处理程序）中。"

数控技术和培训方面的专家通常同意这些条件和要求，他们还增加了一些次要方面的细节要求，如图样识读、不同种类金属的切削性能、车间测量仪器的使用、公差标注和安全操作等方面的知识。

应该到哪里去寻找可以进行培训的学员呢？首先在你自己工厂的车间里。埃德华·F·斯罗斯，一位辛辛那提米拉克龙公司主管销售的副总经理，这样说："我们在将优秀的车工和铣工培训成编程人员方面有很成功的经验。他们没有认识到，但是实际上他们工作的大部分时间都在编程，而且他们有车间所需要的基础数学和三角学知识。你可以很容易地教会他们编程。反之，将一个能力很强的数学家培养成编程人员则很困难。对他来说，刀具轨迹的编程很容易。但是如何实现刀具轨迹的加工——进给量、切削速度等——这可能需要更多的培训工作。"

采用功能较强的计算机辅助编程后，对编程人员在金属切削知识方面的要求就降低了。通过应用这种软件，辛辛那提米拉克龙公司成功地雇用了一些大学刚毕业的学生，其中还包括一些非技术专业毕业的大学生。通过首先对这些学员在工厂中进行机床实际操作培训，然后再对他们进行编程方面的培训，将他们培养成零件数控加工的编程人员。

当然，所有的数控机床供应商都会对他们的产品提供某种编程培训，而且大部分供应商会提供正规的培训计划。例如，米兰克龙公司的销售部门就有 20 名全职的客户培训教师。这个公司要求参加培训的学员应该具有下列必备条件：

"参加人员应该了解机械加工车间的安全规则，能够看懂零件图和剖视图。"

"需要掌握平面几何、直角三角学和公差的基本知识。"

"还需要具有零件手工数控编程、数控机床的调试和操作程序、零件加工、金属切削技术、刀具和夹具等方面的知识。"

把具有以上基础的人送到学校，将能保证数控机床的用户能够从他们所花在培训上的费用（尽管培训费已经包括在购买机床的基本费用中，接受培训的人员还需要花费一周的时间、旅行费和生活费）中获得最大的收益。

第 71 课　工业机器人

关于工业机器人的定义有许多。采用不同的定义，全世界各地工业机器人装置的数量就会发生很大的变化。在制造工厂中使用的许多单用途机器可能会看起来像机器人。这些机器只具有单一的功能，不能通过重复编程的方式去完成不同的工作。这种单一用途的机器不能满足被人们日益广泛接受的关于工业机器人的定义。

国际标准化组织（ISO）对工业机器人的定义为：一种能够自动控制、可重复编程、多用途的操作机，可以对三个或三个以上轴进行编程，它可以是固定式的也可以是移动式的，在工业自动化中使用。

其他的一些协会，如美国机器人协会（RIA）、日本工业机器人协会（JIRA）、英国机器人协会（BRA）等，都对工业机器人提出了各自的定义。由美国机器人协会提出的定义为：

机器人是一种用于移动材料、零件、工具或专用装置的，通过可编程序动作来执行多种任务并具有可重复编程能力的多功能操作机。

在所有的这些定义中有两个共同点，即"可重复编程"和"多功能"这两个词。正是这两个特征将真正的机器人与现代制造工厂中使用的各种单一用途的机器区分开来。

"可重复编程"这个词意味着两件事：机器人根据编写的程序工作，并可以通过重新编写程序来使其适应不同种类的制造工作的需要。

"多功能"这个词意味着机器人可以通过重复编程和使用不同的末端执行器，来完成不同的制造工作。围绕着这两个关键特征所撰写的定义已经变成了被制造业的专业人员所接受的定义。

第一个关节式手臂于 1951 年被研制出来，供美国原子能委员会使用。1954 年，乔治·C·德沃尔设计了第一个工业机器人。它是一个不复杂的、可以编程的物料搬运机器人。

第一个商业化生产的机器人在 1959 年研制成功。1962 年，通用汽车公司安装了第一个用于生产线上的工业机器人。在美国新泽西州的一家汽车厂中，它被用来从压铸机中取出红热的车门拉手和其他类似的汽车零件。它最显著的特点是有一个夹持器，这样就不需要由人去接触那些刚由熔融金属形成的汽车零件。它有五个自由度。这个机器人是由万能自动化公司（Unimation）生产的。

在 1973 年，辛辛那提米拉克龙公司研制出 T^3 工业机器人，在机器人的控制方面取得了重大的进展。T^3 机器人是第一个商业化生产的采用小型计算机控制的机器人。T^3 机器人和它的所有运动如图 71.1 所示。

从那时起，机器人技术在很多方面都得到了发展，这包括焊接、喷漆、装配、机床上下料和检验。

在过去的 30 年中，机器人在许多汽车制造厂中占据了主要地位。在一个工厂中，通常有数以百计的工业机器人在全自动生产线上工作。例如，在一条自动生产线上，车辆底盘装在输送机上，

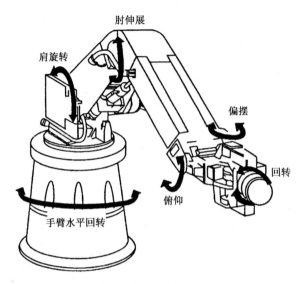

图 71.1 工业机器人

在通过一连串的机器人工作站时进行诸如焊接、喷漆和最后的装配等项工作。

在印制电路板的大批量生产过程中，装配工作几乎完全是采用取放机器人进行的。通常采用 SCARA 机器人抓取微小的电子元件并以非常高的精度将其放置到印制电路板上（见图 71.2）。这类机器人每小时可以放置数以万计的元件，其速度、精度和可靠性都远远超过了人类。

工业机器人成本的降低是促进它们的使用量增长的一个主要原因。从 20 世纪 70 年代开始，工资的快速增长大大增加了制造业中的人工费用。为了生存，制造厂商被迫考虑采用一切能够提高生产率的技术。为了在全球性市场经济的环境中具有竞争能力，制造厂商必须以比较低的成本，生产出质量更好的产品。其他因素，诸如寻找更好的方式来完成带有危险性的制造工作，也促进了工业机器人的发展。但是，其根本原因一直是，而且现在仍然是提高生产率。

机器人的主要优点之一是它们可以在对人类有危险的环境中工作。采用机器人进行焊接和切断工作是比由人工来完成这些工作更为安全的例子。大部分现代机器人被设计用在对人类来说不安全和非常困难的环境中工作。例如，可以设计一个机器人来搬运非常热或非常冷的物体，这些物体如果由人工搬运，则存在不安全因素。

尽管机器人与工作地点的安全密切相关，但是它们本身也可能是危险的。应该精确地计算出机器人的工作空间，并且在这个工作空间的四周清楚地标示出危险区域。可以通过设置障碍物来阻止工人进入机器人的工作空间。即使有了这些预防措施，在使用机器人的场地中设置一个能够自动停止工作的系统仍然不失为一个好主意。这个系统应当具有能够检测出是否有需要自动停止工作的要求的能力。

第 73 课　工业机器人的基本组成部分

为了理解机器人各个组成部分的功能和性能，我们可以同时观察我们的手臂、手腕、手和手指在从货架上取东西、使用手工工具或操纵机器时的各种运动的灵活性和能力。工业机器人的基本组成如下所述。

操作机　操作机是一个能够提供类似人的手臂和手腕运动的机械装置。操作机通常由杆件（它们的功能类似人体中的骨头）和关节（也称为"运动副"）以串联的方式连接而成（如图 72.1 中的那些机器人）。对于一个典型的六自由度机器人（如图 71.1 所示），前三个杆件和关节构成了手臂，而后三个关节构成了手腕。手臂的功能是将一个物体放在三维空间中的某一位置，手腕的功能是确定此物体的方位。

末端执行器 末端执行器是安装在机器人的操作机端部的装置。它相当于人的手。根据其用途，常见的末端执行器有以下几种：

(1) 用于物料搬运的夹持器、电磁铁和真空吸盘；

(2) 用于喷涂的喷枪；

(3) 进行点焊和弧焊的焊接装置；

(4) 诸如电钻等类的电动工具；

(5) 测量仪器。

末端执行器通常是按照客户的特殊要求定制的。最常用的末端执行器是机械夹持器，它们有两根或多根手指。具有两根手指的夹持器（见图73.1）只能抓取形状简单的物体，具有多根手指的夹持器则能够完成更为复杂的工作。为某一特定的用途选择末端执行器时，应该考虑诸如额定负载、环境、可靠性和成本等项因素。

驱动器 驱动器就像机器人的肌肉，它们为操作机和末端执行器提供运动。根据其工作原理，可以将它们分为气动、液压和电气驱动器。

气动驱动器利用活塞或气动马达将由压缩机提供的压缩空气转换为机械能。气动驱动器只有极少数运动部件，这决定了它们具有较高的可靠性和只需要较低的维修费用。在所有的驱动器中，它是价格最便宜的。但是，由于空气的可压缩性，气动驱动器不适合在需要精确控制的场合搬运重物。

液压驱动器利用油等高压流体将力传递到需要的作用点。液压驱动器在外观上很像气动驱动器。二者相比，液压驱动器的工作压力要高很多（通常为7～17MPa）。它们适合用于需要大功率的场合。与电气驱动的机器人相比，液压驱动的机器人具有更强的承受冲击载荷的能力。

作为操作机的驱动器，电动机的使用数量最多。直流（DC）电动机具有很高的转矩体积比。它们还具有高精度、高加速度和高可靠性。尽管它们的功率重量比不如气动驱动器和液压驱动器，但是其可控性使得它们适用于中小型操作机。

交流（AC）电动机和步进电动机在工业机器人中很少使用。前者难以控制和后者只能产生较低的扭矩都限制了它们的应用。

传感器 传感器可以将一种信号转换为另一种信号。例如，人眼可以将光图像转换为电信号。传感器可以分为几种类型：视觉、触觉、位置、力、速度、加速度等。其中的一些传感器将在74课中详细说明。

数字控制器 数字控制器是一种专用的电子装置，在其中有中央处理器（CPU）、存储器，有时还会有硬盘。在机器人系统中，这些元器件被装在一个叫作控制器的密封盒子里。它被用来控制操作机和末端执行器的运动。由于计算机具有与数字控制器相同的特性，所以它也可以被用来作为机器人的控制器。

模数转换器 模数转换器（简称ADC）是一个将模拟信号转换为数字信号的电子器件。这种电子器件与传感器和机器人控制器相连接。通常，模数转换器将输入的模拟电压（或电流）转换成与电压或电流成正比的数值。例如，模数转换器（ADC）可以将由于应变片（见图74.2）的应变而产生的电压信号转换为数字信号，以便机器人的数字控制器能够处理这个信息。

数模转换器（DAC） 数模转换器（DAC）把从机器人控制器中获得的数字信号转换为模拟信号，以启动驱动器。为了使驱动器（例如，一台直流电动机）工作，与数字控制器相连接的数模转换器（DAC）要把它的数字信号转换回模拟信号，也就是，直流电动机的电压。

放大器 一般来说，放大器是一种能够改变，通常是增大信号幅值的装置。由于数字控制器中发出的指令通过数模转换器后生成的模拟信号很弱，因此只有将它们放大之后才能驱动机器人操作机中的电动机。

第 75 课　信息时代的机械工程

在 20 世纪 80 年代初期，工程师们曾经认为要加快产品的研制开发，就必须进行大量的研究工作。结果表明，实际上只进行了较少的研究工作，这是因为产品开发周期的缩短，促使工程师们尽可能地利用现有的技术。研制开发一种创新性的技术并且将其应用在新产品上是有风险的，而且易于招致失败。在产品开发过程中采用较少的步骤是一种安全的和易于成功的方法。

对于资本和劳动力都处于全球性环境中的工程领域而言，缩短产品研制开发周期也是有益的。能够设计和制造各种产品的人可以在世界各地找到。但是，具有创新思想的人则比较难找。如果你已经进行 6 个月的研制开发工作，那么地理上的距离已经不再是使其他人发现它的障碍。如果你的研制周期较短，只要你仍然保持领先，这种情况并不会造成严重后果。如果你正处于一个长达 6 年的项目研制开发过程的中期，一个竞争对手了解到你的研究工作的一些信息，这个项目可能将面临比较大的麻烦。

工程师们在解决任何问题时都需要进行新的设计这种观念很快就过时了。现代设计过程的第一步是浏览互联网或其他信息系统，看看是否有人已经设计了一种类似于你所需要的产品，如变速器或换热器等。通过这些信息系统，你可能发现有些人已经有了制造图样、数控程序和制造你的产品所需要的其他所有东西。这样，工程师们就可以把他们的职业技能集中在尚未解决的问题上。

在解决这些问题时，利用计算机和计算机网络可以极大地提高工程团队的能力和效率。这些信息时代的工具使工程团队可以利用一些大型数据库。这些数据库中有材料性能、标准、技术和成功的设计方案等信息。这些经过验证的设计方案可以通过下载直接应用，或者通过对其进行快速、简单的改进来满足特定的要求。将产品的技术要求通过网络发送出去的远程制造也是可行的。你可以建立一个没有任何加工设备的虚拟公司。在产品加工完成后，你可以指示制造商将其直接发运给你的客户。定期访问你的客户可以保证你设计的产品按照设计要求进行工作。尽管这些产品开发方式不可能对每个公司都完全适用，但是这种可能性是存在的。

用户定制设计业务过去通常是留给小公司的。大公司对此嗤之以鼻，它们讨厌补缺市场或与用户设计的小批量产品打交道。"这是我的产品，"一家大公司这样说，"这是我们能够制造出来的最好产品，你应该喜欢它。如果你不喜欢，那么在这条街上有一家小公司，它会按你的要求去做。"

今天，因为顾客们有较大的选择余地，几乎所有的市场都是补缺市场。如果你不能使产品满足某些特定客户的要求，你就会失去大部分市场份额，甚至失去全部份额。由于这些补缺市场是经常变化的，所以你的公司应该对市场的变化做出快速的反应。

补缺市场和根据客户需求进行设计这种现象的出现改变了工程师们进行研究工作的方式。今天，研究工作通常是针对解决特定问题进行的。许多由政府资助或由大公司出资开发的技术可以在非常低的成本下被自由使用，尽管这种情况可能是暂时的。在对这些技术进行适当改进后，它们通常能够被直接用于产品开发，这使得许多公司可以省去进行大量研究工作的费用。在主要的技术障碍被克服后，研究工作应该专注于产品的商品化，而不是追求新的、有趣的但不确定的替代方案。

从这个角度来看，工程研究应该致力于消除将已知技术快速商品化的障碍。工作的重点是产品的质量和可靠性，这些在当今的顾客的头脑中是最重要的。很明显，一个质量差的声誉是一个不好的企业的同义词。企业应该尽最大努力来保证顾客得到合格的产品，这个努力包括在生产线的终端对产品进行严格的检验和自动更换有缺陷的产品。

研究工作应该关注诸如可靠性等因素带来的成本效益。随着可靠性的提高，制造成本和系统的最终成本将会降低。如果在生产线的终端产生30%的废品，那么这不仅会浪费金钱，也会给你的竞争对手创造一个利用你的想法制造产品，并且将其销售给你的客户的机遇。

提高可靠性和降低成本这个过程的关键是深入、广泛地利用设计软件。设计软件可以使工程师们加快每个阶段的设计工作。然而，仅仅缩短每个阶段的设计时间，可能不会显著地缩短整个设计过程的时间。因而，必须致力于开发并行工程软件，这样可以使设计团队的所有成员都能使用共享数据库。

随着我们更加全面地进入信息时代，要取得成功，工程师们在技术开发和技术管理方面都应该具有一些独特的知识和经验。成功的工程师们不仅应该具有广博的知识和技能，而且还应该是某些关键技术或学科的专家，他们还应该在社会因素和经济因素对市场的影响方面有敏锐的洞察能力。将来，花在解决日常工程问题上的费用将会减少，工程师们将会在一些更富有挑战性、更亟待解决的问题上协同工作，会大大缩短解决这些问题所需要的时间。我们已经开始了工程实践的新阶段。计算机和网络使工程师们具有越来越强的解决问题的能力，这也给他们的工作带来了很大的希望和喜悦。机械工程是一个伟大的行业，在我们尽可能多地利用了信息时代所提供的机遇后，它将变得更加伟大。

第76课　如何撰写科技论文

题目　作者在准备论文题目时，应该记住一个显而易见的事实：论文的题目将被成千上万人读到。如果会有人能够完整地读完整篇论文的话，那么也可能只是少数几个人。大多数读者或者会通过原始期刊，或者会通过某种二次文献（文摘或索引）读到论文的题目。因此，对题目中的每个词都应该仔细地推敲，词与词之间的关系也应该细心处理。论文的题目用词应该仅限于那些既容易理解，又便于检索，还能突出文章的重要内容的词。

摘要　摘要应该是论文的缩写版本。它应该是论文各主要章节的简要总结。一个精心准备的摘要能使读者迅速而又准确地了解论文的基本内容，以决定他们是否对此论文感兴趣，进而决定他们是否要阅读全文。摘要一般不超过250个单词，并应该清楚地反映论文的内容。许多人会阅读原始期刊或《工程索引》(EI)、《科学引文索引》(SCI)或其他种类的二次出版刊物上刊登的摘要。

摘要应该：(1)陈述该项研究工作的主要目的和范围；(2)描述所使用的方法；(3)总结研究成果；(4)陈述主要结论。

摘要中不应该出现任何未在论文中陈述的内容或结论。在摘要中不得引用参考文献（极少数情况除外，例如对以前发表过的方法的改进）。

引言　当然，论文的第一部分应该是引言。引言的目的是提供足够的背景资料，使读者不需要阅读过去已经发表的与此课题有关的出版物，就能够了解和评价目前的研究成果。最重要的是，你应该简单明了地陈述写这篇论文的目的。应该慎重地选择参考文献以提供最重要的背景资料。

实验过程　"实验过程"一节的主要目的是描述实验方案和提供足够的细节，以使有能力的研究人员可以重复这个实验。如果你的方法是新的（从未发表过的），那么你就应该提供所需要的全部细节。然而，如果这个实验方法已经在正规的期刊上发表过，那么只要给出参考文献就可以了。

认真撰写这一节是非常重要的，因为科学方法的基石就是要求你的研究成果不仅有科学价值，而且也必须是能够重复的。为了判断研究成果能否重复，就必须为其他人提供进行重复实验的依据。不能被重复的实验是没有意义的。必须能够产生相同或相似的结果，否则你的论文就没有科学价值。

当对你的论文进行同行评审时，一个好的审稿人会认真地阅读"实验过程"这一节。如果他确实怀疑你的实验能够被重复，那么无论你的研究成果多么令人敬畏，审稿人都会建议退回你的稿件。

结果 现在我们来到论文的核心部分——数据。论文的这部分称为"结果"。"结果"这一节通常由两部分组成。首先，你应该对实验进行全面的叙述，提出一个"大的轮廓"，但不要重复已经在"实验过程"一节中提到的实验细节；其次，你应该提供数据。

当然，这部分的写作不是一件很容易的事。你将如何提供数据呢？通常不能直接将实验笔记本上的数据抄到稿件上。最重要的是，在稿件中你应该提供有代表性的数据，而不是那些无限重复的数据。

"结果"一节应该写得清晰和简练，因为"结果"是由你提供给世界的新知识组成的。论文的前几部分（"引言"和"实验过程"）目的在于告诉人们你为什么和如何得到这些结果；而论文后面的部分（"讨论"）目的在于告诉人们这些结果意味着什么。显然，整篇论文都是以"结果"为基础的。因此，"结果"的表达方式必须是清晰和准确的。

讨论 与其他章节相比，"讨论"一节所写的内容更难于确定。因此，它通常是最难写的一节。而且，无论你知道与否，在许多论文中尽管数据正确，而且能够引起人们的兴趣，但是由于讨论部分写得不好也会遭到期刊编辑的拒绝。甚至更有可能的是，因为在"讨论"中所做的论述使得数据的真正含义变得模糊不清，而使论文遭到退稿。

一个好的"讨论"章节的主要特征如下所述。

1. 设法给出"结果"一节中的原理和相互关系。应该记住，一个好的"讨论"应该对"结果"进行讨论和论述，而不是扼要重述。

2. 指出任何的例外情况或相互关系中有问题的地方，并应该明确指出尚未解决的问题。决不要冒着很大的风险去采取另一方式，即试图对不适合的数据进行掩盖或捏造。

3. 说明和解释你的结果与以前发表过的研究结果有什么相符（或者不符）的地方。

4. 大胆地讨论你的研究工作的理论意义，以及任何可能的实际应用。

结论 在论文的结尾，你应该简要地叙述你的研究成果。要解释为什么你的结果是重要的和有用的，并总结那些能够表明你的观点是对以前的研究工作有所改进的实验证据或理论证明。同时，也要清晰地叙述这项研究工作的局限性。

参 考 文 献

[1] 施平. 机械工程专业英语（21版）. 哈尔滨：哈尔滨工业大学出版社，2022.
[2] 施平. 机电工程专业英语（9版）. 哈尔滨：哈尔滨工业大学出版社，2014.
[3] 施平. 先进制造技术. 哈尔滨：哈尔滨工业大学出版社，2006.
[4] Asfahl, R. *Robots and Manufacturing Automation*. John Wiley & Sons. 1992.
[5] Beer, F. P. Johnston, E. R. and DeWolf, J. T. *Mechanics of Materials*. McGraw-Hill. 2004.
[6] Bertoline, G. R. Wiebe, E. N. *Fundamentals of Graphics Communication*. McGraw-Hill. 2007.
[7] Bradley, D. A., Dawson, D., Burd, N. C. and Loader, A. J. *Mechatronics*. Chapman and Hall. 1991.
[8] Bolton W. C. *Mechanical Science*. Wiley-Blackwell. 2006.
[9] Budynas, R. J. and Nisbett, J. K. *Shigley's Mechanical Engineering Design*. McGraw-Hill. 2014.
[10] Craig, J. J. *Introduction to Robotics*. Pearson Education. 2005.
[11] Cross, N. *Engineering Design Methods*. John Wiley & Sons. 2008.
[12] Darbyshire, A. *Mechanical Engineering*. Elsevier. 2010.
[13] Dieter, G. E. and Schmidt, L. C. *Engineering Design*. McGraw-Hill. 2013.
[14] Dimarogonas, A. D. *Machine Design*. John Wiley & Sons. 2000.
[15] Dorf, R. C. and Kuisiak, A. *Handbook of Design, Manufacturing and Automation*. John Wiley & Sons. 1994.
[16] Eckhartdt, H. D. *Kinematic Design of Mechanics and Mechanisms*. McGraw-Hill. 1998.
[17] El Wakil, S. D. *Processes and Design for Manufacturing*. Prentice-Hall. 1998.
[18] Figliola, R. S. and Beasley, D. E. *Theory and Design for Mechanical Measurements*. John Wiley & Sons. 1999.
[19] Fitzpatrick, M. *Machining and CNC Technology*. McGraw-Hill. 2005.
[20] Gere, J. M. and Goodno, B. J. *Mechanics of Materials*. Cengage learning. 2012.
[21] Groover, M. P. *Principles of Modern Manufacturing*. John Wiley & Sons. 2013.
[22] Hamrock, R. J. *Fundamentals of Fluid Film Lubrication*. McGraw-Hill. 1994.
[23] Harris, T. A. *Rolling Bearing Analysis*. John Wiley & Sons. 2000.
[24] Juvinall, R. C. and Marshek, K. M. *Fundamentals of Machine Component Design*. John Wiley & Sons. 2011.
[25] Kalpakjian, S., Schmid, S. *Manufacturing Engineering and Technology*. Pearson Education. 2006.
[26] Kelly, S. G. *Fundamentals of Mechanical Vibrations*. McGraw-Hill. 1999.
[27] Krar, S. F and Check, A. F. *Technology of Machine Tools*. McGraw Hill. 1997.
[28] Kromacek, S. A., Lawson, A. E. and Horton, A. C. *Manufacturing Technology*. Glencoe/McGraw-Hill. 1990.
[29] Kurfess, P. E. *What Can CMMs Do?* Manufacturing Engineering. 2006.
[30] Kusiak, A. *Intelligent Manufacturing Systems*. Prentice Hall. 1990.
[31] Lynch, M. *Computer Numerical Control*. McGraw-Hill. 1994.
[32] Madsen, D. A., Shumaker, T. M., Turpin, J. L. and Stark, C. *Engineering Drawing and Design*. Delmar Publishing. 1991.
[33] Mattson, M. *CNC Programming: Principles and Applications*. Delmar. 2002.

[34] Niebel, B. W., Draper, A. B. and Wysk, R. A. *Modern Manufacturing Process Engineering*. McGraw-Hill. 1989.
[35] Nikravesh, P. E. *Computer-Aided Analysis of Mechanical Systems*. Prentice-Hall. 1988.
[36] Norton, R. L. *Design of Machinery*. McGraw-Hill. 2011.
[37] Olivo, C. T. and Olivo, T. P. *Applied Mechanics*. Delmar Publisher Inc. 1986.
[38] Ostwald, P. and Munoz, J. *Manufacturing Processes and Systems*. McGraw-Hill. 1997.
[39] Otto, K. N., Wood, K. L. *Product Design*. Pearson Education. 2003.
[40] Pfeifer, T., Eversheim, W., Konig, W., Weck, M. *Manufacturing Excellence*. Chapman & Hall. 1994.
[41] Rao, P. N. *Manufacturing Technology Metal Cutting & Machine Tools*. McGraw-Hill. 2009.
[42] Regh., J. A., Kraebber, H. W. *Computer-Integrated Manufacturing*. Pearson Education. 2002.
[43] Repp, V. E. and McCarthy, W. J. *Machine Tool Technology*. McKnight Publishing. 1984.
[44] Rowe, W. B. *Principles of Modern Grinding Technology*. William Andrew. 2013.
[45] Shaw, M. C. *Metal Cutting Principles*. Oxford University Press. 1984.
[46] Shetty, D., and Kolk, R. A. *Mechatronics System Design*. Thomson Learning. 2011.
[47] Trent, E. M. *Metal Cutting*. Butterworth-Heinemann Ltd. 2000.
[48] Ugural, A. C. *Mechanical Design*. McGraw-Hill. 2004.
[49] Ullman, D. G. *The Mechanical Design Process*. McGraw-Hill. 2010.
[50] Walker, K. M. *Applied Mechanics for Engineering Technology*. Pearson Education. 2000.
[51] Wickert, J. *An Introduction to Mechanical Engineering*. Gengage Learning. 2012.
[52] Wright, P. K. *21st Century manufacture*. Prentice-Hall. 2001.
[53] Youssef, H. A. and Hassan E. *Machining Technology*. CRC Press. 2008.

反侵权盗版声明

电子工业出版社依法对本作品享有专有出版权。任何未经权利人书面许可,复制、销售或通过信息网络传播本作品的行为;歪曲、篡改、剽窃本作品的行为,均违反《中华人民共和国著作权法》,其行为人应承担相应的民事责任和行政责任,构成犯罪的,将被依法追究刑事责任。

为了维护市场秩序,保护权利人的合法权益,我社将依法查处和打击侵权盗版的单位和个人。欢迎社会各界人士积极举报侵权盗版行为,本社将奖励举报有功人员,并保证举报人的信息不被泄露。

举报电话:(010)88254396;(010)88258888
传　　真:(010)88254397
E-mail: dbqq@phei.com.cn
通信地址:北京市海淀区万寿路173信箱
　　　　　电子工业出版社总编办公室
邮　　编:100036